Microwave Imaging

WILEY SERIES IN MICROWAVE AND OPTICAL ENGINEERING

KAI CHANG, Editor
Texas A&M University

A complete list of the titles in this series appears at the end of this volume.

Microwave Imaging

MATTEO PASTORINO

WILEY
JOHN WILEY & SONS, INC., PUBLICATION

Copyright © 2010 by John Wiley & Sons, Inc. All rights reserved

Published by John Wiley & Sons, Inc., Hoboken, New Jersey
Published simultaneously in Canada

No part of this publication may be reproduced, stored in a retrieval system, or transmitted in any form or by any means, electronic, mechanical, photocopying, recording, scanning, or otherwise, except as permitted under Section 107 or 108 of the 1976 United States Copyright Act, without either the prior written permission of the Publisher, or authorization through payment of the appropriate per-copy fee to the Copyright Clearance Center, Inc., 222 Rosewood Drive, Danvers, MA 01923, (978) 750-8400, fax (978) 750-4470, or on the web at www.copyright.com. Requests to the Publisher for permission should be addressed to the Permissions Department, John Wiley & Sons, Inc., 111 River Street, Hoboken, NJ 07030, (201) 748-6011, fax (201) 748-6008, or online at http://www.wiley.com/go/permission.

Limit of Liability/Disclaimer of Warranty: While the publisher and author have used their best efforts in preparing this book, they make no representations or warranties with respect to the accuracy or completeness of the contents of this book and specifically disclaim any implied warranties of merchantability or fitness for a particular purpose. No warranty may be created or extended by sales representatives or written sales materials. The advice and strategies contained herein may not be suitable for your situation. You should consult with a professional where appropriate. Neither the publisher nor author shall be liable for any loss of profit or any other commercial damages, including but not limited to special, incidental, consequential, or other damages.

For general information on our other products and services or for technical support, please contact our Customer Care Department within the United States at (800) 762-2974, outside the United States at (317) 572-3993 or fax (317) 572-4002.

Wiley also publishes its books in a variety of electronic formats. Some content that appears in print may not be available in electronic formats. For more information about Wiley products, visit our web site at www.wiley.com.

Library of Congress Cataloging-in-Publication Data:

Pastorino, Matteo.
 Microwave imaging / Matteo Pastorino.
  p. cm.—(Wiley series in microwave and optical engineering)
 Includes bibliographical references and index.
 ISBN 978-0-470-27800-0 (cloth)
 1. Microwave imaging—Industrial applications. 2. Nondestructive testing. 3. Radiography, Industrial. I. Title.
  TA417.25.P37 2010
  621.36'7—dc22
                         2009041788

Printed in the United States of America

10 9 8 7 6 5 4 3 2 1

Contents

1	**Introduction**	**1**
2	**Electromagnetic Scattering**	**4**
	2.1 Maxwell's Equations	4
	2.2 Interface Conditions	6
	2.3 Constitutive Equations	7
	2.4 Wave Equations and Their Solutions	9
	2.5 Volume Scattering by Dielectric Targets	14
	2.6 Volume Equivalence Principle	16
	2.7 Integral Equations	18
	2.8 Surface Scattering by Perfectly Electric Conducting Targets	19
	References	19
3	**The Electromagnetic Inverse Scattering Problem**	**20**
	3.1 Introduction	20
	3.2 Three-Dimensional Inverse Scattering	22
	3.3 Two-Dimensional Inverse Scattering	24
	3.4 Discretization of the Continuous Model	28
	3.5 Scattering by Canonical Objects: The Case of Multilayer Elliptic Cylinders	41
	References	53

4 Imaging Configurations and Model Approximations — 57

- 4.1 Objectives of the Reconstruction — 57
- 4.2 Multiillumination Approaches — 58
- 4.3 Tomographic Configurations — 59
- 4.4 Scanning Configurations — 63
- 4.5 Configurations for Buried-Object Detection — 65
- 4.6 Born-Type Approximations — 65
- 4.7 Extended Born Approximation — 68
- 4.8 Rytov Approximation — 70
- 4.9 Kirchhoff Approximation — 73
- 4.10 Green's Function for Inhomogeneous Structures — 73
- References — 77

5 Qualitative Reconstruction Methods — 79

- 5.1 Introduction — 79
- 5.2 Generalized Solution of Linear Ill-Posed Problems — 80
- 5.3 Regularization Methods — 82
- 5.4 Singular Value Decomposition — 84
- 5.5 Singular Value Decomposition for Solving Linear Problems — 87
- 5.6 Regularized Solution of a Linear System Using Singular Value Decomposition — 90
- 5.7 Qualitative Methods for Object Localization and Shaping — 91
- 5.8 The Linear Sampling Method — 92
- 5.9 Synthetic Focusing Techniques — 101
- 5.10 Qualitative Methods for Imaging Based on Approximations — 103
- 5.11 Diffraction Tomography — 103
- 5.12 Inversion Approaches Based on Born-Like Approximations — 108
- 5.13 The Born Iterative Method — 117
- 5.14 Reconstruction of Equivalent Current Density — 118
- References — 119

6 Quantitative Deterministic Reconstruction Methods — 123

- 6.1 Introduction — 123
- 6.2 Inexact Newton Methods — 125
- 6.3 The Truncated Landweber Method — 127
- 6.4 Inexact Newton Method for Electric Field Integral Equation Formulation — 129
- 6.5 Inexact Newton Method for Contrast Source Formulation — 136
- 6.6 The Distorted Born Iterative Method — 142
- 6.7 Inverse Scattering as an Optimization Problem — 147
- 6.8 Gradient-Based Methods — 148
- References — 150

7 Quantitative Stochastic Reconstruction Methods — 153

- 7.1 Introduction — 153
- 7.2 Simulated Annealing — 154
- 7.3 The Genetic Algorithm — 158
- 7.4 The Differential Evolution Algorithm — 170
- 7.5 Particle Swarm Optimization — 177
- 7.6 Ant Colony Optimization — 180
- 7.7 Code Parallelization — 187
- References — 189

8 Hybrid Approaches — 193

- 8.1 Introduction — 193
- 8.2 The Memetic Algorithm — 195
- 8.3 Linear Sampling Method and Ant Colony Optimization — 201
- References — 203

9 Microwave Imaging Apparatuses and Systems — 205

- 9.1 Introduction — 205
- 9.2 Scanning Systems for Microwave Tomography — 205
- 9.3 Antennas for Microwave Imaging — 211

	9.4 The Modulated Scattering Technique and Microwave Cameras	214
	References	226
10	**Applications of Microwave Imaging**	**229**
	10.1 Civil and Industrial Applications	229
	10.2 Medical Applications of Microwave Imaging	241
	10.3 Shallow Subsurface Imaging	252
	References	258
11	**Microwave Imaging Strategies, Emerging Techniques, and Future Trends**	**264**
	11.1 Introduction	264
	11.2 Potentialities and Limitations of Three-Dimensional Microwave Imaging	265
	11.3 Amplitude-Only Methods	269
	11.4 Support Vector Machines	269
	11.5 Metamaterials for Imaging Applications	272
	11.6 Through-Wall Imaging	274
	References	274

INDEX   **278**

CHAPTER ONE

Introduction

The present monograph is devoted to microwave imaging for diagnostic applications. As is well known, microwave imaging is a technique aimed at sensing a given scene by means of interrogating microwaves. This active technique—considered for a long time an *emerging technique*—has recently proved capable of providing excellent diagnostic capabilities in several areas, including civil and industrial engineering, nondestructive testing and evaluation (NDT&E), geophysical prospecting, and biomedical engineering.

To *localize*, *shape*, and *reconstruct* an unknown target located in an investigation domain and *surrounded* by measurement probes able to acquire the samples of the *scattered* field, several different approaches can be applied. Those considered in this book are *inverse scattering*–based procedures, which address the data inversion in several different ways depending on the target itself (e.g., strongly or weakly contrasted bodies, conducting objects) or on the imaging configuration and operation conditions.

Conceptually, the starting point for the development of these methods is formulation of the electromagnetic inverse scattering problem. A review of this important formulation constitutes Chapters 2 and 3 of this book and is described with engineering style and notations. In particular, the three-dimensional volume and surface scattering by dielectric and conducting targets is introduced, moving from Maxwell's equations to wave equations. The fundamental Fredholm integral equations, which are solved in most of the inverse scattering–based imaging procedures, are then derived in terms of the proper Green functions and tensors. Then, the two-dimensional scattering problem, in which the presence of infinite cylindrical scatterers is assumed, is addressed in detail, since it is of basic importance for the development of tomographic techniques.

In Chapter 3 the distinction between direct and inverse scattering problems is introduced. Moreover, the nonlinear formulation of the latter problem, in

---

Microwave Imaging, By Matteo Pastorino
Copyright © 2010 John Wiley & Sons, Inc.

terms of the so-called data and state equations, is discussed and a related numerical solving procedure is outlined. Finally, the semianalytical solution for plane-wave scattering by a multilayer elliptic cylinder is derived. The importance of using canonical scatterers for testing numerical methods is well known. In addition, a stratified elliptic cylinder is a quite complex structure for representing a significant test for imaging procedures.

The aim of Chapter 4 is to clarify the *objectives* of the short-range imaging approaches considered and specify the geometric arrangements used for microwave tomography and other inspection modalities. Among them, scanning configurations, as well as the detection of buried objects in half-space domains and in cross-borehole configurations, are discussed. Chapter 4 also introduces the most significant approximations on the scattering model that are used in imaging applications. Essentially, these are the Born and the Rytov approximations for dielectrics and the Kirchhoff approximation for conductors. These approximations are discussed in detail, since they constitute the key assumptions of several qualitative reconstruction methods. Such methods are covered in Chapter 5. In particular, in Section 5.1, the *classification* of microwave imaging algorithms into *quantitative* and *qualitative* methods, which is highly arbitrary, is discussed. The chapter includes remarks on uniqueness, ill-posedness, and stability of the electromagnetic inverse scattering problem and introduces concepts such as generalized solutions, regularization, and other fundamental tools for handling the most critical theoretical and numerical aspects in the development of *inversion* techniques.

Two kinds of qualitative methods are considered in Chapter 5: procedures for object localization and shaping, and methods based on approximations. Concerning the former, the formulation of the linear sampling method is outlined, since this method has been found to be efficient when used not only autonomously but also in conjunction with other quantitative reconstruction techniques in order to devise hybrid inspection methods. As far as qualitative methods based on approximations are concerned, Chapter 5 reviews the *classical* diffraction tomography and iterative approaches based on Born-type approximations.

Chapter 6 is devoted to a description of quantitative deterministic reconstruction methods, which are aimed at retrieving the values of the electromagnetic parameters of the unknown scenario. As they are based on *exact* models, they are theoretically able to inspect any scatterers, even highly contrasted targets. Among the various quantitative methods proposed in the scientific literature, this book focuses on inexact Newton procedures, while the so-called distorted Born iterative method and gradient-based techniques are also discussed.

Significant emphasis is then given to quantitative stochastic reconstruction methods, which are based on probabilistic concepts and are in principle able to find an optimum solution to the inverse scattering problem when it has been recast into a global optimization problem. Simulated annealing, the genetic algorithm, differential evolution, particle swarm optimization, and ant colony optimization are discussed and compared in Chapter 7. In particular, the two main properties of stochastic reconstruction techniques are stressed throughout the chapter. In particular, these techniques allow a simple introduction of a priori information into the model (which is of major importance in practical applications) and can be very efficiently combined with other deterministic methods in order to exploit the specific features of the different techniques, improving the effectiveness of the inspection process. Actually, the possibility of devising hybrid methods (Chapter 8 is devoted to these methods) is of paramount importance in microwave imaging, since the development of specific application-oriented procedures and systems is fundamental to increase the efficiency of microwave imaging for real and practical applications in several advanced technological fields.

Chapter 9 deals with instrumentation for microwave imaging. Some proposed imaging apparatuses, aimed at fast and accurate measurement of the scattered field data, are described. In particular, prototypes of microwave axial tomographs and microwave cameras and scanners are discussed with reference to some significant solutions recently proposed. Insights into the so-called modulated scattering technique and the antennas used in microwave imaging systems are also discussed.

Civil, industrial, and medical applications are covered in Chapter 10, with detailed descriptions of several specific examples. Concerning the field of nondestructive evaluation and testing, the use of inverse scattering-based inspection approaches is analyzed with reference to materials evaluation, crack detection, and inspection of civil and industrial structures. Shallow subsurface detection is also addressed in that chapter. Furthermore, concerning the medical field, where the potentialities of microwave imaging are widely recognized, the chapter includes discussion of the assumptions made and results obtained by applying some of the imaging systems and reconstruction techniques considered.

In Chapter 11, some new directions of imaging strategies are briefly delineated. One of them is represented by the current trend of extending, to three-dimensional imaging, most of the inspection methods originally devoted to two-dimensional configurations. It is very encouraging to see the final utilization—thanks to the increased power of the current generation of computers—of a key property of electromagnetic imaging, namely, the possibility of factoring in the vector nature of the field. Other strategies and applications are also briefly mentioned in the chapter, including the use of amplitude-only input data or metamaterial slabs, as well as the recent proposal of using support vector machines for deriving basic parameters of unknown targets starting from field scattered data.

Finally, the author wishes to profusely thank Dr. Giovanni Bozza and Dr. Andrea Randazzo of the Department of Biophysical and Electronic Engineering at the University of Genoa, Genoa, Italy, for having performed the numerical simulations reported throughout this monograph.

CHAPTER TWO

Electromagnetic Scattering

2.1 MAXWELL'S EQUATIONS

The electromagnetic field is governed by a set of experimental laws known as *Maxwell's equations* (Stratton 1941, Jones 1964, Felsen and Marcuvitz 1973, Van Bladel 2007, Chew 1990), which relate the field vectors to their sources. Maxwell's equations can be expressed in the following local forms:

$$\nabla \times \bar{\mathcal{E}}(\mathbf{r}, t) = -\frac{\partial \bar{\mathcal{B}}(\mathbf{r}, t)}{\partial t}, \qquad (2.1.1)$$

$$\nabla \times \bar{\mathcal{H}}(\mathbf{r}, t) = \frac{\partial \bar{\mathcal{D}}(\mathbf{r}, t)}{\partial t} + \bar{\mathcal{J}}(\mathbf{r}, t), \qquad (2.1.2)$$

$$\nabla \cdot \bar{\mathcal{D}}(\mathbf{r}, t) = \rho(\mathbf{r}, t), \qquad (2.1.3)$$

$$\nabla \cdot \bar{\mathcal{B}}(\mathbf{r}, t) = 0 \qquad (2.1.4)$$

Here $\mathbf{r}$ denotes the position vector [in meters (m)] and t the time [in seconds (s)]. $\bar{\mathcal{E}}, \bar{\mathcal{B}}, \bar{\mathcal{H}}$, and $\bar{\mathcal{D}}$ are the vector fields describing the electromagnetic field and are called the *electric field* [in volts per meter (V/m)], the *magnetic flux density* [in webers per meter (Wb/m^2)], the *magnetic field* [in amperes per meter (A/m)], and the *electric flux density* [in coulombs per square meter (C/m^2)], respectively. The sources are specified by the *volume electric charge density* ρ [in coulombs per cubic meter (C/m^3)] and by the *electric current density* $\bar{\mathcal{J}}$ [in amperes per square meter (A/m^2)].

When the time dependence is of cosinusoidal form (time-harmonic fields), a complex representation for field vectors and sources can be used. More precisely, if

---

Microwave Imaging, By Matteo Pastorino
Copyright © 2010 John Wiley & Sons, Inc.

$$\overline{\mathcal{F}}(\mathbf{r},t) = f_x(\mathbf{r},t)\hat{\mathbf{x}} + f_y(\mathbf{r},t)\hat{\mathbf{y}} + f_z(\mathbf{r},t)\hat{\mathbf{z}} \quad (2.1.5)$$

is any of the vectors given above, and

$$f_p(\mathbf{r},t) = f_{p0}(\mathbf{r})\cos(\omega t + \phi_p(\mathbf{r})), \quad p = x, y, z, \quad (2.1.6)$$

where ω is the *angular pulsation* [in radians per second (rad/s)] [which is related to the frequency f, in hertz (Hz), by $\omega = 2\pi f$], then it is possible to introduce the complex vector field given by

$$\mathbf{F}(\mathbf{r}) = f_x(\mathbf{r})\hat{\mathbf{x}} + f_y(\mathbf{r})\hat{\mathbf{y}} + f_z(\mathbf{r})\hat{\mathbf{z}}, \quad (2.1.7)$$

where $f_p(\mathbf{r}) = f_{p0}(\mathbf{r})e^{j\phi_p(\mathbf{r})}$, $p = x, y, z$, such that

$$\overline{\mathcal{F}}(\mathbf{r},t) = \mathrm{Re}\{\mathbf{F}(\mathbf{r})e^{j\omega t}\}. \quad (2.1.8)$$

In this way the vector field $\overline{\mathcal{F}}(\mathbf{r},t)$ is completely described by $\mathbf{F}(\mathbf{r})$ when ω is known. Using these relations, it follows that

$$\frac{\partial^n \overline{\mathcal{F}}(\mathbf{r},t)}{\partial t^n} = \mathrm{Re}\{(j\omega)^n \mathbf{F}(\mathbf{r})e^{j\omega t}\}. \quad (2.1.9)$$

Taking into account equations (2.1.8) and (2.1.9), the following local form of Maxwell's equations for time-harmonic fields can then be deduced, where $\mathbf{E}$, $\mathbf{B}$, $\mathbf{D}$, $\mathbf{H}$, ρ, and $\mathbf{J}$ denote the complex quantities describing the vector fields and their sources:

$$\nabla \times \mathbf{E}(\mathbf{r}) = -j\omega \mathbf{B}(\mathbf{r}), \quad (2.1.10)$$
$$\nabla \times \mathbf{H}(\mathbf{r}) = j\omega \mathbf{D}(\mathbf{r}) + \mathbf{J}(\mathbf{r}), \quad (2.1.11)$$
$$\nabla \cdot \mathbf{D}(\mathbf{r}) = \rho(\mathbf{r}), \quad (2.1.12)$$
$$\nabla \cdot \mathbf{B}(\mathbf{r}) = 0, \quad (2.1.13)$$

From the differential form of Maxwell's equations we can derive the corresponding integral forms, which describe the relations among sources and field vectors inside a given region of space (Balanis 1989). By integrating equations (2.1.10) and (2.1.11) over an oriented open regular surface S and applying the Stokes theorem, one obtains

$$\oint_C \mathbf{E}(\mathbf{r}) \cdot d\mathbf{l} = -j\omega \int_S \mathbf{B}(\mathbf{r}) \cdot \hat{\mathbf{n}}\, ds, \quad (2.1.14)$$
$$\oint_C \mathbf{H}(\mathbf{r}) \cdot d\mathbf{l} = j\omega \int_S \mathbf{D}(\mathbf{r}) \cdot \hat{\mathbf{n}}\, ds + \int_S \mathbf{J}(\mathbf{r}) \cdot \hat{\mathbf{n}}\, ds, \quad (2.1.15)$$

where $\hat{\mathbf{n}}$ is the normal to S and C is its contour line, oriented according to S. Analogously, by integrating equations (2.1.12) and (2.1.13) over a bounded volume V and applying the divergence theorem, one obtains

$$\oint_S \mathbf{D}(\mathbf{r}) \cdot \hat{\mathbf{n}} \, d\mathbf{r} = \int_V \rho(\mathbf{r}) d\mathbf{r}, \tag{2.1.16}$$

$$\oint_S \mathbf{B}(\mathbf{r}) \cdot \hat{\mathbf{n}} \, d\mathbf{r} = 0, \tag{2.1.17}$$

where S denotes the closed surface with normal $\hat{\mathbf{n}}$ enclosing the volume V.

Equations (2.1.14)–(2.1.17) are referred to as the *global (integral) form* of Maxwell's equations.

2.2 INTERFACE CONDITIONS

The field vectors are continuous functions of spatial coordinates except possibly at the boundaries between different media (or where the sources are present). The behavior of the electromagnetic field at the interface between two different media is governed by the *interface conditions*, which can be simply derived by using the Maxwell's equations in their integral form (Stratton 1941). For time-harmonic fields, the boundary conditions read as

$$\hat{\mathbf{n}} \times [\mathbf{E}_2(\mathbf{r}) - \mathbf{E}_1(\mathbf{r})] = 0, \tag{2.2.1}$$

$$\hat{\mathbf{n}} \times [\mathbf{H}_2(\mathbf{r}) - \mathbf{H}_1(\mathbf{r})] = \mathbf{J}_s(\mathbf{r}), \tag{2.2.2}$$

$$\hat{\mathbf{n}} \cdot [\mathbf{D}_2(\mathbf{r}) - \mathbf{D}_1(\mathbf{r})] = \rho_s(\mathbf{r}), \tag{2.2.3}$$

$$\hat{\mathbf{n}} \cdot [\mathbf{B}_2(\mathbf{r}) - \mathbf{B}_1(\mathbf{r})] = 0, \tag{2.2.4}$$

where the subscripts 1 and 2 denote the fields in media 1 and 2, respectively. Moreover, $\mathbf{J}_s(\mathbf{r})$ and $\rho_s(\mathbf{r})$ are the surface electric current density (A/m) and the surface charge density (C/m^2), possibly lying on the interface between the two media. Finally, $\hat{\mathbf{n}}$ indicates the normal to the boundary surface directed toward medium 2.

An interesting particular case occurs when medium 1 is a *perfect electric conductor* (PEC). In this situation, the time-harmonic field vanishes inside region 1 (Balanis 1989). Consequently, $\mathbf{E}_1(\mathbf{r}) = 0$ and $\mathbf{H}_1(\mathbf{r}) = 0$ and the interface conditions with a PEC region are given by

$$\hat{\mathbf{n}} \times \mathbf{E}_2(\mathbf{r}) = 0, \tag{2.2.5}$$

$$\hat{\mathbf{n}} \times \mathbf{H}_2(\mathbf{r}) = \mathbf{J}_s(\mathbf{r}), \tag{2.2.6}$$

$$\hat{\mathbf{n}} \cdot \mathbf{D}_2(\mathbf{r}) = \rho_s(\mathbf{r}), \tag{2.2.7}$$

$$\hat{\mathbf{n}} \cdot \mathbf{B}_2(\mathbf{r}) = 0. \tag{2.2.8}$$

On the other hand, it is important to note that surface sources can be induced only on a PEC.

2.3 CONSTITUTIVE EQUATIONS

Maxwell's equations state relationships between the vector fields **D**, **E**, **B**, and **H**, and their sources **J** and ρ, which hold true in every electromagnetic phenomenon. However, they are not sufficient for determining the vector fields univocally. In fact, as can be easily proved (Jones 1964), Maxwell's equations [(2.1.10)–(2.1.13)] correspond to six independent scalar equations, whereas the four unknown vector fields can be represented by 12 unknown scalar functions. This fact suggests that some further information on the vector fields is needed. Namely, Maxwell's equations [(2.1.10)–(2.1.13)] contain no information on the media in which the electromagnetic phenomena occur. This kind of information is provided by the *constitutive equations*, which interrelate the vector fields **D**, **E**, **B**, and **H**, and are specific for the medium where the propagation takes place (Jones 1964). In a very general framework, *constitutive equations* can be written as

$$\mathbf{D}(\mathbf{r}) = F(\mathbf{E}, \mathbf{H})(\mathbf{r}), \tag{2.3.1}$$

$$\mathbf{B}(\mathbf{r}) = G(\mathbf{E}, \mathbf{H})(\mathbf{r}), \tag{2.3.2}$$

where F and G are operators that associate with every pair of vector fields **E** and **H** the vector fields **D** and **B**, respectively. As will be shown below, the properties of these operators are used in classification of the constitutive equations.

It is worth noting that vector relations (2.3.1) and (2.3.2) provide six scalar conditions that are to be coupled with Maxwell's equations [(2.1.10)–(2.1.13)]. Equations (2.3.1), (2.3.2), and (2.1.10)–(2.1.13) constitute a system of 12 scalar equations with 12 unknown scalar functions.

An important category of media is that of *linear* materials, for which the operators F and G are linear. In fact, since Maxwell's equations are linear with respect to the vector fields **E**, **D**, **B**, and **H**, an electromagnetic problem involving only linear media is a linear problem.

Relations (2.3.1) and (2.3.2) are very general, but, for most materials (except, e.g., the *bianisotropic* and *biisotropic* ones), the electric flux density **D** depends only on the electric field **E** and the magnetic flux density **B** depends only on the magnetic field **H**; thus

$$\mathbf{D}(\mathbf{r}) = F(\mathbf{E})(\mathbf{r}), \tag{2.3.3}$$

$$\mathbf{B}(\mathbf{r}) = G(\mathbf{H})(\mathbf{r}). \tag{2.3.4}$$

8 ELECTROMAGNETIC SCATTERING

A quite general constitutive relation for linear media can then be written as follows

$$\mathbf{D}(\mathbf{r}) = \bar{\boldsymbol{\varepsilon}}(\mathbf{r}) \cdot \mathbf{E}(\mathbf{r}), \tag{2.3.5}$$

$$\mathbf{B}(\mathbf{r}) = \bar{\boldsymbol{\mu}}(\mathbf{r}) \cdot \mathbf{H}(\mathbf{r}), \tag{2.3.6}$$

where $\bar{\boldsymbol{\varepsilon}}$ and $\bar{\boldsymbol{\mu}}$ are the dielectric permittivity tensor [in farads per meter (F/m)] and the magnetic permeability tensor [in henries per meter (H/m)], respectively. A medium with $\bar{\boldsymbol{\varepsilon}}$ and $\bar{\boldsymbol{\mu}}$, both independent of the position vector, is said to be *homogeneous*, and *inhomogeneous* otherwise. An important subset of this kind of medium consists of *isotropic* media, which are characterized by $\bar{\boldsymbol{\varepsilon}}(\mathbf{r}) = \varepsilon(\mathbf{r})\bar{\mathbf{I}}$ and $\bar{\boldsymbol{\mu}}(\mathbf{r}) = \mu(\mathbf{r})\bar{\mathbf{I}}$, where $\bar{\mathbf{I}}$ is the identity tensor and ε and μ scalar functions called the *dielectric permittivity* and the *magnetic permeability*, respectively. For instance, the vacuum is a linear isotropic medium described by the linear constitutive relations

$$\mathbf{D}(\mathbf{r}) = \varepsilon_0 \mathbf{E}(\mathbf{r}), \tag{2.3.7}$$

$$\mathbf{B}(\mathbf{r}) = \mu_0 \mathbf{H}(\mathbf{r}), \tag{2.3.8}$$

where $\varepsilon_0 \approx 8.85 \times 10^{-12}$ F/m denotes the dielectric permittivity of the vacuum and $\mu_0 = 4\pi \times 10^{-7}$ H/m is the magnetic permeability of the vacuum. Dielectric permittivities and magnetic permeabilities are often replaced by their relative counterparts ε_r and μ_r, defined as follows:

$$\varepsilon_r(\mathbf{r}) = \frac{\varepsilon(\mathbf{r})}{\varepsilon_0}, \tag{2.3.9}$$

$$\mu_r(\mathbf{r}) = \frac{\mu(\mathbf{r})}{\mu_0}. \tag{2.3.10}$$

Another important property of media is the *time dispersiveness*, which is essentially related to dynamics of matter polarization. In the frequency domain, time dispersiveness corresponds to complex-valued dielectric permittivities or magnetic permeabilities depending on the operating frequency.

The different microscopic mechanisms at the base of time dispersiveness can be modeled by several formulas. One of the most common is the Debye one (Balanis 1989), which states that

$$\varepsilon(\omega) = \varepsilon_\infty + \frac{\varepsilon_s - \varepsilon_\infty}{1 + j\omega\tau}, \tag{2.3.11}$$

where ε_s and ε_∞ are the dielectric permittivities for $\omega = 0$ and for $\omega \to \infty$, respectively, whereas $\tau(s)$ is a parameter typical of the medium called the *relaxation time*.

Moreover, in conducting media, an induced current is generated by the field. For a wide category of materials, Ohm's law holds, which states that the induced current density is given by

$$\mathbf{J}_i(\mathbf{r}) = \sigma(\mathbf{r})\mathbf{E}(\mathbf{r}), \tag{2.3.12}$$

where σ (S/m) is the *electric conductivity* of the medium. Consequently, in the presence of a linear, isotropic, and conducting medium for which equation (2.3.12) is valid, Maxwell's equation (2.1.11) can be rewritten as

$$\nabla \times \mathbf{H}(\mathbf{r}) = j\omega\varepsilon(\mathbf{r})\mathbf{E}(\mathbf{r}) + \sigma(\mathbf{r})\mathbf{E}(\mathbf{r}) + \mathbf{J}_0(\mathbf{r}) = j\omega\varepsilon_0\left(\varepsilon_r(\mathbf{r}) - j\frac{\sigma(\mathbf{r})}{\omega\varepsilon_0}\right)\mathbf{E}(\mathbf{r}) + \mathbf{J}_0(\mathbf{r}), \tag{2.3.13}$$

where $\mathbf{J}_0$ is the impressed current density. If the effective dielectric permittivity

$$\varepsilon(\mathbf{r}) = \varepsilon_0\left(\varepsilon_r(\mathbf{r}) - j\frac{\sigma(\mathbf{r})}{\omega\varepsilon_0}\right) \equiv \varepsilon'(\mathbf{r}) - j\varepsilon''(\mathbf{r}) = \varepsilon_0(\varepsilon_r'(\mathbf{r}) - j\varepsilon_r''(\mathbf{r})) \tag{2.3.14}$$

is introduced, equation (2.3.13) can then be written as

$$\nabla \times \mathbf{H}(\mathbf{r}) = j\omega\varepsilon(\mathbf{r})\mathbf{E}(\mathbf{r}) + \mathbf{J}_0(\mathbf{r}). \tag{2.3.15}$$

Accordingly, dispersive and conducting media can be treated in the same way, specifically, by using the complex-valued dielectric permittivity defined by (2.3.14). In this way only the impressed current $\mathbf{J}_0$ explicitly appears in Maxwell's equations.

In several applications of microwave imaging, as will be widely discussed in the following chapters, retrieving the values of the complex dielectric permittivity of a given target or scenario often represents the objective of the reconstruction.

2.4 WAVE EQUATIONS AND THEIR SOLUTIONS

Let us consider a homogeneous medium characterized by a dielectric permittivity ε (possibly, complex-valued) and a magnetic permeability μ. By taking the curl of equation (2.1.10) and substituting the result into equation (2.1.11), one obtains

$$\nabla \times \nabla \times \mathbf{E}(\mathbf{r}) - \omega^2 \mu\varepsilon \mathbf{E}(\mathbf{r}) = -j\omega\mu\mathbf{J}_0(\mathbf{r}), \tag{2.4.1}$$

$$\nabla \times \nabla \times \mathbf{H}(\mathbf{r}) - \omega^2 \mu\varepsilon \mathbf{H}(\mathbf{r}) = \nabla \times \mathbf{J}_0(\mathbf{r}). \tag{2.4.2}$$

These equations, called the *vector wave equations*, necessitate suitable boundary conditions. Several uniqueness theorems can be derived for such a problem. For example, if the region of interest V, bounded by the closed surface S, is lossy, the tangential component $\hat{\mathbf{n}} \times \mathbf{E}(\mathbf{r})$ [or $\hat{\mathbf{n}} \times \mathbf{H}(\mathbf{r})$] assigned to S is sufficient for unique determination of the electromagnetic field inside V. The same uniqueness results hold true even if $\hat{\mathbf{n}} \times \mathbf{E}(\mathbf{r})$ is specified on a part of S and $\hat{\mathbf{n}} \times \mathbf{H}(\mathbf{r})$ on the other one.

The relevant conditions to be satisfied when wave equations (2.4.1) and (2.4.2) need to be solved, in an unbounded medium, are the Silver–Müller radiation conditions, which read, for $r = |\mathbf{r}| \to \infty$, as

$$\hat{\mathbf{r}} \times \mathbf{E}(\mathbf{r}) = \eta \mathbf{H}(\mathbf{r}) + o\left(\frac{1}{r}\right), \quad (2.4.3)$$

$$\mathbf{H}(\mathbf{r}) \times \hat{\mathbf{r}} = \frac{1}{\eta} \mathbf{E}(\mathbf{r}) + o\left(\frac{1}{r}\right), \quad (2.4.4)$$

where $\hat{\mathbf{r}} = \mathbf{r}/r$ is the unit radial vector and $\eta = \sqrt{\mu/\varepsilon}$ is the intrinsic impedance [in ohms (Ω)] of the medium (Van Bladel 2007).

The electromagnetic field generated in an unbounded region (free-space radiation) by the impressed source $\mathbf{J}_0$ satisfying equations (2.4.1) and (2.4.2), along with the radiation conditions (2.4.3) and (2.4.4), can be expressed in integral form as

$$\mathbf{E}(\mathbf{r}) = j\omega\mu \int_V \mathbf{J}_0(\mathbf{r}') \cdot \bar{\mathbf{G}}(\mathbf{r}/\mathbf{r}') d\mathbf{r}', \quad (2.4.5)$$

$$\mathbf{H}(\mathbf{r}) = -\int_V \nabla \times \mathbf{J}_0(\mathbf{r}') \cdot \bar{\mathbf{G}}(\mathbf{r}/\mathbf{r}') d\mathbf{r}', \quad (2.4.6)$$

where $\bar{\mathbf{G}}(\mathbf{r}/\mathbf{r}')$ is the free-space Green dyadic tensor given by (Tai 1971)

$$\bar{\mathbf{G}}(\mathbf{r}/\mathbf{r}') = -\frac{1}{4\pi}\left[\bar{\mathbf{I}} + \frac{1}{k^2}\nabla\nabla\right]\frac{e^{-jk|\mathbf{r}-\mathbf{r}'|}}{|\mathbf{r}-\mathbf{r}'|}, \quad (2.4.7)$$

where $k = \omega\sqrt{\mu\varepsilon}$ is the wavenumber [in reciprocal meters (m^{-1})] of the propagation medium. It is worth noting that Green's dyadic tensor is related to the radiation produced by an elementary source and provides a solution to the following tensor equation:

$$\nabla \times \nabla \times \bar{\mathbf{G}}(\mathbf{r}/\mathbf{r}') - k^2 \bar{\mathbf{G}}(\mathbf{r}/\mathbf{r}') = \bar{\mathbf{I}}\delta(\mathbf{r}-\mathbf{r}'). \quad (2.4.8)$$

It is noteworthy that if $\mathbf{r} \in V$, the integral operator of equation (2.4.5) has to be carefully dealt with because of the involved singularities of the free-space

Green tensor. In fact, the correct meaning of such an integral is as follows (Van Bladel 2007)

$$\int_V \mathbf{J}_0(\mathbf{r}') \cdot \overline{\mathbf{G}}(\mathbf{r}/\mathbf{r}') d\mathbf{r}' = \begin{cases} \int_V \mathbf{J}_0(\mathbf{r}') \cdot \overline{\mathbf{G}}(\mathbf{r}/\mathbf{r}') d\mathbf{r}' & \text{if } \mathbf{r} \notin V \\ PV \int_V \mathbf{J}_0(\mathbf{r}') \cdot \overline{\mathbf{G}}(\mathbf{r}/\mathbf{r}') d\mathbf{r}' - \dfrac{\mathbf{J}_0(\mathbf{r})}{3j\omega\varepsilon} & \text{if } \mathbf{r} \in V \end{cases}, \quad (2.4.9)$$

where PV denotes the principal value of the integral.

Equations (2.4.5) and (2.4.6) provide the field radiated in free space by a given bounded source. Similar expressions hold in different conditions, provided the suitable Green function is used. One of the most relevant situations for diagnostic applications consists of a source radiating in a half-space, which model, for example, the detection of objects buried in soil by using an illumination/measurement system located in air.

Without loss of generality, let us suppose that the two half-space regions are separated by the plane $z = 0$. We assume that the upper region ($z > 0$) is characterized by ε_1 and μ, whereas the lower region ($z \leq 0$) is characterized by ε_2 and μ. If a bounded distribution of electric current density is present in the upper region, the electric fields in the two regions can be written as (Chew 1990)

$$\mathbf{E}_1(\mathbf{r}) = j\omega\mu \int_V \mathbf{J}_0(\mathbf{r}') \cdot \overline{\mathbf{G}}_{11}(\mathbf{r}/\mathbf{r}') d\mathbf{r}', \quad z > 0, \quad (2.4.10)$$

$$\mathbf{E}_2(\mathbf{r}) = j\omega\mu \int_V \mathbf{J}_0(\mathbf{r}') \cdot \overline{\mathbf{G}}_{21}(\mathbf{r}/\mathbf{r}') d\mathbf{r}', \quad z \leq 0, \quad (2.4.11)$$

where

$$\overline{\mathbf{G}}_{11}(\mathbf{r}/\mathbf{r}') = \begin{cases} \dfrac{-j}{4\pi} \int_0^\infty \dfrac{1}{\xi h_1} \sum_{n=0}^\infty (2-\delta_0) \{ \mathbf{M}_{n\xi}^\gamma(h_1) [\mathbf{M}_{n\xi}'^\gamma(-h_1) + a\mathbf{M}_{n\xi}'^\gamma(h_1)] \\ + \mathbf{N}_{n\xi}^\gamma(h_1) [\mathbf{N}_{n\xi}'^\gamma(-h_1) + b\mathbf{N}_{n\xi}'^\gamma(h_1)] \} d\xi, \quad z \geq z' \\ \dfrac{-j}{4\pi} \int_0^\infty \dfrac{1}{\xi h_1} \sum_{n=0}^\infty (2-\delta_0) \{ \mathbf{M}_{n\xi}^\gamma(h_1) [\mathbf{M}_{n\xi}^\gamma(-h_1) + a\mathbf{M}_{n\xi}^\gamma(h_1)] \mathbf{M}_{n\xi}'^\gamma(h_1) \\ + [\mathbf{N}_{n\xi}^\gamma(-h_1) + b\mathbf{N}_{n\xi}^\gamma(h_1)] \mathbf{N}_{n\xi}'^\gamma(h_1) \} d\xi, \quad z' \geq z \geq 0 \end{cases}, \quad (2.4.12)$$

$$\overline{\mathbf{G}}_{21}(\mathbf{r}/\mathbf{r}') = \dfrac{-j}{4\pi} \int_0^\infty \dfrac{1}{\xi h_1} \sum_{n=0}^\infty (2-\delta_0) [c\mathbf{M}_{n\xi}^\gamma(-h_2) \mathbf{M}_{n\xi}'^\gamma(h_1) \\ + d\mathbf{N}_{n\xi}^\gamma(-h_2) \mathbf{N}_{n\xi}'^\gamma(h_1)] d\xi, \quad z \leq 0, \quad (2.4.13)$$

where $h_1 = \sqrt{k_1^2 - \xi^2}$, $h_2 = \sqrt{k_2^2 - \xi^2}$, $k_1 = \omega\sqrt{\varepsilon_1\mu}$, $k_2 = \omega\sqrt{\varepsilon_2\mu}$, and δ_0 is the Kronecker delta function, such that $\delta_0 = 1$ if $n = 0$, and $\delta_0 = 0$ otherwise. In equations (2.4.12) and (2.4.13), $\mathbf{M}_{n\xi}^\gamma$ and $\mathbf{N}_{n\xi}^\gamma$ denote cylindrical vector functions given by (Tai 1971)

$$\mathbf{M}_{n\xi}^\gamma(h) = \left[\mp \frac{nJ_n(\xi r)}{r} \frac{\sin}{\cos} n\varphi \hat{\mathbf{r}} - \frac{\partial J_n(\xi r)}{\partial r} \frac{\cos}{\sin} n\varphi \hat{\boldsymbol{\varphi}} \right] e^{-jhz} \quad (2.4.14)$$

$$\mathbf{N}_{n\xi}^\gamma(h) = \frac{1}{\sqrt{\xi^2 + h^2}} \left[-jh \frac{\partial J_n(\xi r)}{\partial r} \frac{\cos}{\sin} n\varphi \hat{\mathbf{r}} \pm \frac{jhnJ_n(\xi r)}{r} \frac{\sin}{\cos} n\varphi \hat{\boldsymbol{\varphi}} \right.$$
$$\left. + \xi^2 J_n(\xi r) \frac{\cos}{\sin} n\varphi \hat{\mathbf{z}} \right] e^{-jhz} \quad (2.4.15)$$

where the superscript γ indicates even (when $\gamma = e$) and odd (when $\gamma = o$) modes, so that, in equations (2.4.14) and (2.4.15) the upper symbols are valid for $\gamma = e$, whereas the lower ones hold when $\gamma = o$. Finally, the coefficients $a, b, c,$ and d are given by (Tai 1971)

$$a = \frac{h_1 - h_2}{h_1 + h_2} \quad (2.4.16)$$

$$b = \frac{k_2^2 h_1 - k_1^2 h_2}{k_2^2 h_1 + k_1^2 h_2} \quad (2.4.17)$$

$$c = \frac{2h_1}{h_1 + h_2} \quad (2.4.18)$$

$$d = \frac{2k_1 k_2 h_1}{k_2^2 h_1 + k_1^2 h_2}. \quad (2.4.19)$$

Analogous relations hold for the magnetic fields. Obviously, by suitably changing the subscripts related to the two regions, similar expressions can be derived for the field produced by a source located in the lower region. Moreover, for a multilayer structure (a model seldom used for imaging applications), similar relationships can be devised. In particular, the expression for the related Green tensor can be found in the text by Chew (1990).

It should be noted that equations (2.4.5) and (2.4.6) (free space) are also basic equations in antenna theory since an antenna can usually be modeled as a distribution of electric current. Moreover, at a great distance from the source, the following *far-field* approximation can be used (Collin and Zucker 1969):

$$\mathbf{E}(\mathbf{r}) = -\eta \hat{\mathbf{r}} \times \mathbf{H}(\mathbf{r}) = \frac{jk\eta e^{-jkr}}{4\pi r} \int_V [\hat{\mathbf{r}} \cdot \mathbf{J}_0(\mathbf{r}')\hat{\mathbf{r}} - \mathbf{J}_0(\mathbf{r}')] e^{jk\hat{\mathbf{r}} \cdot \mathbf{r}'} d\mathbf{r}', \quad (2.4.20)$$

$$\mathbf{H}(\mathbf{r}) = -\frac{jke^{-jkr}}{4\pi r} \hat{\mathbf{r}} \times \int_V \mathbf{J}_0(\mathbf{r}') e^{jk\hat{\mathbf{r}} \cdot \mathbf{r}'} d\mathbf{r}'. \quad (2.4.21)$$

This approximation is valid in the so-called far-field region, that is, for $r > 10\lambda$, $r > 10d$, and $r > 2d^2\lambda^{-1}$, where d is the diameter of the smallest sphere, including the source and centered at the origin of the frame, and λ is the wavelength in the propagation medium. According to equations (2.4.20) and (2.4.21), in the far-field region the electromagnetic field results to be a transverse electromagnetic spherical wave. Expression (2.4.20) can also be rewritten as

$$\mathbf{E}(\mathbf{r}) = \frac{e^{-jkr}}{r} \mathbf{E}_\infty(\hat{\mathbf{r}}), \tag{2.4.22}$$

Where $\mathbf{E}_\infty$ is a tangential vector (radiation pattern vector) given by

$$\mathbf{E}_\infty(\hat{\mathbf{r}}) = \frac{jk\eta}{4\pi} \int_V [\hat{\mathbf{r}} \cdot \mathbf{J}_0(\mathbf{r}')\hat{\mathbf{r}} - \mathbf{J}_0(\mathbf{r}')] e^{jk\hat{\mathbf{r}} \cdot \mathbf{r}'} d\mathbf{r}'. \tag{2.4.23}$$

Moreover, from equations (2.4.20) and (2.4.21), the electromagnetic field is found to have the *local* characteristics of a plane wave propagating along the radial direction. In fact, the most general form of a plane wave is

$$\mathbf{E}(\mathbf{r}) = E_0 e^{-j\mathbf{k} \cdot \mathbf{r}} \hat{\mathbf{p}}, \tag{2.4.24}$$

$$\mathbf{H}(\mathbf{r}) = \frac{1}{\eta} \hat{\mathbf{k}} \times \mathbf{E}(\mathbf{r}), \tag{2.4.25}$$

where $\mathbf{k}$ is the propagation vector such that $|\mathbf{k}| = k$, $\hat{\mathbf{k}} = \mathbf{k}/k$, and $\hat{\mathbf{p}}$ the complex polarization vector ($|\hat{\mathbf{p}}| = 1$), which must be orthogonal to the propagation direction ($\hat{\mathbf{p}} \cdot \mathbf{k} = 0$). Plane waves are particular solutions of Mawxell's equations, which are usually considered in microwave imaging techniques to model the interrogating fields used for the exposition of the unknown targets. However, since far-field radiation conditions are seldom valid for short-range inspection techniques in the relevant frequency bands, also other source models are often adopted. One of the most common approaches consists in assuming the field generated by the transmitting antenna to be similar to that of an infinite line-current source. In this case, if the source coincides with the z axis (without loss of generality), the electric field it produces is given by (Harrington 1961)

$$\mathbf{E}(\mathbf{r}) = -I \frac{\omega\mu}{4} H_0^{(2)}(k\rho) \hat{\mathbf{z}}, \tag{2.4.26}$$

$$\mathbf{H}(\mathbf{r}) = -jI \frac{k}{4} H_1^{(2)}(k\rho) \hat{\boldsymbol{\varphi}}, \tag{2.4.27}$$

where $H_0^{(2)}(x) = J_0(x) - jY_0(x)$ and $H_1^{(2)}(x) = J_1(x) - jY_1(x)$ denote the zero- and first-order Hankel functions of second kind, where $J_n(x)$ and $Y_n(x)$ are the Bessel functions of the first and second kind of nth order, respectively

(Harrington 1961). Moreover, in equations (2.4.26) and (2.4.27), I denotes the complex amplitude of the current density vector $\mathbf{J}_0(\mathbf{r}) = I\delta(\rho)\hat{\mathbf{z}}$.

It is interesting to observe that, for large arguments, these expressions can be approximated by their asymptotic expressions as

$$\mathbf{E}(\mathbf{r}) \approx -\eta I \sqrt{\frac{jk}{8\pi\rho}} e^{-jk\rho} \hat{\mathbf{z}}, \qquad (2.4.28)$$

$$\mathbf{H}(\mathbf{r}) \approx I \sqrt{\frac{jk}{8\pi\rho}} e^{-jk\rho} \hat{\boldsymbol{\varphi}}, \qquad (2.4.29)$$

which again show that the electromagnetic field locally behaves like a plane wave in the far-field region.

2.5 VOLUME SCATTERING BY DIELECTRIC TARGETS

In the previous sections we recalled some basic concepts of radiation. When an object—which in this context is also referred to as *target* or *scatterer*—is present in the propagation medium, the wave produced by the source interacts with it and the field distribution is affected by the presence of the scatterer. The situation is schematized in Figure 2.1, which describes a *scattering* configuration involving only dielectrics or materials with a finite electric conductivity.

In this context, let us assume the object to be characterized by ε and μ and immersed in a homogeneous and infinite medium (free-space scattering) characterized by ε_b and μ_b. Since the scatterer may be inhomogeneous, its dielectric parameters are in general dependent on $\mathbf{r}$.

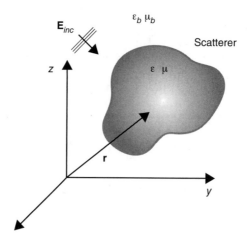

FIGURE 2.1 Electromagnetic scattering by a three-dimensional inhomogeneous target.

The perturbed field (which is the only field that can be *measured* in the presence of the object) is indicated by **E** and **H**. This field is clearly different from the field generated by the source when the object in not present, which is usually indicated as the *unperturbed* or *incident* field and denoted by $\mathbf{E}_{inc}$ and $\mathbf{H}_{inc}$. The incident field is a known quantity if the source is completely characterized, and it can be computed everywhere by using relations (2.4.5) and (2.4.6). Moreover, we can write

$$\mathbf{E}_{scat}(\mathbf{r}) = \mathbf{E}(\mathbf{r}) - \mathbf{E}_{inc}(\mathbf{r}), \quad (2.5.1)$$

$$\mathbf{H}_{scat}(\mathbf{r}) = \mathbf{H}(\mathbf{r}) - \mathbf{H}_{inc}(\mathbf{r}), \quad (2.5.2)$$

where the *difference* between the perturbed field (i.e., the field when the object is present) and the unperturbed incident field (i.e., the field when the object is not present) is called the *scattered* field and can be ascribed to the presence of the object and, in particular, to the interaction between the incident field and the object itself. Since equations (2.5.1) and (2.5.2) can be immediately rewritten as

$$\mathbf{E}(\mathbf{r}) = \mathbf{E}_{inc}(\mathbf{r}) + \mathbf{E}_{scat}(\mathbf{r}), \quad (2.5.3)$$

$$\mathbf{H}(\mathbf{r}) = \mathbf{H}_{inc}(\mathbf{r}) + \mathbf{H}_{scat}(\mathbf{r}), \quad (2.5.4)$$

the perturbed fields **E** and **H** are usually indicated as the *total* fields.

Two situations usually occur. In the first one, the object is completely known and one has to compute the perturbed fields. This is called a *direct scattering problem*. For the free-space configuration in Figure 2.1, the following quantities are assumed as known quantities in the direct scattering problem: the incident fields $\mathbf{E}_{inc}$ and $\mathbf{H}_{inc}$ (for any **r** inside and outside the object); the values of ε_b and μ_b; the space region V_o occupied by the object; and the distributions of the dielectric parameters of the object, $\varepsilon(\mathbf{r})$ and $\mu(\mathbf{r})$, $\mathbf{r} \in V_o$. The goal is the computation of the scattered fields $\mathbf{E}_{scat}$ and $\mathbf{H}_{scat}$ everywhere. From these fields, one can immediately deduce the total fields **E** and **H** by using (2.5.3) and (2.5.4).

In the second situation considered, which is of paramount importance for this book, the object is unknown and one has to deduce information on it from some measurements of the perturbed field generally collected outside the object. This is called an *inverse scattering problem*.

For the free-space configuration considered, in the inverse scattering problem, the following quantities are still assumed known: the incident fields $\mathbf{E}_{inc}$ and $\mathbf{H}_{inc}$ (for any **r** inside and outside the object) and the values of ε_b and μ_b. On the contrary, V_o and the distributions of $\varepsilon(\mathbf{r})$ and $\mu(\mathbf{r})$, $\mathbf{r} \in V_o$, are now unknown quantities. Moreover, it is assumed that the total electric fields, **E** and **H**, are known quantities (e.g., obtained by suitable measurements) only for $\mathbf{r} \notin V_o$. In this case, the objective of the computation is the definition of the object support V_o and the reconstruction of $\varepsilon(\mathbf{r})$ and $\mu(\mathbf{r})$, $\mathbf{r} \in V_o$. It is evident that in practice it is impossible to measure **E** and **H** for any **r** outside the object, and so these vectors are usually available at a discrete set of points.

Furthermore, in certain applicative scenarios, some information about the object, referred to as a priori information, could be available. This information can be useful in limiting the space wherein the solution is searched for and to improve the reconstruction output.

With regard to practical applications, it is very important to observe that reconstruction of the complete distributions of the dielectric parameters of the unknown objects may be a of limited interest, since far fewer details may be required (e.g., in nondestructive testing, retrieving the position and/or shape of a defect in an otherwise known object could be the only information needed). Although these considerations will be discussed in the following chapters, in order to describe both the direct and inverse scattering problems in greater depth, it is necessary to deduce equations relating the measured values of the electromagnetic field to the properties of the scatterer under test. Such relations can be obtained by using the volume equivalence principle, described in the next paragraph.

2.6 VOLUME EQUIVALENCE PRINCIPLE

Let us consider the configuration shown in Figure 2.2a. Since the object actually is a *discontinuity* in the propagation medium, it is opportune to formulate the problem by using the integral form of Maxwell's equations. To this end, by applying equations (2.1.14) and (2.1.15), using an arbitrary open surface S whose contour is represented by a line C, for a linear isotropic medium, one obtains

$$\oint_C \mathbf{E}(\mathbf{r}) \cdot d\mathbf{l} = -j\omega \int_S \mu(\mathbf{r}) \mathbf{H}(\mathbf{r}) \cdot \hat{\mathbf{n}} \, ds, \qquad (2.6.1)$$

$$\oint_C \mathbf{H}(\mathbf{r}) \cdot d\mathbf{l} = j\omega \int_S \varepsilon(\mathbf{r}) \mathbf{E}(\mathbf{r}) \cdot \hat{\mathbf{n}} \, ds + \int_S \mathbf{J}_0(\mathbf{r}) \cdot \hat{\mathbf{n}} \, ds. \qquad (2.6.2)$$

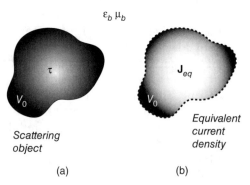

FIGURE 2.2 Volume equivalence principle. (a) real configuration; (b) equivalent problem with the equivalent current density.

VOLUME EQUIVALENCE PRINCIPLE

Analogously, the incident field (i.e., the field when the object is not present) satisfies the following equations

$$\oint_C \mathbf{E}_{inc}(\mathbf{r}) \cdot d\mathbf{l} = -j\omega \int_S \mu_b \mathbf{H}_{inc}(\mathbf{r}) \cdot \hat{\mathbf{n}} \, ds, \qquad (2.6.3)$$

$$\oint_C \mathbf{H}_{inc}(\mathbf{r}) \cdot d\mathbf{l} = j\omega \int_S \varepsilon_b \mathbf{E}_{inc}(\mathbf{r}) \cdot \hat{\mathbf{n}} \, ds + \int_S \mathbf{J}_0(\mathbf{r}) \cdot \hat{\mathbf{n}} \, ds. \qquad (2.6.4)$$

Subtracting (2.6.3) from (2.6.1) and (2.6.4) from (2.6.2), yields

$$\oint_C [\mathbf{E}(\mathbf{r}) - \mathbf{E}_{inc}(\mathbf{r})] \cdot d\mathbf{l} = -j\omega \int_S [\mu(\mathbf{r})\mathbf{H}(\mathbf{r}) - \mu_b \mathbf{H}_{inc}(\mathbf{r})] \cdot \hat{\mathbf{n}} \, ds, \qquad (2.6.5)$$

$$\oint_C [\mathbf{H}(\mathbf{r}) - \mathbf{H}_{inc}(\mathbf{r})] \cdot d\mathbf{l} = j\omega \int_S [\varepsilon(\mathbf{r})\mathbf{E}(\mathbf{r}) - \varepsilon_b \mathbf{E}_{inc}(\mathbf{r})] \cdot \hat{\mathbf{n}} \, ds. \qquad (2.6.6)$$

According to equations (2.5.1) and (2.5.2), we obtain

$$\oint_C \mathbf{E}_{scat}(\mathbf{r}) \cdot d\mathbf{l} = -j\omega \int_S \{\mu(\mathbf{r})\mathbf{H}(\mathbf{r}) - \mu_b [\mathbf{H}(\mathbf{r}) - \mathbf{H}_{scat}(\mathbf{r})]\} \cdot \hat{\mathbf{n}} \, ds, \qquad (2.6.7)$$

$$\oint_C \mathbf{H}_{scat}(\mathbf{r}) \cdot d\mathbf{l} = j\omega \int_S \{\varepsilon(\mathbf{r})\mathbf{E}(\mathbf{r}) - \varepsilon_b [\mathbf{E}(\mathbf{r}) - \mathbf{E}_{scat}(\mathbf{r})]\} \cdot \hat{\mathbf{n}} \, ds. \qquad (2.6.8)$$

By introducing the equivalent sources

$$\mathbf{M}_{eq}(\mathbf{r}) = j\omega[\mu(\mathbf{r}) - \mu_b]\mathbf{H}(\mathbf{r}), \qquad (2.6.9)$$

$$\mathbf{J}_{eq}(\mathbf{r}) = j\omega[\varepsilon(\mathbf{r}) - \varepsilon_b]\mathbf{E}(\mathbf{r}), \qquad (2.6.10)$$

we can rewrite equations (2.6.7) and (2.6.8) as

$$\oint_C \mathbf{E}_{scat}(\mathbf{r}) \cdot d\mathbf{l} = -j\omega \int_S \mu_b \mathbf{H}_{scat}(\mathbf{r}) \cdot \hat{\mathbf{n}} \, ds - \int_S \mathbf{M}_{eq}(\mathbf{r}) \cdot \hat{\mathbf{n}} \, ds, \qquad (2.6.11)$$

$$\oint_C \mathbf{H}_{scat}(\mathbf{r}) \cdot d\mathbf{l} = j\omega \int_S \varepsilon_b \mathbf{E}_{scat}(\mathbf{r}) \cdot \hat{\mathbf{n}} \, ds + \int_S \mathbf{J}_{eq}(\mathbf{r}) \cdot \hat{\mathbf{n}} \, ds. \qquad (2.6.12)$$

The scattered field can then be considered to be generated by an equivalent electric current density and an equivalent magnetic current density, both radiating in free space. According to (2.6.9) and (2.6.10), these sources have supports coinciding with the space region occupied by the object [$\mathbf{M}_{eq}(\mathbf{r}) = 0$ and $\mathbf{J}_{eq}(\mathbf{r}) = 0, \mathbf{r} \notin V_o$].

Equations (2.6.11) and (2.6.12) express the *volume equivalence theorem*, which states that the field scattered by a *real* object is the same as the field produced by the equivalent current densities radiating in free space, provided that such sources are given by (2.6.9) and (2.6.10) (Fig. 2.2b). As expected, $\mathbf{M}_{eq}$ and $\mathbf{J}_{eq}$ depend on the object dielectric properties and the total internal field, which, in turn, depends on the incident field.

In this way, the scattered electric and magnetic fields can be expressed in integral form as

$$\mathbf{E}_{scat}(\mathbf{r}) = j\omega\mu_b \int_{V_o} \mathbf{J}_{eq}(\mathbf{r}') \cdot \bar{\mathbf{G}}(\mathbf{r}/\mathbf{r}') d\mathbf{r}' + \int_{V_o} \nabla \times \mathbf{M}_{eq}(\mathbf{r}') \cdot \bar{\mathbf{G}}(\mathbf{r}/\mathbf{r}') d\mathbf{r}', \qquad (2.6.13)$$

18 ELECTROMAGNETIC SCATTERING

$$\mathbf{H}_{\text{scat}}(\mathbf{r}) = j\omega\varepsilon_b \int_{V_o} \mathbf{M}_{\text{eq}}(\mathbf{r}') \cdot \bar{\mathbf{G}}(\mathbf{r}/\mathbf{r}')d\mathbf{r}' - \int_{V_o} \nabla \times \mathbf{J}_{\text{eq}}(\mathbf{r}') \cdot \bar{\mathbf{G}}(\mathbf{r}/\mathbf{r}')d\mathbf{r}'. \quad (2.6.14)$$

2.7 INTEGRAL EQUATIONS

Let us suppose now that the object of Figure 2.1 is nonmagnetic: $\mu(\mathbf{r}) = \mu_0$, $\mathbf{r} \in V_o$. Accordingly, $\mathbf{M}_{\text{eq}}(\mathbf{r}) = 0$ [equation (2.6.9)] and, on the right-hand sides of equations (2.6.13) and (2.6.14), the terms including the equivalent magnetic current density vanish. Although an expression for the scattered electric field has been deduced, the problem is not solved, since $\mathbf{J}_{\text{eq}}$ is an unknown quantity in both direct and inverse scattering problems. In the former, $\mathbf{J}_{\text{eq}}$ is unknown since $\mathbf{E}$ is to be computed [see equation (2.6.10)], whereas in the latter it is unknown because both $\mathbf{E}$ and ε are unknown functions.

Substituting equation (2.6.13) [with $\mathbf{M}_{\text{eq}}(\mathbf{r}) = 0$] into equation (2.5.3), we have

$$\mathbf{E}(\mathbf{r}) = \mathbf{E}_{\text{inc}}(\mathbf{r}) + j\omega\mu_b \int_{V_o} \mathbf{J}_{\text{eq}}(\mathbf{r}') \cdot \bar{\mathbf{G}}(\mathbf{r}/\mathbf{r}')d\mathbf{r}' \quad (2.7.1)$$

and, by using equation (2.6.10), we obtain

$$\mathbf{E}(\mathbf{r}) = \mathbf{E}_{\text{inc}}(\mathbf{r}) + j\omega\mu_b \int_{V_o} \tau(\mathbf{r}')\mathbf{E}(\mathbf{r}') \cdot \bar{\mathbf{G}}(\mathbf{r}/\mathbf{r}')d\mathbf{r}', \quad (2.7.2)$$

where

$$\tau(\mathbf{r}) = j\omega[\varepsilon(\mathbf{r}) - \varepsilon_b] \quad (2.7.3)$$

is the *object function* or *scattering potential*.

In the direct scattering problem, equation (2.7.2) must be solved for any $\mathbf{r}$. This equation is a Fredholm linear integral equation of the second kind (Morse and Feshbach 1953), in which the only unknown is the total electric field vector $\mathbf{E}$. In practical applications, a solution of this equation can be obtained only using a numerical method. The numerical resolution of the integral equation (2.7.2) consists of two stages. In the first one, equation (2.7.2) is considered for $\mathbf{r} \in V_o$ so that the *internal* total electric field can be computed. Afterward, the *external* total electric field [i.e., $\mathbf{E}(\mathbf{r})$, $\mathbf{r} \notin V_o$] can be easily obtained by computing the integral occurring in equation (2.7.2) by means of a quadrature method, since the right-hand side of (2.7.2) now involves only known functions. A more detailed description of this approach is provided in Chapter 3.

In the inverse scattering problem, $\mathbf{E}(\mathbf{r})$ is assumed to be measurable only for $\mathbf{r} \notin V_o$. In that case, equation (2.7.1) turns out to be a Fredholm linear integral equation of the first kind having the equivalent current density $\mathbf{J}_{\text{eq}}$ as an unknown. Unfortunately, as will be discussed below, knowledge of the equivalent current density generally provides only limited information about the scatterer under test. Consequently, it is usually necessary to reconsider equation (2.7.2). In this case, since the unknown quantities are $\mathbf{E}(\mathbf{r})$ and $\tau(\mathbf{r})$, $\mathbf{r} \in V_o$, the equation to be solved turns out to be nonlinear (Harrington 1961).

2.8 SURFACE SCATTERING BY PERFECTLY ELECTRIC CONDUCTING TARGETS

As recalled from Section 2.2, the electromagnetic field inside a PEC object is zero and a surface current density $\mathbf{J}_S$ is present on the scatterer surface S according to equation (2.2.6). Moreover, when the object is illuminated by an incident field $\mathbf{E}_{\text{inc}}$, the total electric field $\mathbf{E}$ outside the scatterer is still given by equation (2.5.3).

Since the presence of a PEC target does not affect the field radiated by $\mathbf{J}_S$ outside the scatterer (Harrington 1961), the total electric field can be written as

$$\mathbf{E}(\mathbf{r}) = \mathbf{E}_{\text{inc}}(\mathbf{r}) + j\omega\mu_b \int_S \mathbf{J}_S(\mathbf{r}') \cdot \bar{\mathbf{G}}(\mathbf{r}/\mathbf{r}') d\mathbf{r}', \qquad (2.8.1)$$

which, in the inverse scattering problem, represents an integral equation having S and $\mathbf{J}_S$ as the unknowns to be reconstructed.

REFERENCES

Balanis, C. A., *Advanced Engineering Electromagnetics*, Wiley, New York, 1989.

Chew, W. C., *Waves and Fields in Inhomogeneous Media*, Van Nostrand Reinhold, New York, 1990.

Collin, R. E. and F. J. Zucker, *Antenna Theory*, McGraw-Hill, New York, 1969.

Felsen, L. B. and N. Marcuvitz, *Radiation and Scattering of Waves*, Prentice-Hall, Englewood Cliffs, NJ, 1973.

Harrington, R. F., *Time-Harmonic Electromagnetic Fields*, McGraw-Hill, New York, 1961.

Jones, D. S., *The Theory of Electromagnetism*, Macmillan, New York, 1964.

Morse, P. M. and M. Feshbach, *Methods of Theoretical Physics*, McGraw-Hill, New York, 1953.

Stratton, J. A., *Electromagnetic Theory*, McGraw-Hill, New York, 1941.

Tai, C. T., *Dyadic Green's Functions in Electromagnetic Theory*, Intext, Scranton, PA, 1971.

Van Bladel, J., *Electromagnetic Fields*, Wiley, Hoboken, NJ, 2007.

CHAPTER THREE

The Electromagnetic Inverse Scattering Problem

3.1 INTRODUCTION

In the previous chapter, the equations governing the three-dimensional scattering phenomena involving dielectric and conducting materials were derived. A preliminary distinction between *direct* and *inverse* scattering problems has been introduced.

Since this book is devoted to microwave imaging techniques, which are essentially short-range imaging approaches, the scattering equations constitute the foundation for the formulation and the development of the various reconstruction procedures.

The inverse scattering problem considered here belongs to the category of inverse problems (Colton and Kress 1998), which includes many very challenging problems encountered in several applications, including atmospheric sounding, seismology, heat conduction, quantum theory, and medical imaging.

From a strictly mathematical perspective, the definition of a problem as the inverse counterpart of a direct one is completely arbitrary. To this end, it is helpful to recall the following well-known sentence by J. B. Keller (Keller 1976) quoted by Bertero and Boccacci (1998, pp. 1–2): "We call two problems inverses of one another if the formulation of each involves all part of the solution of the other. Often, for historical reasons, one of the two problems has been studied extensively for some time, while the other has never been studied and is not so well understood. In such cases, the former is called a direct problem, while the latter is the inverse problem."

Microwave Imaging, By Matteo Pastorino
Copyright © 2010 John Wiley & Sons, Inc.

On the contrary, from a physical perspective, it is generally recognized that the differences between direct and inverse problems are related to the concepts of cause and effect. In the specific case considered in this book, the direct problem is related to computation of the field that is scattered by a known object (where the interaction between the incident field and the object is the cause of the scattering phenomena), whereas the inverse problem concerns the determination of the object starting from knowledge of the field scattered (the effect of the interaction).

Nowadays, inverse problems, which in the past were considered difficult and "strange" problems, have been widely studied from a mathematical perspective, and several books discuss such problems in depth (e.g., Colton and Kress 1998, Bertero and Boccacci 1998, Engl et al. 1996, Tarantola 1987, Chadan and Sabatier 1992, Herman et al. 1987, Tikhonov and Arsenin 1977).

The most critical aspect of an inverse problem is usually its *ill-posedness*. Following the definition by Hadamard (1902, 1923), a problem is well posed if its solution exists, is unique, and depends continuously on the data. The last property essentially means that a small perturbation of the data results in a small perturbation of the solution. If one of these conditions is not satisfied, the problem is called *ill-posed* or *improperly posed*.

In imaging applications, one measures the scattered field and tries to obtain information on the object subjected to the incident radiation. If, for a given set of measurement values (real data can be affected by noise or measurement errors), there is no object that produce the prescribed field distribution, the problem lacks in existence. Moreover, if two or more different objects produce the same measurement data, the problem solution is not unique. Finally, if two very similar sets of measurements are generated by two significantly different objects, the problem solution does not depend continuously on the data, since small *errors* in the measurements result in large errors in the solution.

The well-posedness of the very general problem of finding $f \in X$, given $g \in Y$, such that

$$Af = g, \qquad (3.1.1)$$

where A is an operator (potentially nonlinear) mapping elements of the normed space X into elements of the normed space Y, depends essentially on the properties of the operator itself.

In particular, the problem turns out to be well posed if the operator A is bijective (i.e., injective and surjective) and the inverse operator A^{-1} such that

$$A^{-1}g = f \qquad (3.1.2)$$

is continuous. *Injectivity* ensures that for any g for which the solution exists, such a solution is unique (uniqueness), whereas *surjectivity* guarantees that there is a solution f for any g (existence). Finally, if A^{-1} is continuous, the solution depends continuously on the data (stability). Obviously, if A does not fulfill all the requirements, described above, the problem is ill-posed.

If A is a completely continuous operator (i.e., an operator that is compact and continuous), then the problem described by equation (3.1.1) can represent an important example of an ill-posed problem. In fact, if A is such an operator, then equation (3.1.1) is ill-posed unless X is of finite dimension.

The proof of this statement (Colton and Kress 1998) can be provided as follows. Assume that A^{-1} exists and is continuous. As a result, $I = A^{-1}A$ (where I is the identity operator in the space X) is compact, since the composition of a continuous and a compact operator is also compact.

Because the identity operator is compact only if the space in which they are defined is of finite dimension, the thesis follows. We note that such a result holds true even if A is a nonlinear operator.

Ill-posedness is a very common property of inverse problems that make them very difficult to solve. *Regularization* procedures are useful tools in controlling ill-posedness. Applying a regularization procedure means replacing the original ill-posed problem with another well-posed problem, in which some additional information can be added. From this new problem one expects to obtain an approximate solution of the original problem. However, adding further information requires some knowledge of the behavior of the solution to the original problem. This information is usually called *a priori information* and can be related, in imaging applications, to the physical nature of the body to be inspected, such as its spatial extension, and/or to the noise level of the measured data.

3.2 THREE-DIMENSIONAL INVERSE SCATTERING

Let us consider Figure 3.1. According to the previous definitions, it is assumed that $\mathbf{E}$ can be measured for $\mathbf{r} \notin V_o$. In particular, $\mathbf{E}$ can be available in an *observation domain* V_m, which corresponds, in practical applications, to the region spanned by the measurement probes (resulting in $V_m \cap V_o = \emptyset$). Moreover, because of the limited information content of the data, equation (2.7.2), for $\mathbf{r} \in V_m$ (hereafter called the *data equation*), is often solved together with the equation that provides the internal field distribution (often called the *state equation*). This relationship is still given by equation (2.7.2), but in this case it is valid for $\mathbf{r} \in V_o$. So the problem solution is reduced to solving the following set of nonlinear integral equations:

$$\mathbf{E}(\mathbf{r}) = \mathbf{E}_{\text{inc}}(\mathbf{r}) + j\omega\mu_b \int_{V_o} \tau(\mathbf{r}') \mathbf{E}(\mathbf{r}') \cdot \bar{\mathbf{G}}(\mathbf{r}/\mathbf{r}') d\mathbf{r}', \quad \mathbf{r} \in V_m \quad \text{(data equation)}, \quad (3.2.1)$$

$$\mathbf{E}(\mathbf{r}) = \mathbf{E}_{\text{inc}}(\mathbf{r}) + j\omega\mu_b \int_{V_o} \tau(\mathbf{r}') \mathbf{E}(\mathbf{r}') \cdot \bar{\mathbf{G}}(\mathbf{r}/\mathbf{r}') d\mathbf{r}', \quad \mathbf{r} \in V_o \quad \text{(state equation)}. \quad (3.2.2)$$

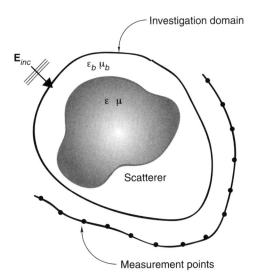

FIGURE 3.1 Imaging configuration for three-dimensional inverse scattering; investigation and observation domains.

Although formally similar, these equations are, of course, very different, since the left-hand side of equation (3.2.1) (which is a Fredholm equation of the first kind) is a known quantity, whereas the corresponding term in equation (3.2.2) (which is a Fredholm equation of the second kind) is unknown.

Formulated in this way, one can view the resolution of the inverse problem as the search for the object that produces a prescribed scattered field distribution (equal to the one that has been measured in the observation domain), but, at the same time, *produces* an internal field distribution consistent with the known incident field inside the object itself. It is simply the presence of the known internal field [and, consequently, the constraint imposed by the requirement of fulfilling equation (3.2.2)] that distinguishes the inverse scattering problem so sharply (also in terms of well-posedness issues) from the inverse source problem (i.e., the retrieval of a source from the field that it radiates). This point is discussed further, for example, by Bleinsten and Cohen (1977), Devaney (1978), Hoenders (1978), and Stone (1987) and in the books cited in Section 3.1.

It should also be noted that the formulation given above is based entirely on electric field integral equations (EFIEs). However, other formulations for describing the scattering phenomena are available. In microwave imaging, one of the most widely applied is the *contrast source formulation* (van den Berg and Kleinman 1997, van den Berg and Abubakar 2001, Abubakar et al. 2006). In such a framework, the inverse problem is still treated in its nonlinear form, but the problem unknowns are the object function and the equivalent current density [which is directly related to the internal field through equation (2.6.10)].

According to this alternative formulation, the following two integral equations are considered:

$$\mathbf{E}(\mathbf{r}) = \mathbf{E}_{inc}(\mathbf{r}) + j\omega\mu_b \int_{V_o} \mathbf{J}_{eq}(\mathbf{r}') \cdot \overline{\mathbf{G}}(\mathbf{r}/\mathbf{r}') d\mathbf{r}', \quad \mathbf{r} \in V_m \quad \text{(data equation)}, \quad (3.2.3)$$

$$\mathbf{J}_{eq}(\mathbf{r}) = \tau(\mathbf{r})\mathbf{E}_{inc}(\mathbf{r}) + j\omega\mu_b\tau(\mathbf{r}) \int_{V_o} \mathbf{J}_{eq}(\mathbf{r}') \cdot \overline{\mathbf{G}}(\mathbf{r}/\mathbf{r}') d\mathbf{r}', \quad \mathbf{r} \in V_o \quad \text{(state equation)}.$$
$$(3.2.4)$$

As mentioned, the problems unknowns are τ and $\mathbf{J}_{eq}$, which vanish outside V_o.

It is also worth noting that, in several applications, the external shape of the object is known and the integration domain is therefore known. In other applications, the shape of the unknown object is itself a problem unknown. In those cases, it is natural to define an *investigation domain* (a *test region*) V_i, which by definition includes the support of the scatterer under test ($V_o \subset V_i$).

In order to retrieve the shape of the scatterer, specific methods can be used (see Chapter 5). However, as a general rule, it is possible to assume that the unknown object to be inspected coincides with the investigation domain; that is, the support of $\tau(\mathbf{r})$ is extended to all $\mathbf{r} \in V_i$ and the integrals in equations (3.2.1) and (3.2.2) [or in equations (3.2.3) and (3.2.4)] are defined over V_i instead of V_o. Clearly, a perfect reconstruction (i.e., a successful solution of the inverse scattering problem) would yield $\tau(\mathbf{r}) = 0$, for $\mathbf{r} \in (V_i \backslash V_o)$, and so it would be possible to precisely define the actual object shape.

3.3 TWO-DIMENSIONAL INVERSE SCATTERING

The scattering formulation reported in the previous section concerns three-dimensional configurations. In fact, although microwave imaging techniques can in principle be applied to three-dimensional configurations without theoretical limitations, most of the approaches proposed so far in the scientific literature are still related to two-dimensional problems. In fact, the imaging of two-dimensional structures can be simplified by means of some assumptions regarding the scatterer under test and the illumination system considered. On the contrary, fully three-dimensional approaches can still be considered as preliminary proposals. Nevertheless, in the following chapters, general formulations will be discussed in a three-dimensional framework, whereas some approaches proposed for two-dimensional configurations will be described with reference to their specific imaging modalities.

If the object to be inspected has an elongate shape with respect to the space region illuminated by the source, it can be approximated as an infinite cylinder (see Fig. 3.2). This is an assumption that should be carefully verified in each application. However, under this approximation, the cross section of the cylinder can be assumed to be independent of one of the spatial coordinates (in Fig. 3.2, the z coordinate), and we obtain

$$\varepsilon(\mathbf{r}) = \varepsilon_0 \varepsilon_r(\mathbf{r}_t), \qquad (3.3.1)$$

$$\mu(\mathbf{r}) = \mu_0 \mu_r(\mathbf{r}_t), \qquad (3.3.2)$$

where $\mathbf{r}_t$ is the transversal component of $\mathbf{r}$, such that (in Cartesian coordinates)

$$\mathbf{r} = x\hat{\mathbf{x}} + y\hat{\mathbf{y}} + z\hat{\mathbf{z}} = \mathbf{r}_t + z\hat{\mathbf{z}}. \qquad (3.3.3)$$

Moreover, if we assume that $\mathbf{E}_{\text{inc}}(\mathbf{r}) = E_{\text{inc}_z}(\mathbf{r}_t)\hat{\mathbf{z}}$, that is, that the incident field is z-polarized and uniform along z [*transverse magnetic* incident field (TM)], for symmetry reasons both the scattered electric field and the total electric field turn out to be independent of z and z-polarized fields [i.e., $\mathbf{E}_{\text{scat}}(\mathbf{r}) = E_{\text{scat}_z}(\mathbf{r}_t)\hat{\mathbf{z}}$ and $\mathbf{E}(\mathbf{r}) = E_z(\mathbf{r}_t)\hat{\mathbf{z}}$]. In this case, equation (2.7.2) can be rewritten as

$$\mathbf{E}(\mathbf{r}_t) = \mathbf{E}_{\text{inc}}(\mathbf{r}_t) + j\omega\mu_b \int_{S_o} \int_{-\infty}^{\infty} \tau(\mathbf{r}'_t) \mathbf{E}(\mathbf{r}'_t) \cdot \bar{\mathbf{G}}(\mathbf{r}_t/\mathbf{r}') dz' d\mathbf{r}'_t, \qquad (3.3.4)$$

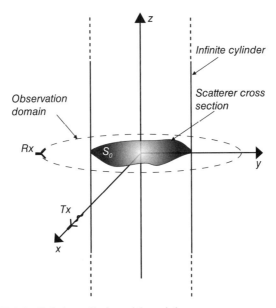

FIGURE 3.2 Infinite cylinder with an inhomogeneous cross section.

where S_o is the cross section of the cylindrical object. By using (2.4.7), we have

$$\mathbf{E}(\mathbf{r}_t) = \mathbf{E}_{\text{inc}}(\mathbf{r}_t) + j\omega\mu_b \int_{S_o} \tau(\mathbf{r}_t') \mathbf{E}(\mathbf{r}_t') \cdot \left(\bar{\mathbf{I}} + \frac{1}{k_b^2}\nabla\nabla\right) \left[\int_{-\infty}^{\infty} \frac{-e^{-jk_b|\mathbf{r}_t-\mathbf{r}'|}}{4\pi|\mathbf{r}_t-\mathbf{r}'|} dz'\right] d\mathbf{r}_t'. \quad (3.3.5)$$

Moreover, since the following relation holds (Balanis 1989)

$$-\frac{1}{4\pi} \int_{-\infty}^{+\infty} \frac{e^{-jk_b|\mathbf{r}_t-\mathbf{r}'|}}{|\mathbf{r}_t-\mathbf{r}'|} dz = \frac{j}{4} H_0^{(2)}(k_b|\mathbf{r}_t-\mathbf{r}_t'|), \quad (3.3.6)$$

from equation (3.3.5), one obtains

$$E_z(\mathbf{r}_t) = E_{\text{inc}_z}(\mathbf{r}_t) - \frac{\omega\mu_b}{4} \int_{S_o} \tau(\mathbf{r}_t') E_z(\mathbf{r}_t') H_0^{(2)}(k_b|\mathbf{r}_t-\mathbf{r}_t'|) d\mathbf{r}_t', \quad (3.3.7)$$

or

$$E_z(\mathbf{r}_t) = E_{\text{inc}_z}(\mathbf{r}_t) + j\omega\mu_b \int_{S_o} \tau(\mathbf{r}_t') E_z(\mathbf{r}_t') G_{2D}(\mathbf{r}_t/\mathbf{r}_t') d\mathbf{r}_t', \quad (3.3.8)$$

where

$$G_{2D}(\mathbf{r}_t/\mathbf{r}_t') = \frac{j}{4} H_0^{(2)}(k_b|\mathbf{r}_t-\mathbf{r}_t'|) \quad (3.3.9)$$

is the free-space Green function for two-dimensional problems. By using the integral representation of the Hankel function, which will be useful in development of the diffraction tomography algorithm (Section 5.2), this can also be written as (Morse and Feshback 1953)

$$G_{2D}(\mathbf{r}_t/\mathbf{r}_t') = -\frac{j}{2} \int_{-\infty}^{+\infty} \frac{e^{-j\sqrt{k_b^2-4\pi^2\lambda^2}|y-y'|}}{\sqrt{k_b^2-4\pi^2\lambda^2}} e^{-j2\pi\lambda(x-x')} d\lambda. \quad (3.3.10)$$

It is important to note that these assumptions result in a greatly simplified formulation, since the inverse scattering problem turns out to be a two-dimensional scalar problem.

Furthermore, as in the three-dimensional case, the two-dimensional inverse scattering problem under transverse magnetic illumination conditions can be formulated in terms of the following two nonlinear scalar integral equations

$$E_z(\mathbf{r}_t) = E_{\text{inc}_z}(\mathbf{r}_t) + j\omega\mu_b \int_{S_o} \tau(\mathbf{r}_t') E_z(\mathbf{r}_t') G_{2D}(\mathbf{r}_t/\mathbf{r}_t') d\mathbf{r}_t', \quad \mathbf{r}_t \in S_m \quad \text{(data equation)}, \quad (3.3.11)$$

$$E_z(\mathbf{r}_t) = E_{\text{inc}_z}(\mathbf{r}_t) + j\omega\mu_b \int_{S_o} \tau(\mathbf{r}_t') E_z(\mathbf{r}_t') G_{2D}(\mathbf{r}_t/\mathbf{r}_t') d\mathbf{r}_t', \quad \mathbf{r}_t \in S_o \quad \text{(state equation)}, \tag{3.3.12}$$

where S_m is the observation domain ($S_m \cap S_o = \emptyset$) where the measurements are performed. This domain is usually a circle or a segment (*probing line*) along which the probes are located or moved (see Chapter 4). Clearly, (3.3.11) and (3.3.12) are the two-dimensional counterparts of (3.2.1) and (3.2.2) of the three-dimensional case. Analogously, if the contrast source approach is used, the two-dimensional counterparts of equations (3.2.3) and (3.2.4) are

$$E_z(\mathbf{r}_t) = E_{\text{inc}_z}(\mathbf{r}_t) + j\omega\mu_b \int_{S_o} J_{\text{eq}_z}(\mathbf{r}_t') G_{2D}(\mathbf{r}_t/\mathbf{r}_t') d\mathbf{r}_t', \quad \mathbf{r}_t \in S_m \quad \text{(data equation)}, \tag{3.3.13}$$

$$J_{\text{eq}_z}(\mathbf{r}_t) = \tau(\mathbf{r}_t) E_{\text{inc}_z}(\mathbf{r}_t) \\ + j\omega\mu_b \tau(\mathbf{r}_t) \int_{S_o} J_{\text{eq}_z}(\mathbf{r}_t) G_{2D}(\mathbf{r}_t/\mathbf{r}_t') d\mathbf{r}_t', \quad \mathbf{r} \in S_o \quad \text{(state equation)}, \tag{3.3.14}$$

where

$$J_{\text{eq}_z}(\mathbf{r}_t) = \tau(\mathbf{r}_t) E_z(\mathbf{r}_t) \tag{3.3.15}$$

is the z component of the equivalent current density.

It should be mentioned that, when the incident field is a *transverse electric* (TE) field [i.e., $\mathbf{E}_{\text{inc}}(\mathbf{r}_t) = E_{\text{inc}_x}(\mathbf{r}_t)\hat{\mathbf{x}} + E_{\text{inc}_y}(\mathbf{r}_t)\hat{\mathbf{y}}$], the problem is still two-dimensional but the vector nature of the equations is conserved. The TE illumination has been considered by some authors (e.g., Chiu and Liu 1996, Cho and Kiang 1999, Ramananjoana et al. 2001, Qing 2002). In certain cases this illumination condition can provide better results, due to the increased information contained in the measured samples of the scattered electric field. However, the simplification inherent to the transverse magnetic illumination is partially lost.

Moreover, in exactly the same way as described for the three-dimensional case in Section 3.2, if the cross section of the object is unknown, an *investigation area* (a test region) S_i can be defined, which includes the cross section of the scatterer under test ($S_o \subset S_i$). In this case, the integrals in (3.3.11)–(3.3.14) are now extended to S_i instead of S_o. In this case, too, a perfect reconstruction would yield $\tau(\mathbf{r}_t) = 0$ for $\mathbf{r}_t \in (S_i \backslash S_o)$, allowing the definition of the object cross section.

Finally, if the infinite cylinder is a PEC one, then, under the *transverse magnetic* (TM) illumination condition, equation (2.8.1) can be rewritten as

$$E_z(\mathbf{r}_t) = E_{\text{inc}_z}(\mathbf{r}_t) + j\omega\mu_b \oint_G J_{S_z}(\mathbf{r}_t') G_{2D}(\mathbf{r}_t/\mathbf{r}_t') d\mathbf{r}', \tag{3.3.16}$$

where G is a closed line that determines the object profile in the transverse plane and is the contour of the object cross section in that plane. Moreover,

J_{S_z} is the z component of the surface current density $\mathbf{J}_S$ (see Section 2.8) and is defined on G.

3.4 DISCRETIZATION OF THE CONTINUOUS MODEL

In practical applications, the continuous model must be discretized. Several numerical methods can be applied to solve the equations involved in scattering problems. The most widely used are the method of moments (MoM), the finite-element method (FEM), and the finite-difference (FD) methods. These methods can be implemented with reference to several different schemes, and a plethora of modified and hybrid techniques can be adopted. Detailed description of these approaches is outside the scope of the present monograph, and the reader is referred to specialized books (e.g., Harrington 1968, Moore and Pizer 1984, Wang 1991, Zienkiewicz 1977, Chari and Silvester 1980, Chew et al. 2001, Mittra 1973, Press et al. 1992, Bossavit 1998, Jin 2002, Monk 2003, Taflove and Hagness 2005, Sullivan 2000, Kunz and Luebbers 1993, and references cited therein). However, for illustration purposes only, and in order to define some quantities used in the following chapters, a straightforward discretization, often used to obtain pixelated representations (images) of the original and reconstructed dielectric distributions, is presented, which is in principle based on application of MoM to a two-dimensional dielectric configuration under TM illumination. With this goal in mind, let us consider equation (3.3.8) and search for an approximate numerical solution. The problem unknowns can be expanded in a set of N basis function $f_n(\mathbf{r}_t)$, such that

$$\tau(\mathbf{r}_t) = \sum_{n=1}^{N} \tau_n f_n(\mathbf{r}_t) \qquad (3.4.1)$$

and

$$E_z(\mathbf{r}_t) = \sum_{n=1}^{N} E_n f_n(\mathbf{r}_t). \qquad (3.4.2)$$

In order to obtain the simplest partitioning of the investigation domain S_i, piecewise constant representations of the unknowns can be used. To this end, one chooses $f_n(\mathbf{r}_t)$ such that $f_n(\mathbf{r}_t) = 1$ if $\mathbf{r}_t \in S_n$, where S_n is the nth subdomain of the partitioned investigation domain (i.e., $\cup_n S_n = S_i$), and $f_n(\mathbf{r}_t) = 0$, elsewhere. Moreover, an inner product must be introduced. Typically, for complex functions, the following product is considered (Harrington 1968)

$$<u(\mathbf{r}_t), v(\mathbf{r}_t)> = \int_S u(\mathbf{r}_t') v^*(\mathbf{r}_t') d\mathbf{r}_t', \qquad (3.4.3)$$

where u and v are two generic functions of $\mathbf{r}_t$ having S as domain, and v^* denotes the complex conjugate of v. Considering the discrete nature of the measurements in imaging applications, one can assume that the values of the scattered field are available in a set of M measurement points where the acquisition probes are located. These conditions suggest the use, as testing functions, of a set of Dirac delta functions, such that

$$w_m(\mathbf{r}_t) = \delta(\mathbf{r}_t - \mathbf{r}_m), \qquad (3.4.4)$$

where $\mathbf{r}_m$, $m = 1,\ldots, M$, is the mth measurement point. By substituting (3.4.1) and (3.4.2) in equation (3.3.8) and sequentially *multiplying* [by means of the inner product of equation (3.4.3)] the resulting equation by each testing function, one obtains

$$\sum_{n=1}^{N} h_{mn} \tau_n E_n = E_m^s, \quad m = 1,\ldots, M, \qquad (3.4.5)$$

where

$$h_{mn} = j\omega\mu_b \int_{S_n} G_{2D}(\mathbf{r}_m/\mathbf{r}_t') d\mathbf{r}_t' \qquad (3.4.6)$$

and

$$E_m^s = E_m - E_m^i \equiv E_z(\mathbf{r}_m) - E_{\text{inc}_z}(\mathbf{r}_m), \qquad (3.4.7)$$

where E_m^s, E_m, and E_m^i respectively are the z components of the scattered, total, and incident fields at the mth measurement point. In order to obtain equation (3.4.5), it has been assumed that the integral in equation (3.4.3) is extended to an infinite domain [this is clearly possible since $\tau(\mathbf{r}_t) = 0$ for $\mathbf{r}_t \notin S$] and the well-known properties of the Dirac delta functions have been exploited.

As a result, the problem solution is reduced to the resolution of a set of M nonlinear algebraic equations given by (3.4.5). Equation (3.4.5) can also be expressed in matrix form as

$$[\mathbf{H}][\mathbf{T}]\mathbf{e} = \mathbf{e}^s, \qquad (3.4.8)$$

where $\mathbf{e} = [E_1,\ldots, E_N]^T$, $\mathbf{e}^s = [E_1^s,\ldots, E_M^s]^T$; $[\mathbf{H}]$ is a rectangular $M \times N$ matrix, whose elements are the coefficients h_{mn}, $m = 1,\ldots, M$, $n = 1,\ldots, N$; and $[\mathbf{T}]$ is a square $N \times N$ diagonal matrix whose diagonal elements are given by τ_n, $n = 1,\ldots, N$.

It should be noted that the coefficients h_{mn} can be numerically computed. A very simple expression, widely used in imaging approaches and sufficiently accurate for two-dimensional TM scattering, is obtained by approximating the subdomains S_n, $n = 1,\ldots, N$, by circles of equivalent areas (Richmond 1965). In this case

$$h_{mn} = \frac{j}{2}\pi k a_n J_1(k_b a_n) H_0^{(2)}(k|\mathbf{r}_m - \mathbf{r}_n|), \qquad (3.4.9)$$

where $a_n = \sqrt{\Delta S_n/\pi}$ is the radius of the equivalent circle, ΔS_n is the area of S_n, and $\mathbf{r}_n$, $n = 1, \ldots, N$, is the barycenter of S_n.

As mentioned previously, the discretization procedure presented above is very simple. Various different approaches can be followed. For example, several different basis and testing functions can be used. The objective is usually to minimize the number of problem unknowns, which has a direct impact on the computational load of the method. It is also worth noting that the selected basis and testing functions are required to allow a simple computation of the coefficients and the known terms of the discretized equations.

It is also evident that the various above mentioned numerical methods can be applied not only to the two-dimensional equations considered here but also to the three-dimensional formulation of Section 3.2. A detailed example is given below. (This procedure has been applied for the numerical simulations presented throughout this book.) For three-dimensional scattering by isotropic dielectric bodies, the integral equation (2.7.2) can be immediately rewritten as follows (Zhang et al. 2003):

$$\mathbf{E}(\mathbf{r}) = \mathbf{E}_{\text{inc}}(\mathbf{r}) + j\omega\mu_b \int_{V_o} \tau(\mathbf{r}')\mathbf{E}(\mathbf{r}')G(\mathbf{r}/\mathbf{r}')\,d\mathbf{r}'$$
$$+ \frac{j\omega\mu_b}{k_b^2} \nabla\nabla \cdot \int_{V_o} \tau(\mathbf{r}')\mathbf{E}(\mathbf{r}')G(\mathbf{r}/\mathbf{r}')\,d\mathbf{r}'. \qquad (3.4.10)$$

This integrodifferential equation relating the electric field and the scattering potential is often preferred to the EFIE since it avoids the problems concerning the singularities of Green's tensor (Van Bladel 2007). For development of the numerical method, it is then useful to introduce the vector field (Zhang et al. 2003)

$$\mathbf{A}(\mathbf{r}) = -\frac{j\omega\mu_b}{k_b^2} \int_{V_o} \tau(\mathbf{r}')\mathbf{E}(\mathbf{r}')G(\mathbf{r}/\mathbf{r}')\,d\mathbf{r}', \qquad (3.4.11)$$

where $V_o = \{(x, y, z) \in \mathfrak{R}^3 : x_1 \leq x \leq x_2,\ y_1 \leq y \leq y_2,\ z_1 \leq z \leq z_2\}$ is assumed for convenience to be a parallelepiped containing the support of the scatterers. As a consequence, equation (3.4.10) can be written as

$$\mathbf{E}(\mathbf{r}) = \mathbf{E}_{\text{inc}}(\mathbf{r}) - k_b^2 \mathbf{A}(\mathbf{r}) - \nabla\nabla \cdot \mathbf{A}(\mathbf{r}). \qquad (3.4.12)$$

The first step in the implementation of the method consists in discretizing region V_0 into $N = I \times J \times K$ parallelepipeds with faces parallel to coordinate directions and centers located at

$$\mathbf{r}_{ijk} = \left[x_1 + \left(i - \frac{1}{2}\right)\Delta x\right]\hat{\mathbf{x}} + \left[y_1 + \left(j - \frac{1}{2}\right)\Delta y\right]\hat{\mathbf{y}} + \left[z_1 + \left(k - \frac{1}{2}\right)\Delta z\right]\hat{\mathbf{z}}, \qquad (3.4.13)$$

for $i = 1,\ldots, I, j = 1,\ldots, J, k = 1,\ldots, K$, where

$$\Delta x = \frac{x_2 - x_1}{I}, \quad \Delta y = \frac{y_2 - y_1}{J}, \quad \Delta z = \frac{z_2 - z_1}{K}, \quad (3.4.14)$$

are the lengths of the sides of the cells. In such a way, a grid of points $G = \{\mathbf{r}_{ijk} \in \Re^3, i = 1,\ldots, I, j = 1,\ldots, J, k = 1,\ldots, K\}$ is introduced.

If equation (3.4.12) is enforced at each point of such a grid [i.e., Dirac delta functions located at the center of each cell are used again as testing functions (Harrington 1968)], it must hold that

$$\mathbf{E}_{ijk} = \mathbf{E}_{ijk}^{\text{inc}} - k_b^2 \mathbf{A}_{ijk} - (\nabla\nabla \cdot \mathbf{A})_{ijk}, \quad (3.4.15)$$

where $\mathbf{E}_{ijk} = \mathbf{E}(\mathbf{r}_{ijk})$, $\mathbf{E}_{ijk}^{\text{inc}} = \mathbf{E}_{\text{inc}}(\mathbf{r}_{ijk})$, $\mathbf{A}_{ijk} = \mathbf{A}(\mathbf{r}_{ijk})$, and $(\nabla\nabla \cdot \mathbf{A})_{ijk}$ is the value of $\nabla\nabla \cdot \mathbf{A}$ evaluated at point $\mathbf{r}_{ijk}$. In order to express $(\nabla\nabla \cdot \mathbf{A})_{ijk}$ in terms of the values of $\mathbf{A}$ computed at $\mathbf{r}_{ijk}$, a finite-difference scheme is used to approximate the vector differential operator $\nabla\nabla \cdot$, which in a Cartesian frame can be represented as

$$\nabla\nabla \cdot \mathbf{A} = (\nabla\nabla \cdot \mathbf{A})_x \hat{\mathbf{x}} + (\nabla\nabla \cdot \mathbf{A})_y \hat{\mathbf{y}} + (\nabla\nabla \cdot \mathbf{A})_z \hat{\mathbf{z}}, \quad (3.4.16)$$

where

$$(\nabla\nabla \cdot \mathbf{A})_x = \frac{\partial^2 A_x}{\partial x^2} + \frac{\partial^2 A_y}{\partial x \partial y} + \frac{\partial^2 A_z}{\partial x \partial z}, \quad (3.4.17)$$

$$(\nabla\nabla \cdot \mathbf{A})_y = \frac{\partial^2 A_x}{\partial y \partial x} + \frac{\partial^2 A_y}{\partial y^2} + \frac{\partial^2 A_z}{\partial y \partial z}, \quad (3.4.18)$$

$$(\nabla\nabla \cdot \mathbf{A})_z = \frac{\partial^2 A_x}{\partial z \partial x} + \frac{\partial^2 A_y}{\partial z \partial y} + \frac{\partial^2 A_z}{\partial z^2}, \quad (3.4.19)$$

where A_x, A_y, and A_z are the Cartesian components of the vector field $\mathbf{A}$.

The adopted finite-difference scheme is based on two different approximations of the first-order derivatives. Namely, for computation of the second order-derivatives $\partial^2/\partial p^2$, $p = x, y, z$, the following expressions are used twice

$$\frac{\partial A_p(x, y, z)}{\partial x} \approx \frac{A_p\left(x + \frac{\Delta x}{2}, y, z\right) - A_p\left(x - \frac{\Delta x}{2}, y, z\right)}{\Delta x}, \quad p = x, y, z, \quad (3.4.20)$$

$$\frac{\partial A_p(x, y, z)}{\partial y} \approx \frac{A_p\left(x, y + \frac{\Delta y}{2}, z\right) - A_p\left(x, y - \frac{\Delta y}{2}, z\right)}{\Delta y}, \quad p = x, y, z, \quad (3.4.21)$$

$$\frac{\partial A_p(x,y,z)}{\partial z} \approx \frac{A_p\left(x,y,z+\frac{\Delta z}{2}\right) - A_p\left(x,y,z-\frac{\Delta z}{2}\right)}{\Delta z}, \quad p = x, y, z, \quad (3.4.22)$$

whereas the approximations used for computing the second-order mixed derivatives are as follows:

$$\frac{\partial A_p(x,y,z)}{\partial x} \approx \frac{A_p(x+\Delta x, y, z) - A_p(x-\Delta x, y, z)}{2\Delta x}, \quad p = x, y, z, \quad (3.4.23)$$

$$\frac{\partial A_p(x,y,z)}{\partial y} \approx \frac{A_p(x, y+\Delta y, z) - A_p(x, y-\Delta y, z)}{2\Delta y}, \quad p = x, y, z, \quad (3.4.24)$$

$$\frac{\partial A_p(x,y,z)}{\partial z} \approx \frac{A_p(x, y, z+\Delta z) - A_p(x, y, z-\Delta z)}{2\Delta z}. \quad p = x, y, z. \quad (3.4.25)$$

As a result, one obtains

$$\begin{aligned}\frac{\partial^2 A_p(x,y,z)}{\partial^2 x} &\approx \frac{\dfrac{\partial A_p\left(x+\frac{\Delta x}{2}, y, z\right)}{\partial x} - \dfrac{\partial A_p\left(x-\frac{\Delta x}{2}, y, z\right)}{\partial x}}{\Delta x} \\ &\approx \frac{\dfrac{A_p(x+\Delta x, y, z) - A_p(x, y, z)}{\Delta x} - \dfrac{A_p(x, y, z) - A_p(x-\Delta x, y, z)}{\Delta x}}{\Delta x} \\ &= \frac{A_p(x+\Delta x, y, z) - 2A_p(x, y, z) + A_p(x-\Delta x, y, z)}{\Delta x^2}, \quad p = x, y, z,\end{aligned}$$
(3.4.26)

and

$$\begin{aligned}\frac{\partial^2 A_p(x,y,z)}{\partial x \partial y} &\approx \frac{\dfrac{\partial A_p\left(x+\frac{\Delta x}{2}, y, z\right)}{\partial y} - \dfrac{\partial A_p\left(x-\frac{\Delta x}{2}, y, z\right)}{\partial y}}{\Delta x} \\ &\approx \frac{\dfrac{A_p(x+\Delta x, y+\Delta y, z) - A_p(x+\Delta x, y-\Delta y, z)}{2\Delta y}}{2\Delta x} \\ &\quad - \frac{\dfrac{A_p(x-\Delta x, y+\Delta y, z) - A_p(x-\Delta x, y-\Delta y, z)}{2\Delta y}}{2\Delta x} \\ &= \frac{A_p(x+\Delta x, y+\Delta y, z) - A_p(x+\Delta x, y-\Delta y, z)}{4\Delta x \Delta y} \\ &\quad - \frac{A_p(x-\Delta x, y+\Delta y, z) - A_p(x-\Delta x, y-\Delta y, z)}{4\Delta x \Delta y}, \quad p = x, y, z,\end{aligned}$$
(3.4.27)

DISCRETIZATION OF THE CONTINUOUS MODEL   33

and analogous relations for the other derivatives. Accordingly, the finite-difference approximation of the components of the differential operator $\nabla\nabla\cdot$ can be written as

$$(\nabla\nabla\cdot\mathbf{A})_x \approx \frac{A_x(x+\Delta x, y, z)-2A_x(x, y, z)+A_x(x-\Delta x, y, z)}{\Delta x^2}$$
$$+\frac{A_y(x+\Delta x, y+\Delta y, z)-A_y(x+\Delta x, y-\Delta y, z)}{4\Delta x\,\Delta y}$$
$$-\frac{A_y(x-\Delta x, y+\Delta y, z)-A_y(x-\Delta x, y-\Delta y, z)}{4\Delta x\,\Delta y}$$
$$+\frac{A_z(x+\Delta x, y, z+\Delta z)-A_z(x+\Delta x, y, z-\Delta z)}{4\Delta x\,\Delta z}$$
$$-\frac{A_z(x-\Delta x, y, z+\Delta z)-A_z(x-\Delta x, y, z-\Delta z)}{4\Delta x\,\Delta z}, \quad (3.4.28)$$

$$(\nabla\nabla\cdot\mathbf{A})_y \approx \frac{A_x(x+\Delta x, y+\Delta y, z)-A_x(x+\Delta x, y-\Delta y, z)}{4\Delta x\,\Delta y}$$
$$-\frac{A_x(x-\Delta x, y+\Delta y, z)-A_x(x-\Delta x, y-\Delta y, z)}{4\Delta x\,\Delta y}$$
$$+\frac{A_y(x, y+\Delta y, z)-2A_y(x, y, z)+A_y(x, y-\Delta y, z)}{\Delta y^2}$$
$$+\frac{A_z(x, y+\Delta y, z+\Delta z)-A_z(x, y+\Delta y, z-\Delta z)}{4\Delta y\,\Delta z}$$
$$-\frac{A_z(x, y-\Delta y, z+\Delta z)-A_z(x, y-\Delta y, z-\Delta z)}{4\Delta y\,\Delta z}, \quad (3.4.29)$$

$$(\nabla\nabla\cdot\mathbf{A})_z \approx \frac{A_x(x+\Delta x, y, z+\Delta z)-A_x(x+\Delta x, y, z-\Delta z)}{4\Delta x\,\Delta z}$$
$$-\frac{A_x(x-\Delta x, y, z+\Delta z)-A_x(x-\Delta x, y, z-\Delta z)}{4\Delta x\,\Delta z}$$
$$+\frac{A_y(x, y+\Delta y, z+\Delta z)-A_y(x, y+\Delta y, z-\Delta z)}{4\Delta y\,\Delta z}$$
$$-\frac{A_y(x, y-\Delta y, z+\Delta z)-A_y(x, y-\Delta y, z-\Delta z)}{4\Delta y\,\Delta z}$$
$$+\frac{A_z(x, y, z+\Delta z)-2A_z(x, y, z)+A_z(x, y, z-\Delta z)}{\Delta z^2}. \quad (3.4.30)$$

Therefore, we obtain

$$(\nabla\nabla \cdot \mathbf{A})_{ijk} \approx \left[\frac{A^x_{(i+1)jk} - 2A^x_{ijk} + A^x_{(i-1)jk}}{\Delta x^2} \right.$$
$$+ \frac{A^y_{(i+1)(j+1)k} - A^y_{(i+1)(j-1)k} - A^y_{(i-1)(j+1)k} + A^y_{(i-1)(j-1)k}}{4\Delta x \Delta y}$$
$$\left. + \frac{A^z_{(i+1)j(k+1)} - A^z_{(i+1)j(k-1)} - A^z_{(i-1)j(k+1)} + A^z_{(i-1)j(k-1)}}{4\Delta x \Delta z} \right] \hat{\mathbf{x}}$$
$$+ \left[\frac{A^x_{(i+1)(j+1)k} - A^x_{(i+1)(j-1)k} - A^x_{(i-1)(j+1)k} + A^x_{(i-1)(j-1)k}}{4\Delta x \Delta y} \right.$$
$$+ \frac{A^y_{i(j+1)k} - 2A^y_{ijk} + A^y_{i(j-1)k}}{\Delta y^2}$$
$$\left. + \frac{A^z_{i(j+1)(k+1)} - A^z_{i(j+1)(k-1)} - A^z_{i(j-1)(k+1)} + A^z_{i(j-1)(k-1)}}{4\Delta y \Delta z} \right] \hat{\mathbf{y}}$$
$$+ \left[\frac{A^x_{(i+1)j(k+1)} - A^x_{(i+1)j(k-1)} - A^x_{(i-1)j(k+1)} + A^x_{(i-1)j(k-1)}}{4\Delta x \Delta z} \right.$$
$$+ \frac{A^y_{i(j+1)(k+1)} - A^y_{i(j+1)(k-1)} - A^y_{i(j-1)(k+1)} + A^y_{i(j-1)(k-1)}}{4\Delta x \Delta y}$$
$$\left. + \frac{A^z_{ij(k+1)} - 2A^z_{ijk} + A^z_{ij(k-1)}}{\Delta z^2} \right] \hat{\mathbf{z}}. \quad (3.4.31)$$

It is worth noting that, as readily follows from (3.4.31), the different finite-difference approximations (3.4.20)–(3.4.22) and (3.4.23)–(3.4.25) make it possible to express $(\nabla\nabla \cdot \mathbf{A})_{ijk}$ in terms of the values of the vector field $\mathbf{A}$ computed on the grid points neighboring $\mathbf{r}_{ijk}$. Equation (3.4.31) also shows that, in order to compute $(\nabla\nabla \cdot \mathbf{A})_{ijk}$ for $i = 1,\ldots, I, j = 1,\ldots, J, i = 1,\ldots, K$, the vector field $\mathbf{A}$ has to be known on a grid $G' = \{\mathbf{r}_{ijk} \in \Re^3 : i = 0,\ldots, I + 1, j = 0,\ldots, J + 1, k = 0,\ldots, K + 1\}$ containing G.

If the cells are so small that the dielectric permittivity and the electric field can be assumed to be constant over each cell, the vector field $\mathbf{A}$ can then be written as

$$\mathbf{A}(\mathbf{r}_{ijk}) = -\frac{j\omega\mu_b}{k_b^2} \sum_{i'=1}^{I} \sum_{j'=1}^{J} \sum_{k'=1}^{K} \tau_{i'j'k'} \mathbf{E}_{i'j'k'} \int_{\Delta V_{i'j'k'}} G(\mathbf{r}_{ijk}/\mathbf{r}') d\mathbf{r}', \quad (3.4.32)$$

for $i = 0,\ldots, I + 1, j = 0,\ldots, J + 1, k = 0,\ldots, K + 1$, where $\Delta V_{i'j'k'} = \{(x, y, z) \in \Re^3 : x_1 + (i' - 1)\Delta x \leq x \leq x_1 + i'\Delta x, y_1 + (j' - 1)\Delta y \leq y \leq y_1 + j'\Delta y, z_1 + (k' - 1)\Delta z \leq z \leq z_1 + k'\Delta z\}$, and $\tau_{i'j'k'} = \tau(\mathbf{r}_{i'j'k'})$.

It can be proved that (Zwamborn and van den Berg 1992)

DISCRETIZATION OF THE CONTINUOUS MODEL

$$\int_{\Delta V_{i'j'k'}} G(\mathbf{r}_{ijk}/\mathbf{r}')d\mathbf{r}' \approx \begin{cases} \dfrac{8}{k_b^2}\left[\left(1+\dfrac{1}{2}jk_b\,\Delta\xi\right)e^{-(1/2)jk_b\Delta\xi}-1\right], & \mathbf{r}_{ijk}\in\Delta V_{i'j'k'}, \\[2ex] \dfrac{4\Delta\xi e^{-jk_b|\mathbf{r}_{ijk}-\mathbf{r}_{i'j'k'}|}}{k_b^2|\mathbf{r}_{ijk}-\mathbf{r}_{i'j'k'}|}\dfrac{\left[-\sinh\left(\dfrac{1}{2}jk_b\,\Delta\xi\right)\right]}{\dfrac{1}{2}jk_b\,\Delta\xi}\\ \qquad\qquad\qquad -\cosh\left(\dfrac{1}{2}jk_b\,\Delta\xi\right)\end{cases}, \quad \mathbf{r}_{ijk}\notin\Delta V_{i'j'k'}, \tag{3.4.33}$$

where $\Delta\xi = \min\{\Delta x, \Delta y, \Delta z\}$. Since the ratio in equation (3.4.33) depends on

$$|\mathbf{r}_{ijk}-\mathbf{r}_{i'j'k'}| = \sqrt{(i-i')^2\,\Delta x^2 + (j-j')^2\,\Delta y^2 + (k-k')^2\,\Delta z^2}, \tag{3.4.34}$$

it can be written as

$$g_{(i-i')(j-j')(k-k')} = \int_{\Delta V_{i'j'k'}} G(\mathbf{r}_{ijk}/\mathbf{r}')d\mathbf{r}'. \tag{3.4.35}$$

As a consequence

$$\mathbf{A}(\mathbf{r}_{ijk}) = -\frac{j\omega\mu_b}{k_b^2}\sum_{i'=1}^{I}\sum_{j'=1}^{J}\sum_{k'=1}^{K}\tau_{i'j'k'}\mathbf{E}_{i'j'k'}g_{(i-i')(j-j')(k-k')}. \tag{3.4.36}$$

It is noteworthy that equation (3.4.36) allows one to compute the vector field **A** at the points of grid G' even if the electric field and the contrast function are known only on grid G.

With the development of a computer code in mind, it is useful to introduce an array **e** of $3N$ elements containing the values of the three Cartesian components of the electric field **E** on grid G. Moreover, it is very easy to check that equations (3.4.15), (3.4.31), and (3.4.36) define a linear system of equations for the elements of **e**. If $\mathbf{e}^{\text{inc}}$ denotes an array whose $3N$ entries are the three Cartesian components of the electric field $\mathbf{E}^{\text{inc}}$ on grid G, then

$$[\mathbf{L}]\mathbf{e} = \mathbf{e}^{\text{inc}}, \tag{3.4.37}$$

where $[\mathbf{L}]$ is a $3N \times 3N$ matrix expressing the relationships given by equations (3.4.15), (3.4.31), and (3.4.36).

The idea underlying the numerical method described above is to solve the linear system (3.4.37) to compute the values of the electric field inside V_o and afterward to use these results to compute the electric field at any point **r** outside V_o. In principle, the linear system (3.4.37) can be solved by any numerical algorithm developed for linear systems, both direct and iterative. However, because of the usually enormous number of unknowns, the iterative approach is much more convenient since the particular structure of the equations involved allows for a very efficient computation of the products between matrix [**L**] and the solution vectors. In order to perform the simulation described, the so-called biconjugate stabilized gradient method has been applied (van der Vorst 1992, Xu et al. 2002). Following the notation by Zhang et al. (2003), the method consists in the following steps:

1. Choose a guess solution $\mathbf{e}_0$ and a tolerance $e > 0$.
2. Set

$$\mathbf{r}_0 = \mathbf{e}^{\text{inc}} - [\mathbf{L}]\mathbf{e}_0, \quad (3.4.38)$$

$$\rho_0 = 1, \quad (3.4.39)$$

$$\alpha_0 = 1, \quad (3.4.40)$$

$$\omega_0 = 1, \quad (3.4.41)$$

$$\mathbf{v}_0 = 0, \quad (3.4.42)$$

$$\mathbf{p}_0 = 0. \quad (3.4.43)$$

3. Choose $\hat{\mathbf{r}}_0$ such that

$$(\hat{\mathbf{r}}_0, \mathbf{r}_0)_{C^{3N}} = \sum_{n=1}^{3N} (\hat{\mathbf{r}}_0)_n (\mathbf{r}_0)_n \neq 0, \quad (3.4.44)$$

and set $i = 0$.

4. Compute

$$\rho_i = (\tilde{\mathbf{r}}_0, \mathbf{r}_{i-1}), \quad (3.4.45)$$

$$\beta_i = \frac{\rho_i}{\rho_{i-1}} \frac{\alpha_i}{\omega_{i-1}}, \quad (3.4.46)$$

$$\mathbf{p}_i = \mathbf{r}_i + \beta_i (\mathbf{p}_{i-1} - \omega_{i-1}\mathbf{v}_{i-1}), \quad (3.4.47)$$

$$\mathbf{v}_i = [\mathbf{L}]\mathbf{p}_i, \quad (3.4.48)$$

$$\alpha_i = \frac{\rho_i}{(\tilde{\mathbf{r}}_0, \mathbf{v})}, \quad (3.4.49)$$

$$\mathbf{s}_i = \mathbf{s}_{i-1} - \alpha_i \mathbf{v}_i, \quad (3.4.50)$$

$$\mathbf{t}_i = [\mathbf{L}]\mathbf{s}_i, \quad (3.4.51)$$

$$\omega_i = \frac{(\mathbf{t}_i, \mathbf{s}_i)_{C^{3N}}}{(\mathbf{t}_i, \mathbf{t}_i)_{C^{3N}}}, \quad (3.4.52)$$

DISCRETIZATION OF THE CONTINUOUS MODEL

$$\mathbf{e}_i = \mathbf{e}_{i-1} - \alpha_i \mathbf{p}_i + \omega_i \mathbf{s}_i, \qquad (3.4.53)$$

$$\mathbf{r}_i = \mathbf{s}_i + \omega_i \mathbf{t}_i. \qquad (3.4.54)$$

5. If

$$\frac{(\mathbf{r}_i, \mathbf{r}_i)_{C^{3N}}}{(\mathbf{e}^{\text{inc}}, \mathbf{e}^{\text{inc}})_{C^{3N}}} < e, \qquad (3.4.55)$$

then terminate and the solution is $\mathbf{e}_i$. Otherwise, increment the counter i and go back to step 3. Once the array $\mathbf{e}$ has been determined, the electric field is known at every point of grid G. In order to compute the electric field for $\mathbf{r} \notin V_0$, the integral relation (2.7.2) can be used, without any problem concerning the proper meaning of the involved integral operator.

As a result, by exploiting the previously used approximations, the p Cartesian component of the total electric field vector, $E_p(\mathbf{r})$, $p = x, y, z$, can be written as (Livesay and Chen 1974)

$$E_p(\mathbf{r}) \approx E_{\text{inc}_p}(\mathbf{r}) + j\omega\mu_b \sum_{i'=1}^{I}\sum_{j'=1}^{J}\sum_{k'=1}^{K} \tau_{i'j'k'} \sum_{q=x,y,z} E_{i'j'k'}^{q} \int_{\Delta V_{i'j'k'}} g_{pq}(\mathbf{r}/\mathbf{r}')d\mathbf{r}', \qquad (3.4.56)$$

where $E_{i'j'k'}^{q} = E_q(\mathbf{r}_{i'j'k'})$, E_{inc_p} is the p Cartesian component of the incident field vector, and g_{pq}, $p, q = x, y, z$ is the pq component of Green's dyadic tensor $\overline{\mathbf{G}}$.

Equation (3.4.56) can be simply implemented in a numerical code, since (Livesay and Chen 1974)

$$\int_{\Delta V_{i'j'k'}} g_{pq}(\mathbf{r}_{ijk}/\mathbf{r}')d\mathbf{r}' = -\frac{j\omega\mu_b k_b \Delta V_{i'j'k'} e^{-jk_b|\mathbf{r}_{ijk} - \mathbf{r}_{i'j'k'}|}}{4\pi k_b^3 |\mathbf{r}_{ijk} - \mathbf{r}_{i'j'k'}|}$$

$$\left[\left(k_b^2 |\mathbf{r}_{ijk} - \mathbf{r}_{i'j'k'}|^2 - 1 - jk_b|\mathbf{r}_{ijk} - \mathbf{r}_{i'j'k'}|\delta_{pq}\right) + \frac{(p_{ijk} - p_{i'j'k'})(q_{ijk} - q_{i'j'k'})}{|\mathbf{r}_{ijk} - \mathbf{r}_{i'j'k'}|^2}\left(3 - k_b^2|\mathbf{r}_{ijk} - \mathbf{r}_{i'j'k'}|^2 + 3jk_b|\mathbf{r}_{ijk} - \mathbf{r}_{i'j'k'}|\right)\right] \qquad (3.4.57)$$

where p_{ijk} and $p_{i'j'k'}$ are the p Cartesian components of $\mathbf{r}_{ijk}$ and $\mathbf{r}_{i'j'k'}$, respectively. It is worth remarking that the entries of the array $\mathbf{e}$ are sufficient to compute the electric field at any point outside V_0, according to (3.4.56) and (3.4.57).

Some simulation results on plane-wave scattering by a homogeneous sphere are illustrated in Figure 3.3. The sphere has a radius equal to $\lambda/2$ and is characterized by $\varepsilon_r = 3.0$ and $\sigma = 0.0166\,\text{S/m}$. The incident field is a unit plane wave [equation (2.4.24)] with $\mathbf{k} = k_0\hat{\mathbf{z}}$ and $\hat{\mathbf{p}} = \hat{\mathbf{x}}$. The amplitude and the phase of the

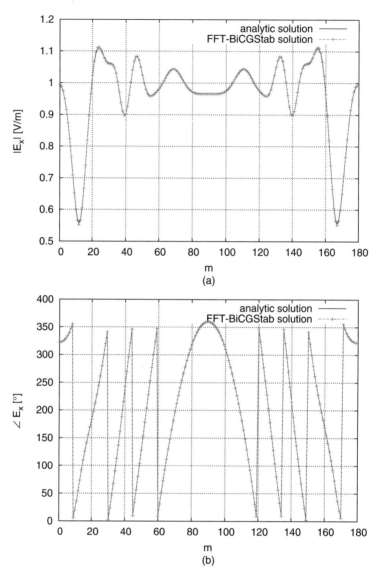

FIGURE 3.3 Scattered electric field produced by the interaction of a plane wave with a sphere ($\varepsilon_r = 3.0$, $\sigma = 0.0166\,\text{S/m}$, radius equal to $\lambda/2$), with comparison between numerical and analytical data; total electric field (amplitude and phase): (a), (b) x component; (c), (d) y component; (e), (f) z component. (Simulations performed by G. Bozza, University of Genoa, Italy.)

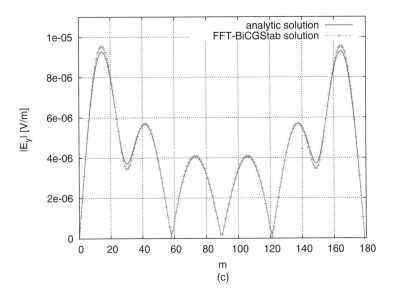

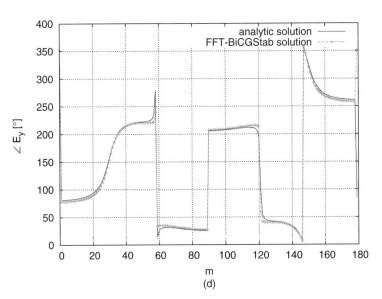

FIGURE 3.3 *Continued*

40 THE ELECTROMAGNETIC INVERSE SCATTERING PROBLEM

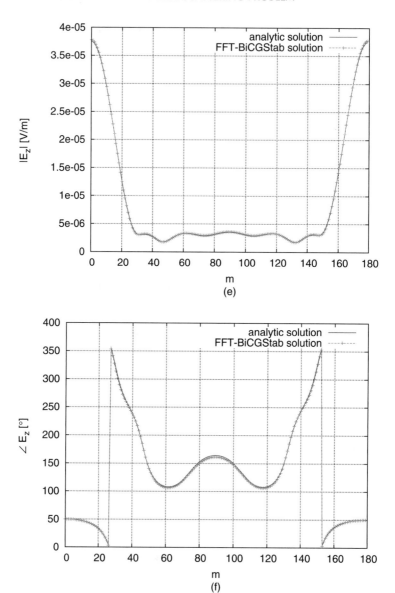

FIGURE 3.3 *Continued*

total electric field have been computed in a set of $M = 180$ points of Cartesian components given by

$$\mathbf{r}_m = (0, 2\lambda \sin\varphi_m, 2\lambda \sin\varphi_m), \varphi_m = m\frac{2\pi}{M}, \quad m = 1, \ldots, M. \quad (3.4.58)$$

Figure 3.3 plots the three Cartesian components of the total electric field. The numerical values are compared, with excellent agreement, with analytical data computed by using the eigenfunction solution for the homogeneous sphere (30 modes are considered) (Stratton 1941).

3.5 SCATTERING BY CANONICAL OBJECTS: THE CASE OF A MULTILAYER ELLIPTIC CYLINDER

As explained above, computation of the electromagnetic field scattered by arbitrarily shaped dielectric objects when they are illuminated by a generic incident field is a complex task that seldom can be performed by using analytic techniques. However, for some canonical geometries of the scatterers and for particular incident fields (e.g., plane or cylindrical waves), series expansions of the solutions for the scattered fields with analytically computable coefficients can be provided.

Among the various canonical objects reported in the literature (e.g., Stratton 1941, Bowman et al. 1969, Wait 1962), we consider here, for the sake of illustration, the scattering by a multilayer dielectric cylinder of elliptic cross section when the incident field is a plane wave polarized along the cylinder axis, that is, under TM illumination conditions (Caorsi et al. 1997a). Similar formulations can be obtained under different illumination conditions (e.g., line-current sources). A stratified elliptic cylinder is sufficiently complex to provide a good test for imaging procedures and, in general, for numerical algorithms. In fact, the assumed configuration is inhomogeneous and may contain both dielectric and conducting materials. Moreover, it does not exhibit a circular symmetry, which is an important aspect in evaluating the capabilities of tomographic imaging configurations, which are essentially based on circular geometries. In addition, elliptic cylinders are often used to approximately model several real structures, such as aircraft fuselage and other cylindrical bodies (Uslenghi 1997).

Let us consider the elliptic coordinates (u, v, z) shown in Figure 3.4. In this coordinate system, $u = $ constant represents a family of elliptic surfaces having the same foci (located at points $x = \pm d$ on the x axis), whereas $v = $ constant represent a family of confocal hyperbolic surfaces. Accordingly, the ith layer of the N-layer (lossless and nonmagnetic) cylinder is bounded by the elliptic surfaces $u = u_{i-1}$ and $u = u_i$, $i = 1, \ldots, N$, with $u_0 = 0$. The external medium is assumed to be the free space and is denoted as the $N + 1$ layer. The semimajor and semiminor axes of the cylinders bounding the various layers are denoted by a_i and b_i.

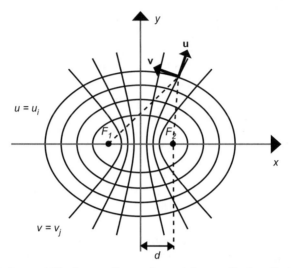

FIGURE 3.4 Elliptic coordinates for multilayer elliptic cylinders.

As mentioned previously, the illuminating field is a transverse magnetic uniform plane wave [equation (2.4.24)] with $\hat{\mathbf{p}} = \hat{\mathbf{z}}$ and

$$\mathbf{k} = -k_b\left(\cos\varphi_{\text{inc}}\,\hat{\mathbf{x}} + \sin\varphi_{\text{inc}}\,\hat{\mathbf{y}}\right), \qquad (3.5.1)$$

where φ_{inc} is the incident direction in the horizontal plane. The total electric field in the ith layer is given by

$$\mathbf{E}^i(u,v) = E_z^i(u,v)\hat{\mathbf{z}}, \qquad (3.5.2)$$

$$\mathbf{H}^i(u,v) = H_u^i(u,v)\hat{\mathbf{u}} + H_v^i(u,v)\hat{\mathbf{v}} = -\frac{j}{\omega\mu_i h}\left[\hat{\mathbf{u}}\frac{\partial}{\partial u} + \hat{\mathbf{v}}\frac{\partial}{\partial v}\right]\mathbf{E}^i(u,v), \qquad (3.5.3)$$

with $i = 1,\ldots,N$, where

$$h = d\sqrt{\cosh^2 u - \cos^2 v}. \qquad (3.5.4)$$

According to Yeh's (1963) solution, the field in the ith layer is expressed in terms of Mathieu functions, which are the eigenfunctions for the elliptic cylinder (Stratton 1941, Morse and Feshback 1953, Blanch 1965)

$$\begin{aligned}E_z^i(u,v) = &\sum_{m=0}^{\infty}\left[e_{m1}^i Mc_m^{(1)}(q_i,u) + e_{m2}^i Mc_m^{(2)}(q_i,u)\right]\cdot ce_m(q_i,v)\\ &+ \sum_{m=1}^{\infty}\left[o_{m1}^i Ms_m^{(1)}(q_i,u) + o_{m2}^i Ms_m^{(2)}(q_i,u)\right]\cdot se_m(q_i,v),\end{aligned} \qquad (3.5.5)$$

where ce_m and se_m denote even and odd angular Mathieu functions, $Mc_m^{(1)}$ and $Ms_m^{(1)}$ denote even and odd radial Mathieu functions of the first kind, $Mc_m^{(2)}$ and $Ms_m^{(2)}$ denote even and odd radial Mathieu functions of the second kind (Blanch 1965), and $q_i = 0.25(k_i d)^2$, where k_i is the wavenumber of the ith layer. Finally, e_{mj}^i and o_{mj}^i, $j = 1, 2$ are the expansion coefficients in the ith layer (to conserve notation, we set $e_{m2}^1 = o_{m2}^1 = 0$).

In the external region, the electric field is the sum of the incident and scattered waves. The scattered wave can be expressed in terms of Mathieu functions of the fourth kind, $Mc_m^{(4)}$ and $Ms_m^{(4)}$, which are analogous to the Hankel functions for the circular cylinder and can be expressed as linear combinations of the corresponding radial Mathieu functions of the first and second kinds [i.e., $Mc_m^{(4)} = Mc_m^{(1)} - jMc_m^{(2)}$ and $Ms_m^{(4)} = Ms_m^{(1)} - jMs_m^{(2)}$]. We then obtain

$$E_{\text{scat}_z}(u,v) = \sum_{m=0}^{\infty} e_m^{N+1} Mc_m^{(4)}(q_{N+1}, u) \cdot ce_m(q_{N+1}, v)$$
$$+ \sum_{m=1}^{\infty} o_m^{N+1} Ms_m^{(4)}(q_{N+1}, u) \cdot se_m(q_{N+1}, v), \quad (3.5.6)$$

where the coefficients e_m^{N+1} and o_m^{N+1} depend on both the amplitude and the direction of the horizontally directed incident plane wave [see equations (2.4.24) and (3.5.1)], which, expanded in Mathieu functions, has the form

$$E_{\text{inc}_z}(u,v) = 2\pi \sum_{m=0}^{\infty} \frac{j^m Mc_m^{(1)}(q_{N+1}, u) ce_m(q_{N+1}, v) ce_m(q_{N+1}, \varphi_{\text{inc}})}{\int_0^{2\pi}[ce_m(q_{N+1}, v)]^2 \, dv}$$
$$+ 2\pi \sum_{m=1}^{\infty} \frac{j^m Ms_m^{(1)}(q_{N+1}, u) se_m(q_{N+1}, v) se_m(q_{N+1}, \varphi_{\text{inc}})}{\int_0^{2\pi}[se_m(q_{N+1}, v)]^2 \, dv}, \quad (3.5.7)$$

where $q_{N+1} = 0.25(k_b d)^2$. The unknown coefficients can be deduced by enforcing the continuity of the tangential components of the electric field across the interfaces between layers. To this end, let us define the following quantities:

$$C_{mn}^{ij} = \int_0^{2\pi} ce_m(q_i, v) ce_n(q_j, v) \, dv, \quad (3.5.8)$$
$$S_{mn}^{ij} = \int_0^{2\pi} se_m(q_i, v) se_n(q_j, v) \, dv, \quad (3.5.9)$$
$$CS_{mn}^{ij} = \int_0^{2\pi} ce_m(q_i, v) se_n(q_j, v) \, dv. \quad (3.5.10)$$

For the ith layer, by applying the Galerkin's method, one obtains

$$\sum_{m=0}^{\infty} C_{mn}^{i(i+1)} \left[e_{m1}^i Mc_m^{(1)}(q_i, u_i) + e_{m2}^i Mc_m^{(2)}(q_i, u_i) \right]$$
$$= \begin{cases} C_{nn}^{(i+1)(i+1)} \left[e_{n1}^{i+1} Mc_n^{(1)}(q_{i+1}, u_i) + e_{n2}^{i+1} Mc_n^{(2)}(q_{i+1}, u_i) \right] & \text{if } i = 1, \ldots, N-1, \\ 2\pi j^n Mc_n^{(1)}(q_{i+1}, u_i) ce_n(q_{i+1}, \varphi_{\text{inc}}) \\ + C_{nn}^{(i+1)(i+1)} e_n^{i+1} Mc_n^{(4)}(q_{i+1}, u_i) & \text{if } i = N, \end{cases} \quad (3.5.11)$$

where $n = 0, 1, 2, \ldots$, and

$$\sum_{m=1}^{\infty} S_{mn}^{i(i+1)} \left[o_{m1}^{i} M s_{m}^{(1)}(q_i, u_i) + o_{m2}^{i} M s_{m}^{(2)}(q_i, u_i) \right]$$
$$= \begin{cases} S_{nn}^{(i+1)(i+1)} \left[o_{n1}^{i+1} M s_{n}^{(1)}(q_{i+1}, u_i) + o_{n2}^{i+1} M s_{n}^{(2)}(q_{i+1}, u_i) \right] & \text{if } i = 1, \ldots, N-1, \\ 2\pi j^n M s_{n}^{(1)}(q_{i+1}, u_i) se_n(q_{i+1}, \varphi_{\text{inc}}) \\ + S_{nn}^{(i+1)(i+1)} o_{n}^{i+1} M s_{n}^{(4)}(q_{i+1}, u_i) & \text{if } i = N, \end{cases} \quad (3.5.12)$$

where $n = 1, 2, \ldots$.

Analogously, the continuity of the $H_v^i(u, v)$ component of $\mathbf{H}^i(u, v)$ gives

$$\sum_{m=0}^{\infty} C_{mn}^{i(i+1)} \left[e_{m1}^{i} DMc_{m}^{(1)}(q_i, u_i) + e_{m2}^{i} DMc_{m}^{(2)}(q_i, u_i) \right]$$
$$= \begin{cases} C_{nn}^{(i+1)(i+1)} \left[e_{n1}^{i+1} DMc_{n}^{(1)}(q_{i+1}, u_i) + e_{n2}^{i+1} DMc_{n}^{(2)}(q_{i+1}, u_i) \right] & \text{if } i = 1, \ldots, N-1, \\ 2\pi j^n DMc_{n}^{(1)}(q_{i+1}, u_i) ce_n(q_{i+1}, \varphi_{\text{inc}}) \\ + C_{nn}^{(i+1)(i+1)} e_{n}^{i+1} DMc_{n}^{(4)}(q_{i+1}, u_i) & \text{if } i = N, \end{cases}$$
$$(3.5.13)$$

where $n = 0, 1, 2, \ldots$, and

$$\sum_{m=1}^{\infty} S_{mn}^{i(i+1)} \left[o_{m1}^{i} DMs_{m}^{(1)}(q_i, u_i) + o_{m2}^{i} DMs_{m}^{(2)}(q_i, u_i) \right]$$
$$= \begin{cases} S_{nn}^{(i+1)(i+1)} \left[o_{n1}^{i+1} DMs_{n}^{(1)}(q_{i+1}, u_i) + o_{n2}^{i+1} DMs_{n}^{(2)}(q_{i+1}, u_i) \right] & \text{if } i = 1, \ldots, N-1, \\ 2\pi j^n DMs_{n}^{(1)}(q_{i+1}, u_i) se_n(q_{i+1}, \varphi_{\text{inc}}) \\ + S_{nn}^{(i+1)(i+1)} o_{n}^{i+1} DMs_{n}^{(4)}(q_{i+1}, u_i) & \text{if } i = N, \end{cases}$$
$$(3.5.14)$$

where $n = 1, 2, \ldots$, and D denotes the derivatives of the Mathieu functions with respect to the u variable. To obtain these relations, we consider the fact that $CS_{mn}^{i(i+1)} = 0$ for any m, n, and i. Moreover, from (3.5.11)–(3.5.14), the following matrix equations can be obtained by imposing series truncations

$$\left[\mathbf{B}^{i+1} \right] \mathbf{w}^{i+1} = \left[\mathbf{A}^{i+1} \right] \mathbf{w}^{i}, \quad (3.5.15)$$

where

$$\mathbf{w}^j = \left[e_{01}^j, e_{02}^j, o_{11}^j, o_{12}^j, e_{11}^j, e_{12}^j, \ldots, e_{(M-1)2}^j \right]^T, \quad (3.5.16)$$

and the matrices involved in (3.5.15) are block matrices given by

$$[\mathbf{B}^j] = \begin{bmatrix} [\mathbf{Be}_0^j] & & & & 0 \\ & [\mathbf{Bo}_1^j] & & & \\ & & [\mathbf{Be}_1^j] & & \\ & & & \ddots & \\ 0 & & & & [\mathbf{Be}_{M-1}^j] \end{bmatrix}, \quad (3.5.17)$$

$$[\mathbf{A}^j] = \begin{bmatrix} [\mathbf{AC}_{00}^j] & [\mathbf{ASC}_{01}^j] & [\mathbf{AC}_{01}^j] & \cdots & [\mathbf{AC}_{0(M-1)}^j] \\ [\mathbf{ACS}_{10}^j] & [\mathbf{AS}_{11}^j] & [\mathbf{ACS}_{11}^j] & \cdots & [\mathbf{ACS}_{1(M-1)}^j] \\ [\mathbf{AC}_{10}^j] & [\mathbf{ASC}_{11}^j] & [\mathbf{AC}_{11}^j] & \cdots & [\mathbf{AC}_{1(M-1)}^j] \\ \vdots & \vdots & \vdots & & \vdots \\ [\mathbf{AC}_{(M-1)0}^j] & [\mathbf{ACS}_{(M-1)1}^j] & [\mathbf{AC}_{(M-1)1}^j] & \cdots & [\mathbf{AC}_{(M-1)(M-1)}^j] \end{bmatrix}, \quad (3.5.18)$$

where

$$[\mathbf{Be}_k^j] = C_{kk}^{jj} \begin{bmatrix} Mc_k^{(1)}(q_j, u_{j-1}) & Mc_k^{(2)}(q_j, u_{j-1}) \\ DMc_k^{(1)}(q_j, u_{j-1}) & DMc_k^{(2)}(q_j, u_{j-1}) \end{bmatrix}, \quad k = 0, 1, \ldots, M-1, \quad (3.5.19)$$

$$[\mathbf{Bo}_k^j] = S_{kk}^{jj} \begin{bmatrix} Ms_k^{(1)}(q_j, u_{j-1}) & Ms_k^{(2)}(q_j, u_{j-1}) \\ DMs_k^{(1)}(q_j, u_{j-1}) & DMs_k^{(2)}(q_j, u_{j-1}) \end{bmatrix}, \quad k = 1, \ldots, M-1, \quad (3.5.20)$$

$$[\mathbf{AC}_{kh}^j] = C_{hk}^{(j-1)j} \begin{bmatrix} Mc_h^{(1)}(q_{j-1}, u_{j-1}) & Mc_h^{(2)}(q_{j-1}, u_{j-1}) \\ DMc_h^{(1)}(q_{j-1}, u_{j-1}) & DMc_h^{(2)}(q_{j-1}, u_{j-1}) \end{bmatrix},$$
$$k = 0, \ldots, M-1, h = 0, \ldots, M-1, \quad (3.5.21)$$

$$[\mathbf{ACS}_{kh}^j] = CS_{hk}^{(j-1)j} \begin{bmatrix} Mc_h^{(1)}(q_{j-1}, u_{j-1}) & Mc_h^{(2)}(q_{j-1}, u_{j-1}) \\ DMc_h^{(1)}(q_{j-1}, u_{j-1}) & DMc_h^{(2)}(q_{j-1}, u_{j-1}) \end{bmatrix},$$
$$k = 1, \ldots, M-1, h = 0, \ldots, M-1, \quad (3.5.22)$$

$$[\mathbf{ASC}_{kh}^j] = CS_{kh}^{j(j-1)} \begin{bmatrix} Ms_h^{(1)}(q_{j-1}, u_{j-1}) & Ms_h^{(2)}(q_{j-1}, u_{j-1}) \\ DMs_h^{(1)}(q_{j-1}, u_{j-1}) & DMs_h^{(2)}(q_{j-1}, u_{j-1}) \end{bmatrix},$$
$$k = 0, \ldots, M-1, h = 1, \ldots, M-1, \quad (3.5.23)$$

$$[\mathbf{AS}_{kh}^j] = S_{hk}^{(j-1)j} \begin{bmatrix} Ms_h^{(1)}(q_{j-1}, u_{j-1}) & Ms_h^{(2)}(q_{j-1}, u_{j-1}) \\ DMs_h^{(1)}(q_{j-1}, u_{j-1}) & DMs_h^{(2)}(q_{j-1}, u_{j-1}) \end{bmatrix},$$
$$k = 1, \ldots, M-1, h = 1, \ldots, M-1, \quad (3.5.24)$$

From equation (3.5.15), the coefficients of the field expansion in the $(i+1)$th layer can be explicitly expressed in terms of the coefficients of the ith layer, yielding

$$\mathbf{w}^{i+1} = [\mathbf{D}^i] \mathbf{w}^i, \quad (3.5.25)$$

where

$$[\mathbf{D}^j] = \begin{bmatrix} [\mathbf{DC}_{00}^j] & [\mathbf{DSC}_{01}^j] & [\mathbf{DC}_{01}^j] & \cdots & [\mathbf{DC}_{0(M-1)}^j] \\ [\mathbf{DCS}_{10}^j] & [\mathbf{DS}_{11}^j] & [\mathbf{DCS}_{11}^j] & \cdots & [\mathbf{DCS}_{1(M-1)}^j] \\ [\mathbf{DC}_{10}^j] & [\mathbf{DSC}_{11}^j] & [\mathbf{DC}_{11}^j] & \cdots & [\mathbf{DC}_{1(M-1)}^j] \\ \vdots & \vdots & \vdots & & \vdots \\ [\mathbf{DC}_{(M-1)0}^j] & [\mathbf{DSC}_{(M-1)1}^j] & [\mathbf{DC}_{(M-1)1}^j] & \cdots & [\mathbf{DC}_{(M-1)(M-1)}^j] \end{bmatrix}, \quad (3.5.26)$$

where

$$[\mathbf{DC}_{kh}^j] = \left(be_{k11}^{j+1}be_{k22}^{j+1} - be_{k12}^{j+1}be_{p21}^{j+1}\right)^{-1}$$
$$\begin{bmatrix} be_{k22}^{j+1}ac_{kh11}^j - be_{k12}^{j+1}ac_{kh21}^i & be_{k22}^{j+1}ac_{kh12}^j - be_{k12}^{j+1}ac_{kh22}^i \\ be_{k11}^{j+1}ac_{kh21}^i - be_{k21}^{j+1}ac_{kh11}^j & be_{k11}^{j+1}ac_{kh22}^i - be_{k21}^{j+1}ac_{kh12}^j \end{bmatrix}, \quad (3.5.27)$$

where $k = 0, 1, \ldots, M - 1, h = 0, 1, \ldots, M - 1$, and

$$[\mathbf{DSC}_{kh}^j] = \left(be_{k11}^{j+1}be_{k22}^{j+1} - be_{k12}^{j+1}be_{p21}^{j+1}\right)^{-1}$$
$$\begin{bmatrix} be_{k22}^{j+1}asc_{kh11}^j - be_{k12}^{j+1}asc_{kh21}^i & be_{k22}^{j+1}asc_{kh12}^j - be_{k12}^{j+1}asc_{kh22}^i \\ be_{k11}^{j+1}asc_{kh21}^i - be_{k21}^{j+1}asc_{kh11}^j & be_{k11}^{j+1}asc_{kh22}^i - be_{k21}^{j+1}asc_{kh12}^j \end{bmatrix}, \quad (3.5.28)$$

where $k = 0, 1, \ldots, M - 1, h = 1, \ldots, M - 1$, and

$$[\mathbf{DCS}_{kh}^j] = \left(bo_{k11}^{j+1}bo_{k22}^{j+1} - bo_{k12}^{j+1}bo_{p21}^{j+1}\right)^{-1}$$
$$\begin{bmatrix} bo_{k22}^{j+1}acs_{kh11}^j - bo_{k12}^{j+1}acs_{kh21}^i & bo_{k22}^{j+1}acs_{kh12}^j - bo_{k12}^{j+1}acs_{kh22}^i \\ bo_{k11}^{j+1}acs_{kh21}^i - bo_{k21}^{j+1}acs_{kh11}^j & bo_{k11}^{j+1}acs_{kh22}^i - boe_{k21}^{j+1}acs_{kh12}^j \end{bmatrix}, \quad (3.5.29)$$

where $k = 1, \ldots, M - 1, h = 0, 1, \ldots, M - 1$, and

$$[\mathbf{DS}_{kh}^j] = \left(bo_{k11}^{j+1}bo_{k22}^{j+1} - bo_{k12}^{j+1}bo_{p21}^{j+1}\right)^{-1}$$
$$\begin{bmatrix} bo_{k22}^{j+1}as_{kh11}^j - bo_{k12}^{j+1}as_{kh21}^i & bo_{k22}^{j+1}as_{kh12}^j - bo_{k12}^{j+1}as_{kh22}^i \\ bo_{k11}^{j+1}as_{kh21}^i - bo_{k21}^{j+1}as_{kh11}^j & bo_{k11}^{j+1}as_{kh22}^i - bo_{k21}^{j+1}as_{kh12}^j \end{bmatrix}, \quad (3.5.30)$$

where $k = 1, \ldots, M - 1, h = 1, \ldots, M - 1$.

In equations (3.5.27)–(3.5.30), be_{klp}^j, bo_{klp}^j, ac_{khlp}^j, as_{khlp}^j, acs_{khlp}^j, asc_{khlp}^j, $l, p = 1, 2$ indicates the elements of the lth row and pth column of matrices $[\mathbf{Be}_k^j]$, $[\mathbf{Bo}_k^j]$, $[\mathbf{AC}_{kh}^j]$, $[\mathbf{AS}_{kh}^j]$, $[\mathbf{ACS}_{kh}^j]$, and $[\mathbf{ASC}_{kh}^j]$, respectively. If equation (3.5.25) is applied recursively, then, for $i = 1, \ldots, N - 1$, we can obtain

$$\mathbf{w}^{i+1} = [\mathbf{F}^i]\mathbf{w}^1, \quad (3.5.31)$$

where

$$[\mathbf{F}^i] = [\mathbf{F}^{i-1}][\mathbf{D}^i] = \prod_{p=1}^{i}[\mathbf{D}^p], \quad (3.5.32)$$

with $[\mathbf{D}^0] = [\mathbf{I}]$.

At the external boundaries, using the same procedure, we obtain analogous relations. In particular

$$[\mathbf{B}]\mathbf{u}^{N+1} = [\mathbf{A}^N]\mathbf{w}^N + [\mathbf{K}], \quad (3.5.33)$$

where

$$\mathbf{u}^{N+1} = \left[e_0^{N+1}, o_0^{N+1}, e_1^{N+1}, \ldots, e_{M-1}^{N+1} \right]^T, \tag{3.5.34}$$

$$[\mathbf{B}] = \begin{bmatrix} [\mathbf{Be}_0] & & & & 0 \\ & [\mathbf{Bo}_1] & & & \\ & & [\mathbf{Be}_1] & & \\ & & & \ddots & \\ 0 & & & & [\mathbf{Be}_{M-1}] \end{bmatrix}, \tag{3.5.35}$$

$$[\mathbf{K}] = \begin{bmatrix} -2\pi j^0 Mc_0^{(1)}(q_{N+1}, u_N) ce_0(q_{N+1}, \varphi_{\text{inc}}) \\ -2\pi j^0 DMc_0^{(1)}(q_{N+1}, u_N) ce_0(q_{N+1}, \varphi_{\text{inc}}) \\ -2\pi j^1 Ms_1^{(1)}(q_{N+1}, u_N) se_1(q_{N+1}, \varphi_{\text{inc}}) \\ -2\pi j^1 DMs_1^{(1)}(q_{N+1}, u_N) se_1(q_{N+1}, \varphi_{\text{inc}}) \\ \vdots \end{bmatrix}, \tag{3.5.36}$$

and $[\mathbf{A}^N]$ is as given by (3.5.18). In equation (3.5.35) the involved matrices are given by

$$[\mathbf{Be}_k] = C_{kk}^{(N+1)(N+1)} \begin{bmatrix} Mc_k^{(4)}(q_{N+1}, u_N) \\ DMc_k^{(4)}(q_{N+1}, u_N) \end{bmatrix}, \quad k = 0, 1, \ldots, M-1, \tag{3.5.37}$$

$$[\mathbf{Bo}_k] = S_{kk}^{(N+1)(N+1)} \begin{bmatrix} Ms_k^{(4)}(q_{N+1}, u_N) \\ DMs_k^{(4)}(q_{N+1}, u_N) \end{bmatrix}, \quad k = 1, \ldots, M-1, \tag{3.5.38}$$

By substituting (3.5.31) in (3.5.33), we obtain

$$[\mathbf{B}]\mathbf{u}^{N+1} - [\mathbf{A}^N][\mathbf{F}^{N-1}]\mathbf{w}^1 = [\mathbf{K}], \tag{3.5.39}$$

and finally

$$\begin{bmatrix} \mathbf{u}^{N+1} \\ \mathbf{w}^1 \end{bmatrix} = \begin{bmatrix} [\mathbf{B}] & 0 \\ 0 & -[\mathbf{A}^N][\mathbf{F}^{N-1}] \end{bmatrix}^{-1} [\mathbf{K}], \tag{3.5.40}$$

which allows us to derive the coefficients of the external and the innermost layers simultaneously. From $\mathbf{u}^{N+1}$ we can also deduce the far-field properties of the scattered field. In particular, the scattering width is defined as

$$W(\varphi) = W(v) = \lim_{\rho \to \infty} 2\pi\rho \left| \frac{E_{\text{scat}_z}}{E_{\text{inc}_z}} \right|^2$$

$$\approx \frac{4}{k_b} \left| \sum_{p=0}^{M} e_M^{N+1} ce_p(q_{N+1}, v) e^{j(p/2)\pi} + \sum_{p=1}^{M} o_M^{N+1} se_p(q_{N+1}, v) e^{j(p/2)\pi} \right|^2, \tag{3.5.41}$$

in which the following asymptotic expression, valid for large values of u, has been applied to equation (3.5.6):

$$Mc_p^{(4)}(q,u) \approx \sqrt{\frac{2}{\pi k_b d \cosh u}} e^{-j\{k_b d \cosh u - [(2p+1)/4]\pi\}} \approx Ms_p^{(4)}(q,u). \quad (3.5.42)$$

Moreover, once $\mathbf{w}^1$ has been obtained from equation (3.5.33), the expansion coefficients in the other internal layers ($i = 2,\ldots, N$) can be immediately obtained by using equation (3.5.25) recursively.

It should be mentioned that the recursive method described above requires the solution of only one matrix equation, exactly as does the Yeh method for a single homogeneous elliptic cylinder (Yeh 1963). Therefore, it is computationally efficient.

This procedure has been checked for several cases. Some examples are described as follows (Caorsi et al. 2000):

1. *A Three-Layer Elliptic Cylinder.* In this case the semimajor axes of the ellipses constituting boundaries in the transversal plane are given by $a_1 = 0.1\,\mathrm{m}$, $a_2 = 0.16\,\mathrm{m}$, and $a_3 = 0.2\,\mathrm{m}$ (external boundary). The semifocal distance is $d = 0.02\,\mathrm{m}$, and the dielectric properties of the three nonmagnetic layers are $\varepsilon_{r_1} = 2.0$, $\varepsilon_{r_2} = 1.3$, and $\varepsilon_{r_3} = 2.5$, respectively. The background is vacuum ($\varepsilon_b = \varepsilon_0$), and the cross section center coincides with the origin of the coordinate system. The incident field is produced by a line-current source ($f = 600\,\mathrm{MHz}$) with unit amplitude and places on the x axis at point $x = 0.505\,\mathrm{m}$, $y = 0.0$. Figure 3.5 shows the computed total electric field (amplitude and phase) along the x and y axes obtained by using $M = 10$ modes for each layer. Since d is very small, the ellipses almost degenerate into circular cylinders. Consequently, the produced field can be compared with the one obtained by the eigenfunction expansion for the multilayer circular cylinder (Bussey and Richmond 1975). As can be seen, the agreement is quite good. Moreover, the same fields have been compared with the one numerically obtained by using a FEM code [with a perfectly matched layer (PML) for truncation of the domain]. Numerical codes can be used to evaluate the reliability of the procedure when the circular cylinder is not a good approximation of the scatterer under test. For the simulation reported in Figure 3.5, a square domain (including the cylinder cross section) has been considered. The side of this domain is $3\,\mathrm{m}$, and the discretization mesh consists of 300×300 equal squares, each of them divided by the positive slope diagonal. The PML used is that described by Caorsi and Raffetto (1998).

2. *A Four-Layer Cylinder.* In this case the simulation parameters are $a_1 = 0.14\,\mathrm{m}$, $a_2 = 0.16\,\mathrm{m}$, $a_3 = 0.9\,\mathrm{m}$, $a_4 = 0.2\,\mathrm{m}$, $d = 0.12\,\mathrm{m}$, $\varepsilon_{r_1} = 2.1$, $\varepsilon_{r_2} = 2.4$, $\varepsilon_{r_3} = 1.8$, and $\varepsilon_{r_4} = 1.4$. The illuminating source is placed at points $x = 0.505\,\mathrm{m}$ and $y = 0.305$. The working frequency is again $f = 600\,\mathrm{MHz}$, and $M = 12$ modes are used. Figure 3.6 provides the results for this simulation and comparison with FEM/PML values (obtained using the same discretization as in case 1). As can be seen, the agreement is quite good,

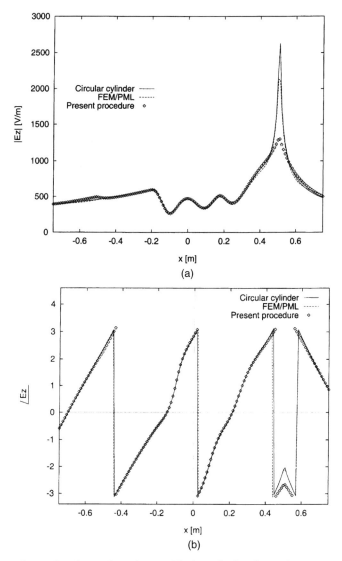

FIGURE 3.5 Scattering by a three-layer elliptic cylinder ($a_1 = 0.1$ m, $a_2 = 0.16$ m, $a_3 = 0.2$ m, $d = 0.02$ m, $\varepsilon_{r_1} = 2.0$, $\varepsilon_{r_2} = 1.3$, $\varepsilon_{r_3} = 2.5$). Line-current source (placed at point $x = 0.505$ m and $y = 0$). Working frequency $f = 600$ MHz. Amplitude (a) and phase (b) of the total electric field computed along the x axis; amplitude (c) and phase (d) of the total electric field computed along the y axis. [Reproduced from S. Caorsi, M. Pastorino, and M. Raffetto, "Electromagnetic scattering by a multilayer elliptic cylinder under line-source illumination," *Microwave Opt. Technol. Lett.* **24**, 322–329 (March 5, 2000), © 2000 Wiley.]

50 THE ELECTROMAGNETIC INVERSE SCATTERING PROBLEM

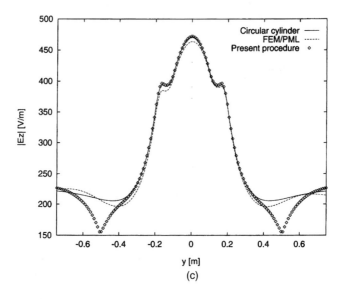

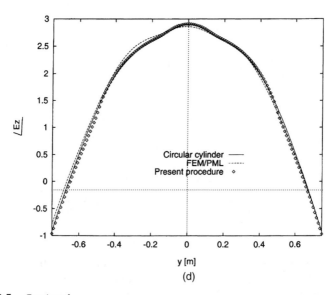

FIGURE 3.5 *Continued*

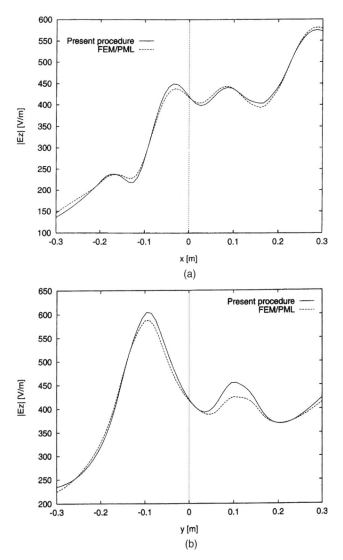

FIGURE 3.6 Scattering by a four-layer elliptic cylinder ($a_1 = 0.14$ m, $a_2 = 0.16$ m, $a_3 = 0.9$ m, $a_4 = 0.2$ m, $d = 0.12$ m, $\varepsilon_{r_1} = 2.1$, $\varepsilon_{r_2} = 2.4$, $\varepsilon_{r_3} = 1.8$, $\varepsilon_{r_4} = 1.4$). Line-current source (placed at point $x = 0.505$ m and $y = 0.305$ m). Working frequency $f = 600$ MHz. Amplitude of the total electric field computed along the x axis (a) and the y axis (b). [Reproduced from S. Caorsi, M. Pastorino, and M. Raffetto, "Electromagnetic scattering by a multilayer elliptic cylinder under line-source illumination," *Microwave Opt. Technol. Lett.* **24**, 322–329 (March 5, 2000), © 2000 Wiley.]

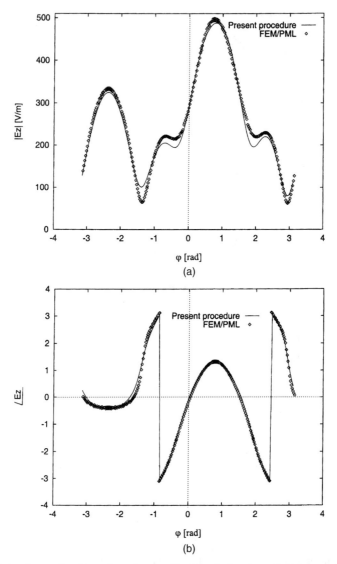

FIGURE 3.7 Scattering by a three-layer elliptic cylinder ($a_1 = 0.14$ m, $a_2 = 0.2$ m, and $a_3 = 0.25$ m, $d = 0.12$ m, $\varepsilon_{r_1} = 1.5$, $\varepsilon_{r_2} = 2.6$, $\varepsilon_{r_3} = 1.9$). Line-current source (placed at point $x = y = 0.505$ m). Working frequency $f = 400$ MHz. Amplitude (a) and phase (b) of the total electric field computed along a circle of radius $R = 0.4$ m. [Reproduced from S. Caorsi, M. Pastorino, and M. Raffetto, "Electromagnetic scattering by a multilayer elliptic cylinder under line-source illumination," *Microwave Opt. Technol. Lett.* **24**, 322–329 (March 5, 2000), © 2000 Wiley.]

although some differences are present between the two solutions, probably due to the quite coarse mesh used for the FEM/PML computation.
3. *A Three-Layer Cylinder with* $a_1 = 0.14$ m, $a_2 = 0.2$ m, *and* $a_3 = 0.25$ m, $d = 0.12$ m, $\varepsilon_{r_1} = 1.5$, $\varepsilon_{r_2} = 2.6$, *and* $\varepsilon_{r_3} = 1.9$. The line current is located at point $x = y = 0.505$ m, the working frequency in this case is $f = 400$ MHz, and $M = 10$ modes are used. Figure 3.7 provides the total electric field (amplitude and phase) computed along a circle of radius $R = 0.4$ m and centered at the origin of the coordinate system (coinciding with the center of the elliptic cross section). In this case, too, there is a very good agreement between the semianalytical data and the data obtained by using the FEM/PML numerical code.

It should be mentioned that other semianalytical solutions involving elliptic cylinders have been considered. Some examples are the simple case in which a PEC core is present (Richmond 1988, Caorsi et al. 1997b), and cases in which the multilayer cylinder consists of isorefractive materials (Caorsi and Pastorino 2004) or metamaterials (see Section 11.5) (Pastorino et al. 2005). Moreover, other similar solutions are available for other geometries, including elliptic cylinders [e.g., the case in which a PEC elliptic cylinder is coated by a circular dielectric layer (Kakogiannos and Roumeliotis 1990)].

REFERENCES

Abubakar, A., T. M. Habashy, and P. M. van den Berg, "Nonlinear inversion of multifrequency microwave Fresnel data using the multiplicative regularized contrast source inversion," *Progress Electromagn. Res.* **62**, 193–201 (2006).

Balanis, C. A., *Advanced Engineering Electromagnetics*, Wiley, New York, 1989.

Bertero, M. and P. Boccacci, *Introduction to Inverse Problems in Imaging*, Institute of Physics, Bristol, UK, 1998.

Blanch, G., "Mathieu functions," in *Handbook of Mathematical Functions*, M. Abramowitz and I. A. Stegun, eds., Dover, New York, 1965.

Bleinsten, N. and J. Cohen, "Nonuniqueness in the inverse source problem in acoustics and electromagnetics," *J. Math. Phys.* **18**, 194–201 (1977).

Bossavit, A., *Computational Electromagnetics: Variational Formulations, Complementarity, Edge Currents*, Academic Press, San Diego, CA, 1998.

Bowman, J. J., T. B. A. Senior, and P. L. E. Uslenghi, *Electromagnetic and Acoustic Scattering by Simple Shapes*, North-Holland, Amsterdam, 1969.

Bussey, H. E. and J. H. Richmond, "Scattering by a lossy dielectric circular cylindrical multilayer, numerical values," *IEEE Trans. Anten. Propag.* **AP-23**, 723–725 (1975).

Caorsi, S. and M. Pastorino, "Scattering by a multilayer isorefractive elliptic cylinder," *IEEE Trans. Anten. Propag.* **52**, 189–196 (2004).

Caorsi, S., M. Pastorino, and M. Raffetto, "Electromagnetic scattering by a multilayer elliptic cylinder: Series solution in terms of Mathieu functions," *IEEE Trans. Anten. Propag.* **45**, 926–935 (1997a).

Caorsi, S., M. Pastorino, and M. Raffetto, "Scattering by a conducting elliptic cylinder with a multilayer dielectric coating," *Radio Sci.* **32**, 2155–2166 (1997b).

Caorsi, S., M. Pastorino, and M. Raffetto, "Electromagnetic scattering by a multilayer elliptic cylinder under line-source illumination," *Microwave Opt. Technol. Lett.* **24**, 322–329 (2000).

Caorsi, S. and M. Raffetto, "Perfectly matched layers for the truncation of finite element meshes in layered half-space geometries and applications to electromagnetic scattering by buried objects," *Microwave Opt. Technol. Lett.* **19**, 427–434 (1998).

Chadan, K. and P. C. Sabatier, *Inverse Problems in Quantum Theory*, Springer, Berlin, 1992.

Chari, M. V. K. and P. P. Silvester, eds., *Finite Elements in Electrical and Magnetic Field Problems*, Wiley, New York, 1980.

Chew, W. C., J. Jin, E. Michielssen, and J. Song, *Fast and Efficient Algorithms in Computational Electromagnetics*, Artech House, New York, 2001.

Chiu, C.-C. and P.-T. Liu, "Image reconstruction of a complex cylinder illuminated by TE waves," *IEEE Trans. Microwave Theory Tech.* **44**, 1921–1927 (1996).

Cho, C.-P. and Y.-W. Kiang, "Inverse scattering of dielectric cylinders by a cascaded TE-TM method," *IEEE Trans. Microwave Theory Tech.* **47**, 1923–1930 (1999).

Colton, D. and R. Kress, *Inverse Acoustic and Electromagnetic Scattering Theory*, Springer, Berlin, 1998.

Devaney, A. J., "Nonuniqueness in the inverse scattering problem," *J. Math. Phys.* **19**, 1526–1535, 1978.

Engl, H. W., M. Hanke, and A. Neubauer, *Regularization of Inverse Problems*. Kluwer Academic Publisher, Dordrecht, 1996.

Hadamard, J., "Sur les problèms aux dérivées partielles et leur significtions physiques," *Univ. Princeton Bull.* **13**, 49–52 (1902).

Hadamard, J., *Lectures on Cauchy's Problem in Linear Partial Differential Equations*, Yale Univ. Press, New Haven, CT, 1923.

Harrington, R. F., *Field Computation by Moment Methods*, Macmillan, New York, 1968.

Herman, G. T. et al., *Basic Methods of Tomography and Inverse Problems*, Adam Hilger, Bristol, UK, 1987.

Hoenders, B. J., "The uniqueness of inverse problems," in H. P. Baltes, ed., *Inverse Problems in Optics*, Springer-Verlag, New York, 1978.

Jin, J., *The Finite Element Method in Electromagnetics*, Wiley, New York, 2002.

Kakogiannos, N. B. and J. A. Roumeliotis, "Electromagnetic scattering from an infinite elliptic metallic cylinder coated by a circular dielectric one," *IEEE Trans. Microwave Theory Tech.* **38**, 1660–1666 (1990).

Keller, J. B., "Inverse problems," *Am. Math. Montly.* **83**, 107–118 (1976).

Kunz, K. S. and R. J. Luebbers, *The Finite Difference Time Domain Method for Electromagnetics*, CRC Press, Boca Raton, FL, 1993.

Livesay, D. E. and K. M. Chen, "Electromagnetic fields induced inside arbitrarily shaped biological bodies," *IEEE Trans. Microwave Theory Tech.* **22**, 1273–1280 (1974).

Mittra, R., *Computer Techniques for Electromagnetics*, Pergamon Press, New York, 1973.

Monk, P., *Finite Element Methods for Maxwell's Equations*, Oxford Univ. Press, Oxford, UK, 2003.

Moore, I. and R. Pizer, eds., *Moment Methods in Electromagnetics*, Wiley, New York, 1984.

Morse, P. M. and H. Feshback, *Methods of Theoretical Physics*, McGraw-Hill, New York, 1953.

Pastorino, M., M. Raffetto, and A. Randazzo, "Interactions between electromagnetic waves and elliptically-shaped metamaterials," *IEEE Anten. Wireless Propagat. Lett.* **4**, 165–168 (2005).

Press, W. H., B. P. Flannery, S. A. Teukolsky, and W. T. Vetterling, *Numerical Recipes in C: The Art of Scientific Computing*, Cambridge Univ. Press, Cambridge, UK, 1992.

Qing, A., "Electromagnetic imaging of two-dimensional perfectly conducting cylinders with transverse electric scattered field," *IEEE Trans. Anten. Propagat.* **50**, 1786–1794 (2002).

Ramananjoana, C., M. Lambert, and D. Lesselier, "Shape inversion from TM and TE real data by controlled evolution of level sets," *Inv. Probl.* **17**, 1585–1595 (2001).

Richmond, J. H., "Scattering by a dielectric cylinder of arbitrary cross section shape," *IEEE Trans. Anten. Propagat.* **AP-13**, 334–341 (1965).

Richmond, J. H., "Scattering by a conducting elliptic cylinder with dielectric coating," *Radio Sci.* **23**, 1061–1066 (1988).

Stone, W. R., "A review and examination of results on uniqueness in inverse problems," *Radio Sci.* **22**, 1026–1030 (1987).

Stratton, J. A., *Electromagnetic Theory*, McGraw-Hill, New York, 1941.

Sullivan, D. M., *Electromagnetic Simulation Using the FDTD Method*, IEEE Press, New York, 2000.

Taflove, A. and S. C. Hagness, *Computational Electrodynamics: The Finite-Difference Time-Domain Method*, Artech House, Norwood, MA, 2005.

Tarantola, A. *Inverse Problem Theory*, Elsevier, Amsterdam, 1987.

Tikhonov, A. N. and V. Y. Arsenin, *Solutions of Ill-Posed Problems*, Winston/Wiley, Washington, DC, 1977.

Uslenghi, P. L. E., "Exact scattering by isorefractive bodies," *IEEE Trans. Anten. Propag.* **45**, 1382–1385 (1997).

Van Bladel, J., *Electromagnetic Fields*, Wiley, Hoboken, NJ, 2007.

van den Berg, P. M. and R. E. Kleinman, "A contrast source inversion method," *Inverse Problems* **13**, 1607–1620 (1997)

van den Berg, P. M. and A. Abubakar, "Contrast source inversion method: State of art," *Progress Electromagn. Res.* **34**, 189–218 (2001).

van der Vorst, H. A., "Bi-CGSTAB: A fast and smoothly converging variant of Bi-CG for the solution of nonsymmetric linear systems," *SIAM J. Sci. Statist. Comput.* **13**, 631–644 (1992).

Wait, J. R., *Electromagnetic Waves in Stratified Media*, Pergamon Press, Oxford, UK, 1962.

Wang, J. J. H., *Generalized Moment Methods in Electromagnetics*, Wiley, New York, 1991.

Xu, X., Q. H. Liu, and Z. Q. Zhang, "The stabilized biconjugate gradient fast Fourier transform method for electromagnetic scattering," *Proc. 2002 Antennas and Propagation Society Int. Symp.*, San Antonio, Texas, USA, 614–617 (2002).

Yeh, C., "The diffraction of waves by a penetrable ribbon," *J. Math. Phys.* **4**, 65–71 (1963).

Zhang, Z. Q., Q. H. Liu, C. Xiao, E. Ward, G. Ybarra, and W. T. Joines, "Microwave breast imaging: 3D forward scattering simulation," *IEEE Trans. Biomed. Eng.* **50**, 1180–1189 (2003).

Zienkiewicz, C., *The Finite Element Method*, McGraw-Hill, New York, 1977.

Zwamborn, P. and P. M. van den Berg, "The three dimensional weak form of the conjugate gradient FFT method for solving scattering problems," *IEEE Trans. Microwave Theory Tech.* **40**, 1757–1766 (1992).

CHAPTER FOUR

Imaging Configurations and Model Approximations

4.1 OBJECTIVES OF THE RECONSTRUCTION

In practical applications, the *objectives* of the inspection can be very different. In certain cases, it can be sufficient to locate the unknown scatterers by retrieving their spatial positions starting from measurement of the scattered field performed outside an investigation domain. Methods aimed at solving this problem are actually not imaging methods, but mainly localization methods (e.g., those based on radar concepts and on smart antennas that can estimate the directions of arrival of signals related to the presence of the objects).

In other cases, the aim of the inspection is to retrieve the shape of the object, which in certain applications (discussed in the following sections) is sufficient information, as the values of the object's dielectric parameters are unnecessary or obvious. An example is represented by nondestructive testing, evaluations, and quality control of materials and products (see Chapter 10), where knowledge of the position and shape of a *defect* is often sufficient for deciding whether the structure of apparatus under test can be still used. Moreover, the information on the shape and, consequently, on the dimensions of a defect can also be an indication for determining the residual life of the apparatus or system.

However, the most *ambitious* goal of the inspection is to obtain complete images of the objects under test. Such images, when using microwaves, are directly related to the spatial distributions of the dielectric parameters of the scattering targets.

Microwave Imaging, By Matteo Pastorino
Copyright © 2010 John Wiley & Sons, Inc.

In the following sections, the most common configurations used in microwave imaging systems are presented, with particular focus on the assumption made regarding measurement points and illumination conditions. Tomographic configurations are presented in detail for their significant practical relevance. However, scanning configurations are also considered. The chapter also describes the most frequently used approximations for simplyfing, when possible, the scattering equations. Finally, the problem of numerical computation of Green's function for arbitrary inhomogeneous background is discussed.

4.2 MULTIILLUMINATION APPROACHES

In general, the object to be inspected is sequentially illuminated by using a set of incident fields that are generated by several transmitting antennas or by a single source moving around the target. For each illuminating field, the measurements are performed in an observation domain that can be the same for each incident field or can change at any illumination. If the investigation domain V_i remains unchanged during the measurement phase, the following set of integral equations [see (3.2.1) and (3.2.2)] is obtained:

$$\mathbf{E}^i(\mathbf{r}) = \mathbf{E}^i_{\text{inc}}(\mathbf{r}) + j\omega\mu_b \int_{V_i} \tau(\mathbf{r}')\mathbf{E}^i(\mathbf{r}')\cdot\bar{\mathbf{G}}(\mathbf{r}/\mathbf{r}')d\mathbf{r}', \quad \mathbf{r}\in V_m^i \quad \text{(data equation)},$$
(4.2.1)

$$\mathbf{E}^i(\mathbf{r}) = \mathbf{E}^i_{\text{inc}}(\mathbf{r}) + j\omega\mu_b \int_{V_i} \tau(\mathbf{r}')\mathbf{E}^i(\mathbf{r}')\cdot\bar{\mathbf{G}}(\mathbf{r}/\mathbf{r}')d\mathbf{r}', \quad \mathbf{r}\in V_i \quad \text{(state equation)}.$$
(4.2.2)

Here, the superscript i denotes quantities related to the ith illumination (except, of course, in the symbol for the investigation domain V_i) and V_m^i indicates the observation domain where the samples of the field are collected when the ith incident field is applied. On the other hand, if the observation domain does not change during the measurement phase, then $V_m^i = V_m$ for any i. By considering equations (4.2.1) and (4.2.2), one can observe that the object function does not depend on i, whereas the total internal field changes for any illumination.

Analogous relations hold if the contrast source formulation is adopted [see (3.2.3) and (3.2.4)]

$$\mathbf{E}^i(\mathbf{r}) = \mathbf{E}^i_{\text{inc}}(\mathbf{r}) + j\omega\mu_b \int_{V_i} \mathbf{J}^i_{\text{eq}}(\mathbf{r}')\cdot\bar{\mathbf{G}}(\mathbf{r}/\mathbf{r}')d\mathbf{r}', \quad \mathbf{r}\in V_m^i \text{ (data equation)}, \quad (4.2.3)$$

$$\mathbf{J}^i_{\text{eq}}(\mathbf{r}) = \tau(\mathbf{r})\mathbf{E}^i_{\text{inc}}(\mathbf{r}) + j\omega\mu_b\tau(\mathbf{r})\int_{V_i} \mathbf{J}^i_{\text{eq}}(\mathbf{r}')\cdot\bar{\mathbf{G}}(\mathbf{r}/\mathbf{r}')d\mathbf{r}', \quad \mathbf{r}\in V_i \text{ (state equation)},$$
(4.2.4)

where the contrast sources are dependent on the considered incident field.

4.3 TOMOGRAPHIC CONFIGURATIONS

A classical example of the multiillumination process is represented by *microwave tomography*, which is used for two-dimensional imaging (Kak and Slaney 1988). With reference to Figure 4.1, in tomographic applications, the observation domain is usually a circle including the cross section of the target to be inspected or a rotating straight line always tangent to a circle surrounding the target. Under transverse magnetic (TM) illumination conditions, the two-dimensional counterparts of equations (4.2.1) and (4.2.2) are

$$E_z^i(\mathbf{r}_t) = E_{\text{inc}_z}^i(\mathbf{r}_t) + j\omega\mu_b \int_{S_i} \tau(\mathbf{r}_t') E_z^i(\mathbf{r}_t') G_{2D}(\mathbf{r}_t/\mathbf{r}_t') d\mathbf{r}_t', \quad \mathbf{r}_t \in S_m^i \quad \text{(data equation)},$$
(4.3.1)

$$E_z^i(\mathbf{r}_t) = E_{\text{inc}_z}^i(\mathbf{r}_t) + j\omega\mu_b \int_{S_i} \tau(\mathbf{r}_t') E_z^i(\mathbf{r}_t') G_{2D}(\mathbf{r}_t/\mathbf{r}_t') d\mathbf{r}_t', \quad \mathbf{r}_t \in S_i \quad \text{(state equation)},$$
(4.3.2)

where S_m^i indicates the observation domain where the measured values related to the *i*th illumination are collected. Analogous relationships hold if the contrast source formulation is used:

$$E_z^i(\mathbf{r}_t) = E_{\text{inc}_z}^i(\mathbf{r}_t) + j\omega\mu_b \int_{S_i} J_z^i(\mathbf{r}_t') G_{2D}(\mathbf{r}_t/\mathbf{r}_t') d\mathbf{r}_t', \quad \mathbf{r}_t \in S_m^i \quad \text{(data equation)}, \quad (4.3.3)$$

$$J_z^i(\mathbf{r}_t) = \tau(\mathbf{r}_t) E_{\text{inc}_z}^i(\mathbf{r}_t) + j\omega\mu_b \tau(\mathbf{r}_t) \int_{S_i} J_z^i(\mathbf{r}_t') G_{2D}(\mathbf{r}_t/\mathbf{r}_t') d\mathbf{r}_t', \quad \mathbf{r}_t \in S_i \quad \text{(state equation)}.$$
(4.3.4)

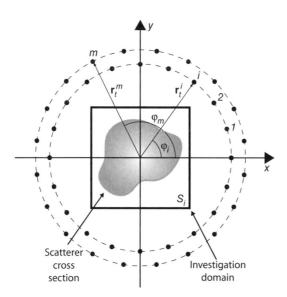

FIGURE 4.1 Two-dimensional imaging configuration, showing object cross section and investigation area and positions of the sources and measurement points.

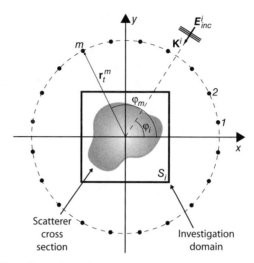

FIGURE 4.2 Two-dimensional imaging configuration; plane-wave illumination.

The target is successively illuminated by S incident fields incoming from different directions. If plane-wave illumination can be assumed (Fig. 4.2), then, according to equation (2.4.24), the incident field is given by

$$\mathbf{E}^i(\mathbf{r}_t) = E^i_{\text{inc}_z}(\mathbf{r}_t)\hat{\mathbf{z}} = E^i_0 e^{-j\mathbf{k}_i \cdot \mathbf{r}}\hat{\mathbf{z}}, \quad i = 1, \ldots, S, \qquad (4.3.5)$$

where $\mathbf{k}^i = k_b(-\cos\varphi_i\hat{\mathbf{x}} - \sin\varphi_i\hat{\mathbf{y}})$. If the incident waves are equally angularly spaced, the incident angle φ_{inc} assumes the following values:

$$\varphi_i = (i-1)\frac{2\pi}{S} \quad i = 1, \ldots, S. \qquad (4.3.6)$$

In other cases, as mentioned in Section 2.4, the illuminating field is modeled as the one produced by a line-current source (Fig. 4.3). In that case, taking into account equation (2.4.26), we obtain

$$\mathbf{E}^i(\mathbf{r}_t) = -I^i \frac{\omega\mu_b}{4} H_0^{(2)}(k_b \rho_t^i)\hat{\mathbf{z}}, \quad i = 1, \ldots, S, \qquad (4.3.7)$$

where $\rho_t^i = |\mathbf{r}_t - \mathbf{r}_t^i|$, with I^i and $\mathbf{r}_t^i$ such that $\mathbf{J}^i(\mathbf{r}_t) = I^i \delta(\mathbf{r}_t - \mathbf{r}_t^i)\hat{\mathbf{z}}$. Obviously, in the plane $z = 0$, the ith line-current source passes through point $\mathbf{r}_t^i$. If the sources are located on a circumference or radius ρ_s, (with polar coordinates), then $\mathbf{r}_t^i = \rho_s \hat{\rho}(\varphi_i)$. Moreover, if points $\mathbf{r}_t^i, i = 1, \ldots, S$, are equally spaced on this circle, φ_i is again as given by equation (4.3.6).

It should be noted that the source can be modeled more realistically by factoring in the radiation pattern of the transmitting antenna. However, in

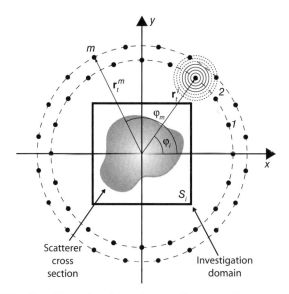

FIGURE 4.3 Two-dimensional imaging configuration; line-current sources.

most cases, the radiation pattern is sufficiently wide (in the azimuth plane $z = 0$) to illuminate the entire investigation domain, so that the above models for the incident field are usually adopted.

A simple method for shaping the incident field to better approximate the radiation properties of the transmitting antenna is to use a relation similar to Silver's formula (Balanis 2005), which is valid for the main lobe of the antenna. In this case (Fig. 4.4), for an antenna located at $\mathbf{r}_t^i$ and radiating toward the center of the coordinate system (which coincides with the center of the investigation domain), the following relation holds

$$\begin{cases} \mathbf{E}^i(\mathbf{r}_t) = -I^i \dfrac{\omega \mu_b}{4} H_0^{(2)}(k_b \rho_t^i) \cos^q\!\left(\tilde{\varphi}_i \dfrac{\pi}{\alpha_i}\right)\hat{\mathbf{z}}, & |\tilde{\varphi}_i| \leq \dfrac{\alpha_i}{2}, \\ \mathbf{E}^i(\mathbf{r}_t) = 0, & |\tilde{\varphi}_i| > \dfrac{\alpha_i}{2}, \end{cases} \quad (4.3.8)$$

where α_i is the beamwidth of the main lobe of the ith antenna in the $z = 0$ plane and $\tilde{\varphi}_i$ is the azimuth angle in the local coordinate system of the ith antenna, which can be expressed as follows:

$$\tilde{\varphi}_i = \cos^{-1}\!\left(\dfrac{(\rho_t^i)^2 + |\mathbf{r}_t^i|^2 - |\mathbf{r}_t|^2}{2|\mathbf{r}_t^i|\rho_t^i}\right). \quad (4.3.9)$$

In equation (4.3.8), q is an exponent controlling the directivity of the antenna; the more directive is the antenna, the higher is q. For isotropic (in the transverse plane) antennas, the result is $q = 0$.

62 IMAGING CONFIGURATIONS AND MODEL APPROXIMATIONS

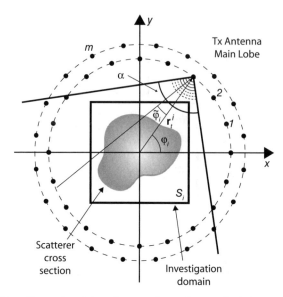

FIGURE 4.4 Two-dimensional imaging configuration; main lobe of the transmitting antenna.

As mentioned in Section 3.4, the measurements are usually performed in a discrete set of points. These measurement points can completely surround the object cross section; that is, they can be located on a circumference of radius ρ_M. This circumference may ($\rho_M = \rho_S$) or may not ($\rho_M \neq \rho_S$) coincide with the one in which the sources are located. In this case, the measurement points (assumed to be the same for any illumination) are located at $\mathbf{r}_t^m = \rho_M \hat{\rho}(\varphi_m)$, $m = 1, \ldots, M$, where

$$\varphi_m = (m-1)\frac{2\pi}{M}, \quad m = 1, \ldots, M. \tag{4.3.10}$$

Sometimes, one tries to avoid placing measurement points near the sources. This can be achieved by locating the probes, for the ith illumination, at points $\mathbf{r}_t^{mi} = \rho_M \hat{\rho}(\varphi_{mi})$, $m = 1, \ldots, M$, $i = 1, \ldots, S$, such that

$$\varphi_{mi} = \frac{2\pi - \beta}{M-1}(m-1) + \varphi_i + \frac{\beta}{2}, \quad m = 1, \ldots, M, i = 1, \ldots, S, \tag{4.3.11}$$

where β is an angular sector *around* the incident direction φ_i, where no measurement points are present.

In other cases, the measurement points can be located on a straight line on the opposite side of the object from the source of the incident field (Fig. 4.5). This measurement line (*probing line*) should ideally be infinite in order to

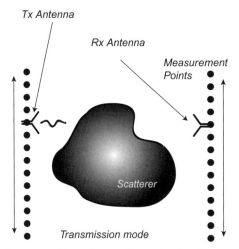

FIGURE 4.5 Two-dimensional imaging configuration for data acquisition in transmission mode; linear probing line.

measure the entire scattered wave. In real applications, the probing line should be long enough to capture most of the energy of the scattered wave. In this case, a set of M measurement points of coordinates $\mathbf{r}_t^{mi} = x^{mi}\hat{\mathbf{x}} + y^{mi}\hat{\mathbf{y}}$, $m = 1,\ldots, M, i = 1,\ldots, S$, such that

$$y^{mi} - y_0^i = \kappa^i\left(x^{mi} - x_0^i\right), \quad (4.3.12)$$

where $\mathbf{r}_0^i = x_0^i\hat{\mathbf{x}} + y_0^i\hat{\mathbf{y}}, i = 1,\ldots, S$ is the center of the linear probing line for the ith view and κ^i is the related angular coefficient.

4.4 SCANNING CONFIGURATIONS

With reference to Figure 4.6, the inspection can be performed on a single side of the object. In this case, the reflection is the main measured contribution to the input data. A set of measurement M points, $\mathbf{r}_t^m = x^m\hat{\mathbf{x}} + y^m\hat{\mathbf{y}}, m = 1,\ldots, M$, are usually located on a straight line (although other configurations are possible), for which equation (4.3.12) still holds. A single probe can be used (Fig. 4.6), which is moved to occupy successively the various positions $\mathbf{r}_t^m$. In this case, the probe location $\mathbf{r}_t^s$ coincides with a given measurement point. At each position of the probe, the scattered field is measured (monostatic configuration). An image of the target based only on the measured scattered field can be constructed.

Otherwise, the measured values can be stored and postprocessed in order to combine the reflected signals to obtain a synthetic aperture focusing (see Chapter 5).

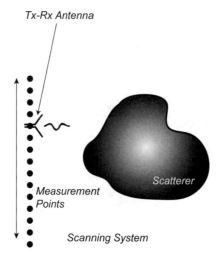

FIGURE 4.6 Two-dimensional imaging configuration for data acquisition in reflection mode; scanning systems, with single probe and linear probing line.

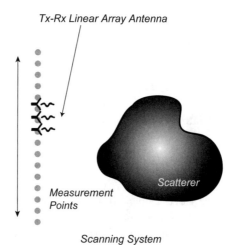

FIGURE 4.7 Two-dimensional imaging configuration for data acquisition in reflection mode; scanning systems, with linear array probe.

An array of probes can also be used (Fig. 4.7). Consequently, for each position of the transmitting element $\mathbf{r}_t^s$, the field scattered can be collected in a set of points $\mathbf{r}_t^m$, $m = 1,\ldots, Q$, in which Q can coincide with M or can be a subset of the scanned domain ($Q \leq M$).

4.5 CONFIGURATIONS FOR BURIED-OBJECT DETECTION

To detect a shallow buried object, the source and the probes are usually located in the upper region (Fig. 4.8). In this case, the measurement points are $\mathbf{r}_t^m = x^m \hat{\mathbf{x}} + y_0 \hat{\mathbf{y}}$, where y_0 is the distance of the probing line from the interface between air and the medium in which the target is located (e.g., the soil), assumed to be the line $y = 0$. Accordingly, in equation (4.3.12), $\kappa^i = 0$. Moreover, the same scanning approaches discussed in Section 4.4 can be adopted.

When possible, a borehole configuration can be used. With reference to Figure 4.9, the source(s) and the probe(s) can be located inside holes. Actually, this technique is used mainly at lower frequencies and usually results in better reconstructions, due to the possibility of accounting for more *contributions* to the scattering mechanism. In the borehole configuration, the points in which the sources are positioned are denoted as $\mathbf{r}_t^s = x_1 \hat{\mathbf{x}} + y^s \hat{\mathbf{y}}, s = 1, \ldots, S$, with $y^s < 0$ and x_1 denoting the x coordinate of the center of the source hole, whereas the measurement points are $\mathbf{r}_t^m = x_2 \hat{\mathbf{x}} + y^m \hat{\mathbf{y}}, m = 1, \ldots, M, y^m < 0$, where x_2 is the x coordinate of the center of the measurement hole.

4.6 BORN-TYPE APPROXIMATIONS

In some cases, information on the target under test are available. Consequently, approximations of the model can be introduced. When the scatterer to be inspected is weak with respect to the propagation medium, Born-type

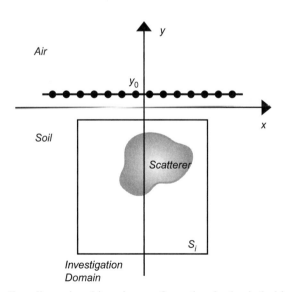

FIGURE 4.8 Two-dimensional imaging configuration for buried-object detection.

66 IMAGING CONFIGURATIONS AND MODEL APPROXIMATIONS

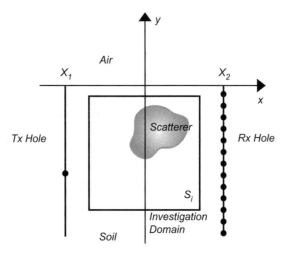

FIGURE 4.9 Two-dimensional configuration for bore-hole imaging of buried objects.

approximations can be used (Morse and Feshbach 1953, Born and Wolf 1965). The simplest approximation is the first-order Born approximation, which assumes that the scattered electric field in equation (2.7.2) can be written only in terms of the incident field:

$$\mathbf{E}(\mathbf{r}) = \mathbf{E}_{inc}(\mathbf{r}) + j\omega\mu_b \int_{V_i} \tau(\mathbf{r}')\mathbf{E}(\mathbf{r}') \cdot \bar{\mathbf{G}}(\mathbf{r}/\mathbf{r}')d\mathbf{r}'$$
$$\approx \mathbf{E}_{inc}(\mathbf{r}) + j\omega\mu_b \int_{V_i} \tau(\mathbf{r}')\mathbf{E}_{inc}(\mathbf{r}') \cdot \bar{\mathbf{G}}(\mathbf{r}/\mathbf{r}')d\mathbf{r}'. \quad (4.6.1)$$

Accordingly, we shall write

$$\mathbf{E}^{1B}(\mathbf{r}) = \mathbf{E}_{inc}(\mathbf{r}) + j\omega\mu_b \int_{V_i} \tau(\mathbf{r}')\mathbf{E}_{inc}(\mathbf{r}') \cdot \bar{\mathbf{G}}(\mathbf{r}/\mathbf{r}')d\mathbf{r}', \quad (4.6.2)$$

where the superscript 1B indicates the approximation presented above.

The Born approximation can be used for both the direct scattering problem (i.e., to compute the field scattered by a known weakly scattering object) and the inverse scattering problem (i.e., to reconstruct an unknown target). In particular, in the inverse scattering problem under the approximation given above, the scattering equation (2.7.2) is *linearized*, since the only problem unknown is the object function $\tau(\mathbf{r})$, $\mathbf{r} \in V_i$, where V_i is the investigation domain defined in Section 3.2. More precisely, the first-order Born approximation is valid for weak scatterers for which (Natterer 2004)

$$k_b a \sup_{|\mathbf{r}|<a} |\varepsilon_r(\mathbf{r}) - 1| < 2\pi\varsigma, \qquad (4.6.3)$$

where a is the radius of the minimum circle that can include the object cross section and ς is a constant that Slaney et al. (1984) set to 0.25, even if different estimations have been proposed (Natterer 2004, Chen and Stamnes 1998). The choice $\varsigma = 0.25$ is obtained in the Slaney et al. (1984) study by considering a plane wave impinging onto a cylindrical scatterer and requiring that the phase change between the incident wave and the field traveling through the scatterer be less than π.

Higher-order Born approximations can be obtained in a recursive way. In particular, a Born series can be constructed by using the following recursive relations

$$\mathbf{E}^{nB}(\mathbf{r}) = \mathbf{E}_{\text{inc}}(\mathbf{r}) + j\omega\mu_b \int_{V_i} \tau(\mathbf{r}') \mathbf{E}^{(n-1)B}(\mathbf{r}') \cdot \bar{\mathbf{G}}(\mathbf{r}/\mathbf{r}') d\mathbf{r}', \quad n \geq 1, \qquad (4.6.4)$$

$$\mathbf{E}^{0B}(\mathbf{r}) = \mathbf{E}_{\text{inc}}(\mathbf{r}), \qquad (4.6.5)$$

where the superscript nB denotes the nth-order approximation. Although this series can be used for a fast solution of the direct scattering problem, it converges only for weak scatterers. Moreover, in microwave imaging, besides the first-order approximation, the second-order approximation has been considered in some studies (Pierri et al. 2000, Estatico et al. 2005). In this case ($n = 2$), and according to (4.6.4), we have

$$\begin{aligned}\mathbf{E}^{2B}(\mathbf{r}) &= \mathbf{E}_{\text{inc}}(\mathbf{r}) + j\omega\mu_b \int_{V_i} \tau(\mathbf{r}') \mathbf{E}^{1B}(\mathbf{r}') \cdot \bar{\mathbf{G}}(\mathbf{r}/\mathbf{r}') d\mathbf{r}' \\ &= \mathbf{E}_{\text{inc}}(\mathbf{r}) + j\omega\mu_b \int_{V_i} \tau(\mathbf{r}') \Big[\mathbf{E}_{\text{inc}}(\mathbf{r}') + j\omega\mu_b \int_{V_i} \tau(\mathbf{r}'') \mathbf{E}_{\text{inc}}(\mathbf{r}'') \cdot \bar{\mathbf{G}}(\mathbf{r}'/\mathbf{r}'') d\mathbf{r}'' \Big] \\ &\quad \cdot \bar{\mathbf{G}}(\mathbf{r}/\mathbf{r}') d\mathbf{r}' \end{aligned} \qquad (4.6.6)$$

which is a quadratic equation with respect to the object function. Moreover, it seems that it can limitedly extend, in practice, the range of applicability of the first-order approximation. The key point in applying this approximation is that the resulting inverse problem is still nonlinear, but only the object function is unknown. This represents a significant savings in computation in comparison to the approaches for which the total electric field inside the target is kept as an additional unknown.

As has been shown (Estatico et al. 2005), the second-order Born approximation provides better reconstructions when the Born series converges [essentially, under condition (4.6.3)].

In two-dimensional problems (see Section 3.3), equation (4.6.2) becomes

$$\begin{aligned} E_z^{1B}(\mathbf{r}_t) &= E_{\text{inc}_z}(\mathbf{r}_t) + E_{\text{scat}_z}^{1B}(\mathbf{r}_t) = E_{\text{inc}_z}(\mathbf{r}_t) \\ &\quad + j\omega\mu_b \int_{S_i} \tau(\mathbf{r}_t') E_{\text{inc}_z}(\mathbf{r}_t') G_{2D}(\mathbf{r}_t/\mathbf{r}_t') d\mathbf{r}_t' \end{aligned} \qquad (4.6.7)$$

where $E_{\text{inc}_z}(\mathbf{r}_t)$ and $E_{\text{scat}_z}^{1B}(\mathbf{r}_t)$ turn out to be the solutions of the scalar Helmholtz equations

$$(\nabla^2 + k_b^2)E_{\text{inc}_z}(\mathbf{r}_t) = j\omega\mu_b J_z(\mathbf{r}_t), \qquad (4.6.8)$$

$$(\nabla^2 + k_b^2)E_{\text{scat}_z}^{1B}(\mathbf{r}_t) = j\omega\mu_b J_{\text{eq}_z}^{1B}(\mathbf{r}_t), \qquad (4.6.9)$$

where J_z and $J_{\text{eq}_z}^{1B}$ are z components of the impressed source [equation (2.3.13)] and of the equivalent current density [equation (3.3.15)] written under the first-order Born approximation:

$$J_{\text{eq}_z}^{1B}(\mathbf{r}_t) = \tau(\mathbf{r}_t)E_{\text{inc}_z}(\mathbf{r}_t). \qquad (4.6.10)$$

Obviously, equations (4.6.8) and (4.6.9) can be deduced with some trivial mathematical steps from the corresponding vector wave equations (in the form of equation (2.4.1)) by using the transverse-magnetic assumptions stated in Section 3.3 (Chew 1990).

4.7 EXTENDED BORN APPROXIMATION

Let us consider a point $\mathbf{r} \in V_o$ located inside the object. By factoring in the definition of the equivalent source in terms of scattering potential, we can rewrite equation (2.7.1) as

$$\mathbf{E}(\mathbf{r}) = \mathbf{E}_{\text{inc}}(\mathbf{r}) + j\omega\mu_b\left(\bar{\mathbf{I}} + \frac{1}{k_b^2}\nabla\nabla\right) \cdot \int_{V_o} \tau(\mathbf{r}')\mathbf{E}(\mathbf{r}')G\mathbf{r}/\mathbf{r}'d\mathbf{r}'. \qquad (4.7.1)$$

If the quantity $j\omega\mu_b \int_{V_o} \tau(\mathbf{r}')\bar{G}(\mathbf{r}/\mathbf{r}') \cdot \mathbf{E}(\mathbf{r})d\mathbf{r}'$ is added and subtracted from equation (4.7.1), we obtain (Habashy et al. 1993)

$$\mathbf{E}(\mathbf{r}) = \mathbf{E}_{\text{inc}}(\mathbf{r}) + j\omega\mu_b\left[\left(\bar{\mathbf{I}} + \frac{1}{k_b^2}\nabla\nabla\right)\int_{V_o} \tau(\mathbf{r}')G(\mathbf{r}/\mathbf{r}')d\mathbf{r}'\right]$$
$$\cdot \mathbf{E}(\mathbf{r}) + j\omega\mu_b \int_{V_o} \tau(\mathbf{r}')\bar{G}\mathbf{r}/\mathbf{r}' \cdot (\mathbf{E}(\mathbf{r}') - \mathbf{E}(\mathbf{r}))d\mathbf{r}', \qquad (4.7.2)$$

or, equivalently

$$\mathbf{E}(\mathbf{r}) = \left(\bar{\mathbf{I}} - j\omega\mu_b\left[\left(\bar{\mathbf{I}} + \frac{1}{k_b^2}\nabla\nabla\right)\int_{V_o} \tau(\mathbf{r}')G(\mathbf{r}/\mathbf{r}')d\mathbf{r}'\right]\right)^{-1}$$
$$\cdot \left(\mathbf{E}_{\text{inc}}(\mathbf{r}) + j\omega\mu_b \int_{V_o} \tau(\mathbf{r}')\bar{G}(\mathbf{r}/\mathbf{r}') \cdot (\mathbf{E}(\mathbf{r}') - \mathbf{E}(\mathbf{r}))d\mathbf{r}'\right). \qquad (4.7.3)$$

If one now considers the fact that Green's function is highly peaked when $\mathbf{r} \to \mathbf{r}'$ (due to its singularity) and, on the contrary, tends to zero when $|\mathbf{r} - \mathbf{r}'| \to \infty$, the following approximate relation can be obtained:

$$\mathbf{E}(\mathbf{r}) \approx \left(\overline{\mathbf{I}} - j\omega\mu_b \left[\left(\overline{\mathbf{I}} + \frac{1}{k_b^2} \nabla\nabla \right) \int_{V_o} \tau(\mathbf{r}') G(\mathbf{r}/\mathbf{r}') d\mathbf{r}' \right] \right)^{-1} \cdot \mathbf{E}_{inc}(\mathbf{r}). \quad (4.7.4).$$

As a consequence, we have

$$\mathbf{E}(\mathbf{r}) = \mathbf{E}_{inc}(\mathbf{r}) + j\omega\mu_b \int_{V_o} \tau(\mathbf{r}')$$
$$\left(\left[\overline{\mathbf{I}} - j\omega\mu_b \left[\left(\overline{\mathbf{I}} + \frac{1}{k_b^2} \nabla\nabla \right) \int_{V_o} \tau(\mathbf{r}'') G(\mathbf{r}'/\mathbf{r}'') d\mathbf{r}'' \right] \right]^{-1} \cdot \mathbf{E}_{inc}(\mathbf{r}') \right) \cdot \overline{\mathbf{G}}(\mathbf{r}/\mathbf{r}') d\mathbf{r}'.$$
(4.7.5)

Equation (4.7.5) is a nonlinear integral equation in the only unknown τ, which can be shown to extend the range of validity of the first-order Born approximation (Habashy et al. 1993). The approximation adopted to derive such a scattering equation is referred to as the *extended Born approximation*, which can be also used in two-dimensional scattering problem with TM illumination (see Section 3.3). To this end, let us consider equation (3.3.8), for $\mathbf{r}_t \in S_o$. An equivalent integral equation can be derived from that equation if the quantity $j\omega\mu_b \int_{S_o} \tau(\mathbf{r}_t') E_z(\mathbf{r}_t) G_{2D}(\mathbf{r}_t/\mathbf{r}_t') d\mathbf{r}_t'$ is added and subtracted on the right-hand side:

$$E_z(\mathbf{r}_t) = E_{inc_z}(\mathbf{r}_t) + j\omega\mu_b \int_{S_o} \tau(\mathbf{r}_t') E_z(\mathbf{r}_t) G_{2D}(\mathbf{r}_t/\mathbf{r}_t') d\mathbf{r}_t'$$
$$+ j\omega\mu_b \int_{S_o} \tau(\mathbf{r}_t')[E_z(\mathbf{r}_t') - E_z(\mathbf{r}_t)] G_{2D}(\mathbf{r}_t/\mathbf{r}_t') d\mathbf{r}_t'. \quad (4.7.6)$$

By rearranging terms, we obtain

$$\left[1 - j\omega\mu_b \int_{S_o} \tau(\mathbf{r}_t') G_{2D}(\mathbf{r}_t/\mathbf{r}_t') d\mathbf{r}_t' \right] E_z(\mathbf{r}_t)$$
$$= E_{inc_z}(\mathbf{r}_t) + j\omega\mu_b \int_{S_o} \tau(\mathbf{r}_t')[E_z(\mathbf{r}_t') - E_z(\mathbf{r}_t)] G_{2D}(\mathbf{r}_t/\mathbf{r}_t') d\mathbf{r}_t' \quad (4.7.7)$$

As a consequence, the internal electric field can be expressed as

$$E_z(\mathbf{r}_t) = \frac{E_{inc_z}(\mathbf{r}_t)}{1 - j\omega\mu_b \int_{S_o} \tau(\mathbf{r}_t') G_{2D}(\mathbf{r}_t/\mathbf{r}_t') d\mathbf{r}_t'} + \frac{j\omega\mu_b \int_{S_o} \tau(\mathbf{r}_t')[E_z(\mathbf{r}_t') - E_z(\mathbf{r}_t)] G_{2D}(\mathbf{r}_t/\mathbf{r}_t') d\mathbf{r}_t'}{1 - j\omega\mu_b \int_{S_o} \tau(\mathbf{r}_t') G_{2D}(\mathbf{r}_t/\mathbf{r}_t') d\mathbf{r}_t'}.$$
(4.7.8)

It is worth noting that no approximation has been introduced so far. However, since Green's function in this case is also a highly peaked function when $\mathbf{r}_t \to \mathbf{r}'_t$, because of its singularity but tends to zero for $|\mathbf{r}_t - \mathbf{r}'_t| \to \infty$, the second term on the right-hand side can be neglected and the extended Born approximation is obtained for the internal field in two-dimensional TM scattering as follows:

$$E_z(\mathbf{r}_t) \approx \left[1 - j\omega\mu_b \int_{S_o} \tau(\mathbf{r}'_t) G_{2D}(\mathbf{r}_t/\mathbf{r}'_t) d\mathbf{r}'_t \right]^{-1} E_{\text{inc}_z}(\mathbf{r}_t). \quad (4.7.9)$$

The corresponding integral equation describing the scattering phenomena then reads as

$$E_z(\mathbf{r}_t) = E_{\text{inc}_z}(\mathbf{r}_t) + j\omega\mu_b \int_{S_o} \tau(\mathbf{r}'_t) \left[1 - j\omega\mu_b \int_{S_o} \tau(\mathbf{r}''_t) G_{2D}(\mathbf{r}'_t/\mathbf{r}''_t) d\mathbf{r}''_t \right]^{-1} E_{\text{inc}_z}(\mathbf{r}'_t) G_{2D}(\mathbf{r}_t/\mathbf{r}'_t) d\mathbf{r}'_t. \quad (4.7.10)$$

As already observed for three-dimensional configurations, the equation obtained is still nonlinear, but involves as unknown only the scattering potential. Moreover, the range of validity of such an approximation has been shown to be wider than the classical Born one (Zhang and Liu 2001, Habashy et al. 1994, Song and Liu 2005, Torres-Verdin and Habashy 2001).

For the sake of completeness, it is worth remarking that the extended Born approximation presented above can be generalized and extended to higher-order approximations, as described in by Gao and Torres-Verdin (2006) and in Cui et al. (2004).

4.8 RYTOV APPROXIMATION

The Rytov approximation is another first-order approximation, which is based on the phase of the electromagnetic field. In this section, for simplicity, only the scattering by two-dimensional configurations under TM illumination conditions is considered.

The first step for applying the Rytov approximation, consists in expressing the incident field as (Slaney et al. 1984)

$$E_{\text{inc}_z}(\mathbf{r}_t) = e^{\Phi_{\text{inc}}(\mathbf{r}_t)}, \quad (4.8.1)$$

where $\Phi_{\text{inc}}(\mathbf{r}_t)$ is the incident complex phase function. Analogously, the total field is written as

$$E_z(\mathbf{r}_t) = e^{\Phi(\mathbf{r}_t)}, \quad (4.8.2)$$

where $\Phi(\mathbf{r}_t)$ is the total complex phase function. By using $\Phi_{\text{inc}}(\mathbf{r}_t)$ and $\Phi(\mathbf{r}_t)$, one can define the scattered phase:

$$\Phi_{\text{scat}}(\mathbf{r}_t) = \Phi(\mathbf{r}_t) - \Phi_{\text{inc}}(\mathbf{r}_t). \quad (4.8.3)$$

RYTOV APPROXIMATION

Since

$$\nabla E_z(\mathbf{r}_t) = e^{\Phi(\mathbf{r}_t)} \nabla \Phi(\mathbf{r}_t) \tag{4.8.4}$$

and

$$\nabla^2 E_z(\mathbf{r}_t) = \nabla \cdot [\nabla E_z(\mathbf{r}_t)], \tag{4.8.5}$$

we obtain

$$\nabla^2 E_z(\mathbf{r}_t) = \nabla \cdot \left[e^{\Phi(\mathbf{r}_t)} \nabla \Phi(\mathbf{r}_t)\right]. \tag{4.8.6}$$

Moreover, because

$$\nabla \cdot (\psi \mathbf{F}) = \psi \nabla \cdot \mathbf{F} + \nabla \psi \cdot \mathbf{F} \tag{4.8.7}$$

holds for a scalar field and a vector field, ψ and $\mathbf{F}$, respectively, it follows that

$$\nabla^2 E_z(\mathbf{r}_t) = e^{\Phi(\mathbf{r}_t)} \nabla \cdot \nabla \Phi(\mathbf{r}_t) + \nabla e^{\Phi(\mathbf{r}_t)} \cdot \nabla \Phi(\mathbf{r}_t) = e^{\Phi(\mathbf{r}_t)} \{\nabla^2 \Phi(\mathbf{r}_t) + [\nabla \Phi(\mathbf{r}_t)]^2\}. \tag{4.8.8}$$

Let us now consider the scalar Helmholtz equation for the scattered field produced by the equivalent current density

$$\left(\nabla^2 + k_b^2\right) E_{\text{scat}_z}(\mathbf{r}_t) = j\omega\mu_b J_{\text{eq}_z}(\mathbf{r}_t) = j\omega\mu_b \tau(\mathbf{r}_t) E_z(\mathbf{r}_t), \tag{4.8.9}$$

which is analogous to equation (4.6.9) with the exact equivalent current density $J_{\text{eq}_z}(\mathbf{r}_t)$ instead of $J_{\text{eq}_z}^{1B}(\mathbf{r}_t)$. Since the incident field satisfies the Helmholtz equation (4.6.8), for points inside the investigation domain (i.e., $\mathbf{r}_t \in S_i$), where the impressed source is equal to zero, from (4.8.9) and (4.6.8) it follows that

$$\left(\nabla^2 + k_b^2\right) E_z(\mathbf{r}_t) = j\omega\mu_b \tau(\mathbf{r}_t) E_z(\mathbf{r}_t), \quad \mathbf{r}_t \in S_i. \tag{4.8.10}$$

By substituting (4.8.8) into (4.8.10), and taking into account (4.8.2) we obtain the following Riccati equation

$$\nabla^2 \Phi(\mathbf{r}_t) + [\nabla \Phi(\mathbf{r}_t)]^2 + k_b^2 = j\omega\mu_b \tau(\mathbf{r}_t) \tag{4.8.11}$$

and, using (4.8.3), we have

$$\nabla^2 \Phi_{\text{inc}} + \nabla^2 \Phi_{\text{scat}} + [\nabla \Phi_{\text{inc}}(\mathbf{r}_t)]^2 + [\nabla \Phi_{\text{scat}}(\mathbf{r}_t)]^2 + 2\nabla \Phi_{\text{inc}}(\mathbf{r}_t) \cdot \nabla \Phi_{\text{scat}}(\mathbf{r}_t) + k_b^2 \\ = j\omega\mu_b \tau(\mathbf{r}_t). \tag{4.8.12}$$

With similar mathematical steps, starting from the equation for the incident field [equation (4.6.8), which is a homogeneous equation for $\mathbf{r}_t \in S_i$], we obtain

$$\nabla^2 \Phi_{\text{inc}} + [\nabla \Phi_{\text{inc}}(\mathbf{r}_t)]^2 + k_b^2 = 0, \quad \mathbf{r}_t \in S_i. \tag{4.8.13}$$

Using (4.8.12) and (4.8.13), we then obtain

$$2\nabla \Phi_{\text{inc}}(\mathbf{r}_t) \cdot \nabla \Phi_{\text{scat}}(\mathbf{r}_t) + \nabla^2 \Phi_{\text{scat}} = -[\nabla \Phi_{\text{scat}}(\mathbf{r}_t)]^2 + j\omega\mu_b \tau(\mathbf{r}_t). \tag{4.8.14}$$

Moreover, from the following identity

$$\nabla^2 [E_{\text{inc}}(\mathbf{r}_t) \Phi_{\text{scat}}(\mathbf{r}_t)]$$
$$= \Phi_{\text{scat}}(\mathbf{r}_t) \nabla^2 E_{\text{inc}}(\mathbf{r}_t) + 2\nabla E_{\text{inc}}(\mathbf{r}_t) \cdot \nabla \Phi_{\text{scat}}(\mathbf{r}_t) + E_{\text{inc}}(\mathbf{r}_t) \nabla^2 \Phi_{\text{scat}}(\mathbf{r}_t), \tag{4.8.15}$$

we deduce that

$$\nabla^2 [E_{\text{inc}}(\mathbf{r}_t) \Phi_{\text{scat}}(\mathbf{r}_t)] + k_b^2 E_{\text{inc}}(\mathbf{r}_t) \Phi_{\text{scat}}(\mathbf{r}_t)$$
$$= E_{\text{inc}}(\mathbf{r}_t) [2\nabla \Phi_{\text{inc}}(\mathbf{r}_t) \cdot \nabla \Phi_{\text{scat}}(\mathbf{r}_t) + E_{\text{inc}}(\mathbf{r}_t) \nabla^2 \Phi_{\text{scat}}(\mathbf{r}_t)]. \tag{4.8.16}$$

In deriving equation (4.8.16), we have accounted for

$$\nabla E_{\text{inc}_z}(\mathbf{r}_t) = e^{\Phi_{\text{inc}}(\mathbf{r}_t)} \nabla \Phi_{\text{inc}}(\mathbf{r}_t) \tag{4.8.17}$$

and

$$\nabla^2 E_{\text{inc}_z}(\mathbf{r}_t) = -k_b^2 E_{\text{inc}_z}(\mathbf{r}_t). \tag{4.8.18}$$

Finally, comparing (4.8.16) with (4.8.14), we obtain

$$(\nabla^2 + k_b^2) E_{\text{inc}_z}(\mathbf{r}_t) \Phi_{\text{scat}}(\mathbf{r}_t) = E_{\text{inc}_z}(\mathbf{r}_t) \{j\omega\mu_b \tau(\mathbf{r}_t) - [\nabla \Phi_{\text{scat}}(\mathbf{r}_t)]^2\}. \tag{4.8.19}$$

The Rytov approximation requires that the variations of the complex scattered phase be neglected for $\mathbf{r}_t \in S_i$ (Ishimaru 1978). Accordingly, the solution of equation (4.8.19) can be written in terms of Green's function for two-dimensional configurations as

$$\Phi_{\text{scat}}(\mathbf{r}_t) \approx \Phi_{\text{scat}}^R(\mathbf{r}_t) = \frac{j\omega\mu_b}{E_{\text{inc}_z}(\mathbf{r}_t)} \int_{S_i} \tau(\mathbf{r}_t') E_{\text{inc}_z}(\mathbf{r}_t') G_{2D}(\mathbf{r}_t/\mathbf{r}_t') d\mathbf{r}_t', \tag{4.8.20}$$

which is a linear integral equation having as unknown term only the object function τ, which includes the dielectric properties of the cross sections of the scatterers inside the investigation area S_i [equation (2.7.3)].

It has been argued that the Rytov approximation, as compared with the first-order Born approximation (Section 4.6), is better for larger objects with small changes in the dielectric permittivity distribution (Slaney et al. 1984). However, for small scatterers, the two approximations tend to coincide, in fact, from equation (4.8.3) we have

$$E_z(\mathbf{r}_t) = e^{\Phi_{\text{inc}}(\mathbf{r}_t) + \Phi_{\text{scat}}(\mathbf{r}_t)} = E_{\text{inc}_z}(\mathbf{r}_t) e^{\Phi_{\text{scat}}(\mathbf{r}_t)} = E_{\text{inc}_z}(\mathbf{r}_t) [1 + E_{\text{scat}_z}(\mathbf{r}_t) e^{-\Phi_{\text{inc}}(\mathbf{r}_t)}]. \tag{4.8.21}$$

If the amplitude of the scattered field inside the scatterer is very small, the second term in square brackets can be neglected and $E_z(\mathbf{r}_t) \approx E_{\text{inc}_z}(\mathbf{r}_t)$ results. The Rytov approximation simplifies to the Born approximation.

However, it has been shown that a condition for the validity of the Rytov approximation is (Slaney et al. 1984, Keller 1969)

$$|\tau(\mathbf{r}_t)| \gg \frac{(\nabla \Phi_{\text{scat}}(\mathbf{r}_t))^2}{j\omega\mu_b}. \tag{4.8.22}$$

It has also been observed that "unlike the Born approximation, the size of the object is not a factor in the Rytov approximation" (Slaney et al. 1984). Since the approximation concerns phase changes, the Rytov approximation is expected to work well with smoothly varying dielectric profiles for which the phase change is small (Dunlop et al. 1976).

4.9 KIRCHHOFF APPROXIMATION

In the inspection of PEC scatterers, at high frequencies, use of the Kirchhoff approximation (or physical optics (PO) approximation) has been proposed even for short-range imaging (Pierri et al. 2001). With reference to equation (2.8.1), the unknown current density on the surface S is approximated as

$$\bar{\mathbf{J}}_S(\mathbf{r}) \approx \bar{\mathbf{J}}_S^{PO}(\mathbf{r}) = \begin{cases} 2\hat{\mathbf{n}} \times \bar{\mathbf{H}}_{\text{inc}}(\mathbf{r}), & \mathbf{r} \in S_1 \\ 0, & \mathbf{r} \in S_2 \end{cases} \tag{4.9.1}$$

where S_1 and S_2 are the illuminated and shadowed parts of S, with $S = S_1 \cup S_2$. Clearly, the only problem unknown is now the support of the scatterer, in particular the surface S_1.

It should be noted that imaging methods developed in the high-frequency range (i.e., methods for inspecting objects whose linear dimensions are much higher than the wavelength of the incident field) are not considered in the present book. These techniques are usually based on geometric optics approximations and ray propagation. The reader is referred to the article by Chu and Farhat (1988) and the references therein.

4.10 GREEN'S FUNCTION FOR INHOMOGENEOUS STRUCTURES

So far it has been assumed that the object to be inspected is immersed in free space (Figs. 4.1–4.7) or in a half-space domain (Figs. 4.8 and 4.9). In both cases, the Green functions/tensors involved in the computation are known quantities given by relations (2.4.7), (3.3.9), (2.4.12), and (2.4.13). In some imaging applications, however, the object can be immersed in an inhomogeneous medium, for which a closed-form relation for Green's function is seldom available.

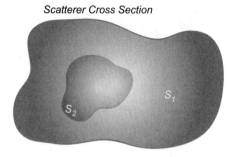

FIGURE 4.10 Cross section of an infinite cylinder with two separated regions.

However, this Green function can be computed numerically. There are two motivations for studying the inhomogeneous Green function for a given structure. The first one concerns the practical case in which the *object* to be detected is a defect inside an otherwise known target. If the Green's function for the *propagation* medium constituted by the *unperturbed* target is known, the inspection can be limited to a reduced investigation domain inside the target, avoiding the need for inspecting the whole body. The second reason for studying the inhomogeneous Green's function is related to development of the distorted Born iterative method, discussed in Section 6.6.

With reference to two-dimensional scattering, let us consider Figure 4.10, in which an object of support S_1 includes a region of interest $S_2 \subset S_1$. In some cases, the two domains can coincide. Given an incident field, the total field inside and outside the object can be computed by using equation (3.3.8). Alternatively, we can express the complex dielectric permittivity of the inner part of the object S_1 in this way

$$\varepsilon_2(\mathbf{r}_t) = \varepsilon_1(\mathbf{r}_t) + \Delta\varepsilon(\mathbf{r}_t), \quad (4.10.1)$$

where $\varepsilon_1(\mathbf{r}_t) = \varepsilon(\mathbf{r}_t)$ for $\mathbf{r}_t \in S_1$ and $\varepsilon_2(\mathbf{r}_t) = \varepsilon(\mathbf{r}_t)$ for $\mathbf{r}_t \in S_2$. For example, region S_2 can be a defect in an otherwise known object. In this case $\varepsilon_1(\mathbf{r}_t)$ is known for any $\mathbf{r}_t \in S_1$. In other cases, the values of $\varepsilon_1(\mathbf{r}_t)$ for $\mathbf{r}_t \in S_2$ could be fixed arbitrarily (it is also not a theoretical requirement that $S_2 \subset S_1$, although this is a situation that can be encountered in practical imaging problems).

According to equation (4.10.1), the object function of equation (3.3.8) can be rewritten as

$$\tau(\mathbf{r}_t) = \tau_1(\mathbf{r}_t) + \tau_2(\mathbf{r}_t), \quad (4.10.2)$$

where

$$\tau_1(\mathbf{r}_t) = j\omega[\varepsilon_1(\mathbf{r}_t) - \varepsilon_b], \quad \mathbf{r}_t \in S_1, \quad (4.10.3)$$

and

$$\tau_2(\mathbf{r}_t) = \begin{cases} j\omega \Delta\varepsilon(\mathbf{r}_t), & \mathbf{r}_t \in S_2, \\ 0, & \mathbf{r}_t \notin S_2. \end{cases} \quad (4.10.4)$$

By considering (4.10.3) and (4.10.3), from equation (3.3.8) we obtain

$$E_z(\mathbf{r}_t) = E_{\text{inc}_z}(\mathbf{r}_t) + j\omega\mu_b \int_{S_1} \tau_1(\mathbf{r}_t') E_z(\mathbf{r}_t') G_{2D}(\mathbf{r}_t/\mathbf{r}_t') d\mathbf{r}_t' \\ + j\omega\mu_b \int_{S_1} \tau_2(\mathbf{r}_t') E_z(\mathbf{r}_t') G_{2D}(\mathbf{r}_t/\mathbf{r}_t') d\mathbf{r}_t'. \quad (4.10.5)$$

Alternatively (Fig. 4.10), the field E_z can be viewed as the sum of the field scattered by the *excess* of permittivity of region S_2 when the *incident* field is the field that *propagates* in an inhomogeneous medium that is characterized by ε_b for $\mathbf{r}_t \notin S_1$ and by $\varepsilon_1(\mathbf{r}_t)$ for $\mathbf{r}_t \in S_1$, specifically,

$$E_z(\mathbf{r}_t) = E_{1_z}(\mathbf{r}_t) + j\omega\mu_b \int_{S_2} \tau_2(\mathbf{r}_t') E_z(\mathbf{r}_t') G_{\text{in}}(\mathbf{r}_t/\mathbf{r}_t') d\mathbf{r}_t', \quad (4.10.6)$$

where E_{1_z} is given by

$$E_{1_z}(\mathbf{r}_t) = E_{\text{inc}_z}(\mathbf{r}_t) + j\omega\mu_b \int_{S_1} \tau_1(\mathbf{r}_t') E_{1_z}(\mathbf{r}_t') G_{2D}(\mathbf{r}_t/\mathbf{r}_t') d\mathbf{r}_t' \quad (4.10.7)$$

and G_{in} is Green's function for the inhomogeneous structure S_1. By definition (Tai 1971), it satisfies equation

$$G_{\text{in}}(\mathbf{r}_t/\mathbf{r}_t') = G_{2D}(\mathbf{r}_t/\mathbf{r}_t') + j\omega\mu_b \int_{S_1} \tau_1(\mathbf{r}_t'') G_{\text{in}}(\mathbf{r}_t''/\mathbf{r}_t') G_{2D}(\mathbf{r}_t/\mathbf{r}_t'') d\mathbf{r}_t''. \quad (4.10.8)$$

Let us consider again the case in which one searches for a defect in a known structure. Green's function for this structure can be computed offline and stored. Then, an imaging method (e.g., one of those described in the following chapters) can be applied with reference to a reduced investigation domain coinciding with S_2. This essentially allows the imaging system to focus on a certain region of the original configuration.

Moreover, even in the case in which $S_1 \equiv S_2$, one can search for a small perturbation over a prescribed dielectric configuration. This is essentially the approach followed in the development of the distorted-wave Born approximation (Section 6.6), which is a very effective microwave imaging technique.

In order to numerically solve equation (4.10.6) according to the approach described in Section 3.4, it is necessary to calculate the following integrals [see equation (3.4.6)]

$$h_{mn}^{\text{in}} = j\omega\mu_b \int_{S_n} G_{\text{in}}(\mathbf{r}_m/\mathbf{r}_t') d\mathbf{r}_t', \quad (4.10.9)$$

which can be approximated by

$$h_{mn}^{\text{in}} = j\omega\mu_b G_{\text{in}}(\mathbf{r}_m/\mathbf{r}_n)\Delta S \qquad (4.10.10)$$

where $\mathbf{r}_n$ is the center of the nth subdomain. To calculate the values of $G_{\text{in}}(\mathbf{r}_m/\mathbf{r}_n)$, used in (4.10.10), the following set of integral equations must be solved [see (4.10.8)]:

$$G_{\text{in}}(\mathbf{r}_m/\mathbf{r}_n) = G_{2D}(\mathbf{r}_m/\mathbf{r}_n) + j\omega\mu_b \int_{S_1} \tau_1(\mathbf{r}_t'') G_{\text{in}}(\mathbf{r}_t''/\mathbf{r}_n) G_{2D}(\mathbf{r}_m/\mathbf{r}_t'') d\mathbf{r}_t'',$$
$$m = 1, \ldots, M, n = 1, \ldots, N. \qquad (4.10.11)$$

By exploiting the reciprocity of Green's function, we have

$$G_{\text{in}}(\mathbf{r}_n/\mathbf{r}_m) = G_{2D}(\mathbf{r}_n/\mathbf{r}_m) + j\omega\mu_b \int_{S_1} \tau_1(\mathbf{r}_t'') G_{\text{in}}(\mathbf{r}_t''/\mathbf{r}_m) G_{2D}(\mathbf{r}_n/\mathbf{r}_t'') d\mathbf{r}_t'',$$
$$m = 1, \ldots, M, n = 1, \ldots, N, \qquad (4.10.12)$$

and by following a procedure similar to that reported in Section 3.4, we obtain

$$G_{\text{in}}(\mathbf{r}_n/\mathbf{r}_m) = G_{2D}(\mathbf{r}_n/\mathbf{r}_m) + \sum_{k=1}^{N} \tau_{1_k} G_{\text{in}}(\mathbf{r}_k/\mathbf{r}_m)\left(j\omega\mu_b \int_{S_k} G_{2D}(\mathbf{r}_n''/\mathbf{r}')d\mathbf{r}'\right)$$
$$= G_{2D}(\mathbf{r}_n/\mathbf{r}_m) + \sum_{k=1}^{N} \tau_{1_k} G_{\text{in}}(\mathbf{r}_k/\mathbf{r}_m) h_{nk}, \ m = 1, \ldots, M, n = 1, \ldots, N.$$
$$(4.10.13)$$

The set of equations in (4.10.13) can also be expressed in matrix form as

$$[\mathbf{G}] = [\mathbf{G}_{2D}] + [\mathbf{H}][\mathbf{T}_1][\mathbf{G}], \qquad (4.10.14)$$

where $[\mathbf{G}]$ is a rectangular $N \times M$ matrix, whose elements are the values of the inhomogeneous Green function $G_{\text{in}}(\mathbf{r}_n/\mathbf{r}_m)$, $[\mathbf{G}_{2D}]$ is a rectangular $N \times M$ matrix, whose elements are the values of the free-space Green function $G_{2D}(\mathbf{r}_n/\mathbf{r}_m)$, and $[\mathbf{T}_1]$ is a square $N \times N$ diagonal matrix whose diagonal elements are given by $\tau_{1_n}, n = 1, \ldots, N$.

Consequently, computation of the inhomogeneous Green function is reduced to resolution of the matrix equation given by (4.10.14), whose solution is

$$[\mathbf{G}] = ([\mathbf{I}] - [\mathbf{H}][\mathbf{T}_1])^{-1}[\mathbf{G}_{2D}], \qquad (4.10.15)$$

where $[\mathbf{I}]$ is the $N \times N$ identity matrix.

REFERENCES

Balanis, C. A., *Antenna Theory: Analysis and Design*, Wiley, Hoboken, NJ, 2005.

Born, M. and E. Wolf, *Principles of Optics*, Pergamon Press, New York, 1965.

Chen, B. and J. J. Stamnes, "Validity of diffraction tomography based on the first Born and the first Rytov approximations," *Appl. Opt.* **37**, 2996–3006 (1998).

Chew, W. C., *Waves and Fields in Inhomogeneous Media*, Van Nostrand Reinhold, New York, 1990.

Chu, T. H. and N. H. Farhat, "Frequency-swept microwave imaging of dielectric objects," *IEEE Trans. Microwave Theory Tech.* **36**, 489–493 (1988).

Cui, T. J., Y. Qin, G.-L. Wang, and W. C. Chew, "Low-frequency detection of two-dimensional buried objects using high-order extended Born approximations," *Inverse Problems* **20**, 41–62 (2004).

Dunlop, G., W. Boerner, and R. Bates, "On an extended Rytov approximation and its comparison with the Born approximation," *Proc. Antennas Propagation Society Int. Symp.* 1976, pp. 587–591.

Estatico, C., M. Pastorino, and A. Randazzo, "An inexact Newton method for short-range microwave imaging within the second order Born approximation," *IEEE Trans. Geosci. Remote Sens.* **43**, 2593–2605 (2005).

Gao, G. and C. Torres-Verdín, "High-order generalized extended Born approximation for electromagnetic scattering," *IEEE Trans. Anten. Propag.* **54**, 1243–1256 (2006).

Habashy, T. M., R. W. Groom, and B. R. Spies, "Beyond the Born and Rytov approximations: A nonlinear approach to electromagnetic scattering," *J. Geophys. Res.* **98**, 1759–1775 (1993).

Habashy, T. M., M. L. Oristaglio, and A. T. de Hoop, "Simultaneous nonlinear reconstruction of two-dimensional permittivity and conductivity," *Radio Sci.* **29**, 1101–1118 (1994).

Ishimaru, A., *Wave Propagation and Scattering in Random Media*, Academic Press, New York, 1978.

Kak, A. C. and M. Slaney, *Principles of Computerized Tomographic Imaging*, IEEE Press, New York, 1988.

Keller, J. B., "Accuracy and validity of the Born and Rytov approximations," *J. Opt. Soc. Am.* **59**, 1003–1004 (1969).

Morse, P. M. and M. Feshbach, *Methods of Theoretical Physics*, McGraw-Hill, New York, 1953.

Natterer, F., "An error bound for the Born approximation," *Inverse Problems* **20**, 447–452 (2004).

Pierri, R., G. Leone, and R. Persico, "Second-order iterative approach to inverse scattering: Numerical results," *J. Opt. Soc. Am. A* **17**, 874–880 (2000).

Pierri, R., A. Liseno, and F. Soldovieri, "Shape reconstruction from PO multifrequency scattered fields via the singular value decomposition approach," *IEEE Trans. Anten. Propag.* **49**, 1331–1343 (2001).

Slaney, M., A. C. Kak, and L. E. Larsen, "Limitations of imaging with first-order diffraction tomography," *IEEE Trans. Microwave Theory Tech.* **32**, 860–874 (1984).

Song, L.-P. and Q. H. Liu, "A new approximation to three-dimensional electromagnetic scattering," *IEEE Geosci. Remote Sens. Lett.* **2**, 238–242 (2005).

Tai, C. T., *Dyadic Green's Functions in Electromagnetic Theory*, Intext, Scranton, PA, 1971.

Torres-Verdin, C. and T. M. Habashy, "Rapid numerical simulation of axisymmetric single-well induction data using the extended Born approximation," *Radio Sci.* **36**, 1287–1306 (2001).

Zhang, Z. Q. and Q. H. Liu, "Two nonlinear inverse methods for electromagnetic induction measurements," *IEEE Trans. Geosci. Remote Sens.* **39**, 1331–1339 (2001).

CHAPTER FIVE

Qualitative Reconstruction Methods

5.1 INTRODUCTION

In the scientific literature the term *qualitative* is used to denote essentially two families of reconstruction methods. In the first case it refers to approaches aimed at obtaining only some information about the scatterers under test. Such methods are generally unable to retrieve the distributions of the unknown electromagnetic parameters (e.g., dielectric permittivity, electric conductivity, magnetic permeability), but they provide information concerning only the shapes and locations of the scatterers.

In the second case, the same term is used to indicate the reconstruction methods that are based on certain approximations in scattering models (e.g., those based on Born- and Rytov-type approximations) and can then be successfully applied only if some a priori information on the unknown object is available and if the inspected objects fulfill the conditions making the approximate models valid. For example, in order to apply the Born approximation, one needs to know that the object is a weak scatterer, as discussed in detail in Chapter 4.

Despite their limited applicability, qualitative methods are of significant interest for their computational efficiency, enabling fast and robust reconstructions possible in a quite short time period. Such considerations suggest a classification of microwave imaging algorithms into two groups, namely, *quantitative* and *qualitative* methods, in order to distinguish qualitative approaches from those aimed at providing the values of the electromagnetic parameters (usually dielectric permittivity and electric conductivity) of the inspected investigation domain, without approximations on the electromagnetic model.

Microwave Imaging, By Matteo Pastorino
Copyright © 2010 John Wiley & Sons, Inc.

As will appear clear from the subsequent developments discussed below, quantitative methods provide very complete information on the inspected targets, but are usually quite computationally expensive and often require a suitable starting guess of the solution in order to converge. The reader is referred to Chapter 6 for a description of this class of methods. It is worth mentioning that the above classification is highly arbitrary and other choices are possible. In particular, we note that mathematicians usually restrict use of the term *qualitative* only in reference to methods providing support of the scatterers (Cakoni and Colton 2006).

In the following text, we shall focus only on qualitative approaches, which are less ambitious because they either attempt to provide less complete characterization of the scatterers or can be applied only to a particular class of targets. Both the approaches are discussed. However, since many qualitative methods lead to resolution of linear ill-posed problems, in the following sections, some preliminary theory and tools useful in dealing with this class of problems are presented. In particular, the concepts of generalized inverse operators and generalized solutions are addressed, along with one of the most commonly used and powerful tools in linear inverse scattering, the singular value decomposition. Some basic concepts regarding regularization theory are also briefly introduced, without cumbersome mathematical details.

5.2 GENERALIZED SOLUTION OF LINEAR ILL-POSED PROBLEMS

As mentioned above, qualitative methods very often require the resolution of ill-posed linear problems. Some examples are represented by the linear sampling method (Section 5.8) and by the approaches based on first-order Born and Rytov approximations, described in Sections 4.6 and 4.8, respectively. However, even nonlinear methods aimed at inspecting strong scatterers can require the resolution of linear problems. An example is represented by the Newton method described in Chapter 6, in which a linearized equation is to be solved at each iteration.

In this section, we shall analyze the problem of solving the linear problem

$$Lf = g \qquad (5.2.1)$$

in a quite abstract framework, since applications are described in the sections devoted to the specific inversion methods. More precisely, we shall assume that L is a linear operator between the Hilbert spaces X and Y.

In particular, we shall discuss issues related to existence, uniqueness, and continuous dependence on the data of the solution f of equation (5.2.1). With regard to existence, equation (5.2.1) admits a solution if and only if $g \in R(L)$, where $R(L)$ is the *range* of the operator L. If, for certain reasons (noise on the data, errors due to the measurement system, etc), $g \notin R(L)$, a solution does not exist in the standard sense.

On the other hand, if at least a nontrivial solution $\tilde{f}$ exists such that

$$L\tilde{f} = 0, \tag{5.2.2}$$

that is, the null space (also referred to as the "kernel") of the operator $N(L)$ does not contain only the null element, then the solution of the equation (5.2.1) is never unique. In fact, if f is a solution, then the function $f' = f + \tilde{f}$ is still a solution (for the same known term g).

As a consequence, either if $R(L)$ is a proper subset of Y (i.e., there exists $g \notin R(L)$) or if $N(L) \neq \{0\}$ [i.e., for any $g \in R(L)$ there exists more than one solution of equation (5.2.1)], the problem of finding f is ill-posed, according to the definition given in Section 3.1.

Nonetheless, as will be discussed below, the standard concept of solution can be extended by introducing the *generalized solution*, usually denoted as f^+, in order to resolve some nonuniqueness and nonexistence issues.

In order to define the generalized solution, it is first useful to introduce the *least-squares solution* (or *pseudosolution*) of the linear problem $Lf = g$. An element $\bar{f}$ of X is called a *least-squares solution* of (5.2.1) if

$$\|L\bar{f} - g\| \leq \|Lf - g\|, \quad \forall f \in X \tag{5.2.3}$$

or, equivalently, if it satisfies the *normal form* of the equation of interest, specifically

$$L^*L\bar{f} = L^*g, \tag{5.2.4}$$

where L^* is the adjoint operator of L.

In other words, a least-squares solution minimizes the norm of the residual

$$r = Lf - g. \tag{5.2.5}$$

For this reason, it is clear that any solution of equation (5.2.1) in classical sense is a least-squares solution, too, since the residual of a classical solution is always exactly zero. However, the opposite statement does not hold true, since a least-squares solution may exist even when a standard solution cannot be found, that is, if $g \notin R(L)$.

On the other hand, it is worth remarking that more than one least-squares solution may exist, as the abovementioned reasoning on the kernel of L suggests. However, the uniqueness can be ensured by choosing the *minimum norm* least-squares solution. In fact, it can be proved that the set of least-squares solutions is a closed and convex set in X and thus always admits a *unique* minimum norm element (Rudin, 1970). For this reason, the generalized solution f^+ is introduced as the least-squares solution of *minimum norm*.

It can also be proved that the generalized solution exists for any datum $g \in R(L) \oplus R(L)^\perp$, where $\oplus$ is the direct sum operator and $R(L)^\perp$ is the orthogonal complement of $R(L)$ (Tikhonov et al. 1995). As a consequence, the generalized solution exists for any datum $g \in Y$ if and only if the range of L is a closed set. Accordingly, the *generalized inverse operator* L^+ can be introduced as an operator mapping the elements of $R(L) \oplus R(L)^\perp$ to the corresponding generalized solutions (Tikhonov et al. 1995) and thus it can be written

$$f^+ = L^+ g. \tag{5.2.6}$$

It is quite easy to verify that the generalized inverse operator is a linear operator and the generalized solution is the unique least-squares solution with no component belonging to the kernel of the operator L (Tikhonov et al. 1995). Moreover, it can be proved that the operator L^+ is continuous when $R(L)$ is closed in Y (Tikhonov et al. 1995). As a consequence, the problem of finding the generalized solution of a linear problem is well posed when $R(L)$ is closed, since, in such a case, the generalized solution exists, is unique, and depends continuously on the data.

On the other hand, when $R(L)$ is not closed, determination of the generalized solution is an ill-posed problem and *regularization methods* need to be introduced. Regularization theory is a wide topic that would necessitate devoting too much space in the present context for a precise description. For this reason, we shall provide only some very basic ideas.

5.3 REGULARIZATION METHODS

A regularization method consists in replacing an ill-posed problem with a well-posed problem, whose solutions approximate the one of the ill-posed problem (Sarkar et al. 1981). More precisely, the set of operators R_α, with $\alpha > 0$, is said to be a regularization scheme for operator L if the following two conditions are fulfilled (Colton and Kress 1998) (where α is the *regularization parameter*):

1. R_α is continuous for any $\alpha > 0$.
2. For every $g \in R(L) \oplus R(L)^\perp$, $\|R_\alpha g - f^+\| \to 0$ when $\alpha \to 0^+$.

In other words, regularization operators provide stable (continuous) approximations of the (unbounded) generalized inverse. It is worth noting that such a definition of regularization scheme does not involve the presence of noise.

However, the practical importance of the regularization methods is due largely to their ability to provide stable solutions even for noisy data, that is, when the known term g or the operator L are not exactly known.

In the case of noisy environments, the previously provided definition of regularization scheme needs to be slightly modified to account for the presence of noise and its level. In particular, a regularization strategy for noisy data/operators has to be equipped with a rule for the choice of the optimal regularization in dependence on noise level.

In order to clarify this point, let us assume that only the datum g is affected by noise. Then let g^δ be the noisy datum, where is δ the noise level (i.e., $\|g - g^\delta\| \leq \delta$); it is interesting to evaluate the norm of the difference between the generalized solution f^+ corresponding to the noiseless datum g and the quantity $f_\alpha^\delta = R_\alpha g^\delta$:

$$\|f_\alpha^\delta - f^+\| = \|R_\alpha g^\delta - f^+\| = \|R_\alpha g^\delta - R_\alpha g + R_\alpha g - f^+\| \leq \delta\|R_\alpha\| + \|R_\alpha g - f^+\|. \quad (5.3.1)$$

Such an inequality is very useful in understanding the effect of noise on a regularized solution. In fact, it provides an estimate of the error on the generalized solution as the sum of two terms: $\delta\|R_\alpha\|$, which is a measure of the error on the regularized approximated solution $f_\alpha^\delta = R_\alpha g^\delta$ due to the error on g; and $\|R_\alpha g - f^+\|$, which represents the approximation error due to the use of R_α instead of the generalized inverse operator. According to a definition of the regularization method, because $\|R_\alpha g - f^+\| \to 0$ as $\alpha \to 0^+$ and, in ill-posed problems, $\|R_\alpha\| \to \infty$ as $\alpha \to 0^+$, a tradeoff between the two errors is required, and it can be obtained by a suitable choice of the regularization parameter α. So it is clear that every regularization scheme requires a strategy for choosing the parameter α in dependence on the error level in order to achieve an acceptable approximation of the generalized solution.

In the scientific literature, many regularization methods have been proposed and studied. Among them, we cite the Tikhonov method, the truncated Landweber algorithm, the truncated conjugate gradient and the truncated singular value decomposition. For example, the Tikhonov method consists in minimizing the residual (5.2.5) with a penalty term. In particular, the Tikhonov regularization method is based on minimization of the functional

$$F_\alpha(f) = \|Lf - g\|^2 + \alpha\|f\|^2, \quad (5.3.2)$$

where the residual on the datum g is minimized along with the norm of the solution. For numerical implementation of the Tikhonov method, it is very useful to know that the function minimizing (5.3.2) is also the solution of the Euler equation

$$(L^*L + \alpha I)f = L^*g. \quad (5.3.3)$$

Therefore, in this case, we obtain

$$R_\alpha = (L^*L + \alpha I)^{-1} L^*. \quad (5.3.4)$$

Other regularization methods (e.g., the truncated Landweber method) will be used in the following text and described when necessary.

Although the previously developed ideas hold true in quite general frameworks, provided that the suitable technical hypotheses are assumed, in the following text we shall address the case of finite-dimensional spaces, by introducing the singular value decomposition (SVD) of a matrix and discussing the resolution of linear problems in terms of this powerful mathematical tool.

5.4 SINGULAR VALUE DECOMPOSITION

Let us consider a generic $N \times M$ matrix $[\mathbf{L}]$ of complex numbers of rank $p \leq \min\{N, M\}$.

In order to introduce the singular value decomposition, let us consider, for $i = 1, \ldots, p$, the vectors $\mathbf{u}_i$ and $\mathbf{v}_i$ of N and M elements, respectively, and the numbers σ_i such that

$$[\mathbf{L}]\mathbf{v}_i = \sigma_i \mathbf{u}_i, \quad (5.4.1)$$

$$[\mathbf{L}]^* \mathbf{u}_i = \sigma_i \mathbf{v}_i. \quad (5.4.2)$$

It is very easy to deduce that

$$[\mathbf{L}]^*[\mathbf{L}]\mathbf{v}_i = \sigma_i [\mathbf{L}]^* \mathbf{u}_i = \sigma_i^2 \mathbf{v}_i, \quad (5.4.3)$$

$$[\mathbf{L}][\mathbf{L}]^* \mathbf{u}_i = \sigma_i [\mathbf{L}]\mathbf{v}_i = \sigma_i^2 \mathbf{u}_i. \quad (5.4.4)$$

As a consequence, $\mathbf{v}_i$ is an eigenvector of the matrix $[\mathbf{L}]^*[\mathbf{L}]$ with eigenvalue σ_i^2 and $\mathbf{u}_i$ is an eigenvector of the matrix $[\mathbf{L}][\mathbf{L}]^*$, again with eigenvalue σ_i^2. Since $[\mathbf{L}]^*[\mathbf{L}]$ and $[\mathbf{L}][\mathbf{L}]^*$ are both self-adjoint matrices [i.e., $([\mathbf{L}]^*[\mathbf{L}])^* = [\mathbf{L}]^*[\mathbf{L}]$ and $([\mathbf{L}][\mathbf{L}]^*)^* = [\mathbf{L}][\mathbf{L}]^*$], as a consequence of well-known results in linear algebra, they are diagonalizable and all their eigenvalues are real (Bertero and Boccacci 1998). Moreover, the eigenvectors of $[\mathbf{L}]^*[\mathbf{L}]$ and $[\mathbf{L}][\mathbf{L}]^*$ can be made orthonormal in C^M and C^N, respectively. Accordingly, there exist two sequences of vectors $\{\mathbf{u}_i\}_{i=1}^N$ and $\{\mathbf{v}_i\}_{i=1}^M$ such that

$$[\mathbf{L}]^*[\mathbf{L}]\mathbf{v}_i = \begin{cases} \sigma_i^2 \mathbf{v}_i, & i = 1, \ldots, p, \\ 0 & i = p+1, \ldots, M, \end{cases} \quad (5.4.5)$$

$$[\mathbf{L}][\mathbf{L}]^* \mathbf{u}_i = \begin{cases} \sigma_i^2 \mathbf{u}_i, & i = 1, \ldots, p, \\ 0 & i = p+1, \ldots, N, \end{cases} \quad (5.4.6)$$

Let us then sort the vectors $\{\mathbf{u}_i\}_{i=1}^{N}$ and $\{\mathbf{v}_i\}_{i=1}^{M}$ such that the corresponding eigenvalues are in decreasing order, and let us build the $N \times N$ matrix

$$[\mathbf{U}] = [\mathbf{u}_1, \mathbf{u}_2, \ldots, \mathbf{u}_N] \tag{5.4.7}$$

and the $M \times M$ matrix

$$[\mathbf{V}] = [\mathbf{v}_1, \mathbf{v}_2, \ldots, \mathbf{v}_M], \tag{5.4.8}$$

with the mutually orthonormal eigenvectors as columns, so that they result in unitary matrices [i.e., $[\mathbf{U}]^* = [\mathbf{U}]^{-1}$ and $[\mathbf{V}]^* = [\mathbf{V}]^{-1}$]. It is worth noting that, since it can be proved that the kernels of $[\mathbf{L}]^*[\mathbf{L}]$ and $[\mathbf{L}]$ coincide (Oden and Demkowicz 1996), the number of non-zero eigenvalues of $[\mathbf{L}]^*[\mathbf{L}]$ is equal to the rank p of $[\mathbf{L}]$. On the basis of the pointed out properties and equations (5.4.5) and (5.4.6), we can then write

$$[\mathbf{L}]\mathbf{v}_i = \begin{cases} \sigma_i \mathbf{u}_i, & i = 1, \ldots, p, \\ 0, & i = p+1, \ldots, M, \end{cases} \tag{5.4.9}$$

$$[\mathbf{L}]^* \mathbf{u}_i = \begin{cases} \sigma_i \mathbf{v}_i, & i = 1, \ldots, p, \\ 0 & i = p+1, \ldots, N. \end{cases} \tag{5.4.10}$$

Equations (5.4.9) and (5.4.10) can be expressed in a compact form by using matrices $[\mathbf{U}]$ and $[\mathbf{V}]$ as

$$[\mathbf{L}][\mathbf{V}] = [\mathbf{U}][\boldsymbol{\Sigma}], \tag{5.4.11}$$

$$[\mathbf{L}]^*[\mathbf{U}] = [\mathbf{V}][\boldsymbol{\Sigma}]^T, \tag{5.4.12}$$

where $[\boldsymbol{\Sigma}]$ is the $N \times M$ real matrix given by

$$[\boldsymbol{\Sigma}] = \begin{bmatrix} \sigma_1 & 0 & \cdots & \cdots & \cdots & \cdots & 0 \\ 0 & \sigma_2 & 0 & \ddots & \ddots & \ddots & \vdots \\ \vdots & 0 & \ddots & \ddots & \ddots & \ddots & \vdots \\ \vdots & \ddots & \ddots & \sigma_p & \ddots & \ddots & \vdots \\ \vdots & \ddots & \ddots & \ddots & 0 & \ddots & \vdots \\ \vdots & \ddots & \ddots & \ddots & \ddots & \ddots & \vdots \\ 0 & \cdots & \cdots & \cdots & \cdots & \cdots & 0 \end{bmatrix}. \tag{5.4.13}$$

In the following paragraphs, it will be useful to define $\sigma_i = 0$ for $i = p + 1, \ldots, M$.

Since $[\mathbf{V}]^* = [\mathbf{V}]^{-1}$, from equation (5.4.11) it immediately follows that

$$[\mathbf{L}] = [\mathbf{U}][\boldsymbol{\Sigma}][\mathbf{V}]^*, \tag{5.4.14}$$

which is called the *singular value decomposition* of the complex matrix $[\mathbf{L}]$. The diagonal elements of $[\boldsymbol{\Sigma}]$ are called the *singular values* of the matrix $[\mathbf{L}]$, and there are exactly p nonzero singular values. Moreover, the columns of $[\mathbf{U}]$ and $[\mathbf{V}]$ are called the *singular vectors* of the matrix $[\mathbf{L}]$. The whole composite of singular values and singular vectors of $[\mathbf{L}]$ is referred to as the *singular system* of $[\mathbf{L}]$.

Of course, as $[\Sigma]$ is a matrix of real numbers, a similar decomposition for the adjoint matrix $[\mathbf{L}]^*$ can be provided as follows:

$$[\mathbf{L}]^* = [\mathbf{V}][\Sigma]^T [\mathbf{U}]^*. \tag{5.4.15}$$

Equations (5.4.14) and (5.4.15) are rich in consequences. In fact, since the singular vectors $\{\mathbf{u}_i\}_{i=1}^N$ and $\{\mathbf{v}_i\}_{i=1}^M$ are mutually orthonormal in C^N and C^M, respectively, $\{\mathbf{u}_i\}_{i=1}^N$ is a orthonormal basis in C^N and $\{\mathbf{v}_i\}_{i=1}^M$ is a orthonormal basis in C^M. More precisely, $\{\mathbf{v}_i\}_{i=1}^p$ is an orthonormal basis for the orthogonal complement of the kernel of $[\mathbf{L}]$, whereas $\{\mathbf{v}_i\}_{i=p+1}^M$ is an orthonormal basis for the kernel of $[\mathbf{L}]$ (Oden and Demkowicz 1996). Analogously, $\{\mathbf{u}_i\}_{i=1}^p$ is an orthonormal basis for the orthogonal complement of the range of $[\mathbf{L}]$, whereas $\{\mathbf{u}_i\}_{i=p+1}^N$ is an orthonormal basis for the orthogonal complement of the range of $[\mathbf{L}]$.

As a consequence of this important property, the image of any element of C^M through $[\mathbf{L}]$ can be expressed by means of its singular system. In fact, as $\mathbf{f}$ is a generic element of C^M, it follows that

$$\mathbf{f} = \sum_{i=1}^{M} \langle \mathbf{f}, \mathbf{v}_i \rangle \mathbf{v}_i, \tag{5.4.16}$$

and, since $\{\mathbf{v}_i\}_{i=1}^M$ are mutually orthonormal in C^M, we obtain

$$\|\mathbf{f}\| = \sqrt{\sum_{i=1}^{M} |\langle \mathbf{f}, \mathbf{v}_i \rangle|^2}. \tag{5.4.17}$$

As a consequence

$$[\mathbf{L}]\mathbf{f} = \sum_{i=1}^{M} \langle \mathbf{f}, \mathbf{v}_i \rangle [\mathbf{L}]\mathbf{v}_i = \sum_{i=1}^{M} \sigma_i \langle \mathbf{f}, \mathbf{v}_i \rangle \mathbf{u}_i = \sum_{i=1}^{p} \sigma_i \langle \mathbf{f}, \mathbf{v}_i \rangle \mathbf{u}_i. \tag{5.4.18}$$

Moreover, using the mutual orthonormality of the elements of $\{\mathbf{u}_i\}_{i=1}^N$, we obtain

$$\|[\mathbf{L}]\mathbf{f}\| = \sqrt{\sum_{i=1}^{p} \sigma_i^2 |\langle \mathbf{f}, \mathbf{v}_i \rangle|^2}. \tag{5.4.19}$$

Because $\sigma_1 \geq \sigma_2 \geq \cdots \geq \sigma_p > 0$, from equation (5.4.19) and (5.4.17), where $p \leq M$, we deduce that

$$\|[\mathbf{L}]\mathbf{f}\| \leq \sigma_1 \sqrt{\sum_{i=1}^{p} |\langle \mathbf{f}, \mathbf{v}_i \rangle|^2} \leq \sigma_1 \sqrt{\sum_{i=1}^{M} |\langle \mathbf{f}, \mathbf{v}_i \rangle|^2} = \sigma_1 \|\mathbf{f}\|. \tag{5.4.20}$$

It is worth remarking that the equality can hold true in (5.4.20) when $\mathbf{f}$ has a nonzero component only along the first singular vector $\mathbf{v}_1$. Accordingly, the standard norm $\|[\mathbf{L}]\|$ of matrix $[\mathbf{L}]$ can be easily expressed in terms of its maximum singular value, by observing that

$$\|[\mathbf{L}]\| = \sup_{\|\mathbf{f}\| \neq 0} \left\{ \frac{\|[\mathbf{L}]\mathbf{f}\|}{\|\mathbf{f}\|} \right\} = \sigma_1. \quad (5.4.21)$$

5.5 SINGULAR VALUE DECOMPOSITION FOR SOLVING LINEAR PROBLEMS

Singular value decomposition (SVD) also plays an important role in the resolution of linear problems, and this is the main reason why it is presented in this book. In order to explain this point, let us consider the system of equations

$$[\mathbf{L}]\mathbf{f} = \mathbf{g}. \quad (5.5.1)$$

and let us look for its generalized solution, that is, for a vector $\mathbf{f}$ such that

$$[\mathbf{L}]^*[\mathbf{L}]\mathbf{f} = [\mathbf{L}]^*\mathbf{g}. \quad (5.5.2)$$

To this end, it is useful to observe that, by using (5.4.18), we can write

$$[\mathbf{L}]^*[\mathbf{L}]\mathbf{f} = \sum_{i=1}^{p} \sigma_i \langle \mathbf{f}, \mathbf{v}_i \rangle [\mathbf{L}]^* \mathbf{u}_i = \sum_{i=1}^{p} \sigma_i^2 \langle \mathbf{f}, \mathbf{v}_i \rangle \mathbf{v}_i. \quad (5.5.3)$$

Moreover

$$\mathbf{g} = \sum_{i=1}^{N} \langle \mathbf{g}, \mathbf{u}_i \rangle \mathbf{u}_i, \quad (5.5.4)$$

and hence

$$[\mathbf{L}]^*\mathbf{g} = \sum_{i=1}^{N} \langle \mathbf{g}, \mathbf{u}_i \rangle [\mathbf{L}]^* \mathbf{u}_i = \sum_{i=1}^{N} \sigma_i \langle \mathbf{g}, \mathbf{u}_i \rangle \mathbf{v}_i = \sum_{i=1}^{p} \sigma_i \langle \mathbf{g}, \mathbf{u}_i \rangle \mathbf{v}_i. \quad (5.5.5)$$

Using (5.5.3) and (5.5.5), we can rewrite equation (5.5.2) as

$$\sum_{i=1}^{p} \sigma_i^2 \langle \mathbf{f}, \mathbf{v}_i \rangle \mathbf{v}_i = \sum_{i=1}^{p} \sigma_i \langle \mathbf{g}, \mathbf{u}_i \rangle \mathbf{v}_i, \quad (5.5.6)$$

or, equivalently, because the singular vectors are mutually orthonormal, as

$$\langle \mathbf{f}, \mathbf{v}_i \rangle = \frac{1}{\sigma_i} \langle \mathbf{g}, \mathbf{u}_i \rangle, \quad i = 1, \ldots, p. \quad (5.5.7)$$

As a consequence, a solution of (5.5.2) is given by

$$\mathbf{f} = \sum_{i=1}^{p} \frac{1}{\sigma_i} \langle \mathbf{g}, \mathbf{u}_i \rangle \mathbf{v}_i. \quad (5.5.8)$$

Since the vectors $\{\mathbf{v}_i\}_{i=1}^p$ are an orthonormal basis for the orthogonal complement of the kernel of $[\mathbf{L}]$, the solution (5.5.8) has no component in the null space of $[\mathbf{L}]$ and so, as observed above in a more general setting, it is the generalized solution of the linear problem (5.5.1):

$$\mathbf{f}^+ = \sum_{i=1}^{p} \frac{1}{\sigma_i} \langle \mathbf{g}, \mathbf{u}_i \rangle \mathbf{v}_i. \quad (5.5.9)$$

In such a way, we have shown that the generalized solution of the generic linear system (5.5.1) can be explicitly expressed by means of the SVD of the matrix $[\mathbf{L}]$.

Equation (5.5.9) is also very useful, since it readily provides the generalized inverse matrix $[\mathbf{L}]^+$, referred to also as the *pseudoinverse* matrix. In fact, on the basis of the general theory described in the previous section, we can write

$$\mathbf{f}^+ = [\mathbf{L}]^+ \mathbf{g} = \sum_{i=1}^{p} \frac{1}{\sigma_i} \langle \mathbf{g}, \mathbf{u}_i \rangle \mathbf{v}_i. \quad (5.5.10)$$

As a consequence

$$[\mathbf{L}]^+ = [\mathbf{V}][\mathbf{\Sigma}]^{+T}[\mathbf{U}]^*, \quad (5.5.11)$$

where $[\mathbf{\Sigma}]^+$ is a $N \times M$ diagonal matrix, with diagonal elements given by

$$\left([\mathbf{\Sigma}]^+\right)_{ii} = \begin{cases} \dfrac{1}{\sigma_i}, & i=1,\ldots,p, \\ 0, & i=p+1,\ldots,\min\{N,M\}. \end{cases} \quad (5.5.12)$$

From (5.5.11) it easily follows that

$$\|[\mathbf{L}]^+\| = \frac{1}{\sigma_p}. \quad (5.5.13)$$

It should be noted that the use of the pseudoinverse solution can solve uniqueness and nonexistence problems, but numerical instability may still affect the computation of $\mathbf{f}^+$. In practice, we mean that, although (5.5.10) always holds true, large numerical errors may occur on $\mathbf{f}^+$ even if the datum $\mathbf{g}$ is affected by small errors.

In order to make these considerations more precise, let us consider the datum $\mathbf{g}$ to be affected by an error $\delta\mathbf{g}$, that is, only the "noisy" version

$$\mathbf{g}^e = \mathbf{g} + \delta\mathbf{g} \quad (5.5.14)$$

of $\mathbf{g}$ is disposable, and let us evaluate the impact of this error on the corresponding generalized solution $\mathbf{f}^{+,e}$. Since

$$\mathbf{f}^{+,e} = [\mathbf{L}]^+ \mathbf{g}^e = \sum_{i=1}^{p} \frac{1}{\sigma_i} \langle \mathbf{g}^e, \mathbf{u}_i \rangle \mathbf{v}_i, \quad (5.5.15)$$

from (5.5.10), it follows that

$$\delta \mathbf{f}^+ = \mathbf{f}^{+,e} - \mathbf{f}^+ = [\mathbf{L}]^+ \mathbf{g}^e - [\mathbf{L}]^+ \mathbf{g} = [\mathbf{L}]^+ \delta \mathbf{g}. \tag{5.5.16}$$

As a consequence, we have

$$\|\delta \mathbf{f}^+\| = \|[\mathbf{L}]^+ \delta \mathbf{g}\| \le \|[\mathbf{L}]^+\| \|\delta \mathbf{g}\| = \frac{\|\delta \mathbf{g}\|}{\sigma_p}. \tag{5.5.17}$$

Moreover, from (5.5.10) it follows that

$$\|\mathbf{f}^+\| = \|[\mathbf{L}]^+ \mathbf{g}\| = \sqrt{\sum_{i=1}^{p} \frac{1}{\sigma_i^2} |\langle \mathbf{g}, \mathbf{u}_i \rangle|^2} \ge \frac{1}{\sigma_1} \sqrt{\sum_{i=1}^{p} |\langle \mathbf{g}, \mathbf{u}_i \rangle|^2} = \frac{1}{\sigma_1} \|\mathbf{g}_{R([\mathbf{L}])}\|, \tag{5.5.18}$$

where $\mathbf{g}_{r([\mathbf{L}])}$ denotes the projection of $\mathbf{g}$ onto the range of the matrix $[\mathbf{L}]$. Therefore, from (5.5.18) and (5.5.17) it follows (for $\|\mathbf{f}^+\| \ne 0$) that

$$\frac{\|\delta \mathbf{f}^+\|}{\|\mathbf{f}^+\|} \le \|[\mathbf{L}]\| \|[\mathbf{L}]^+\| \frac{\|\delta \mathbf{g}\|}{\|\mathbf{g}_{R([\mathbf{L}])}\|} = \frac{\sigma_1}{\sigma_p} \frac{\|\delta \mathbf{g}\|}{\|\mathbf{g}_{R([\mathbf{L}])}\|}. \tag{5.5.19}$$

From equation (5.5.19), it is easily deduced that the relative error on the known term corresponds to a relative error on the generalized solution that is controlled by the quantity

$$m = \|[\mathbf{L}]\| \|[\mathbf{L}]^+\| = \frac{\sigma_1}{\sigma_p}, \tag{5.5.20}$$

which is called the *condition number* of the matrix $[\mathbf{L}]$. It is worth observing that, since $\sigma_1 \ge \sigma_p$, it follows that $m \ge 1$.

The condition number indicates the sensitivity of the solution with respect to the error on the known term and depends on the ratio between the largest and the smallest singular values of $[\mathbf{L}]$. From equation (5.5.20), we see immediately that if m is small, small changes in the known term will result in small variations in the solution, otherwise, small changes in the data will result in large variations in the solution. In the former case, the discrete problem (and the related matrix) is said to be *well-conditioned*, in the latter case, it is said to be *ill-conditioned*. Unfortunately, the discretization of ill-posed inverse problems usually leads to systems of equations characterized by very high condition numbers. Moreover, the finer is the discretization of the continuous ill-posed problem, typically the higher is the condition number of the resulting matrix. This can be explained by observing that a finer discretization produces a better approximation of the continuous operator, whose inverse is usually unbounded. This should be carefully accounted for in practical applications, where a tradeoff between resolution (high resolution would require a fine discretization) and stability (coarse discretization) must be pursued.

Consequently, the linear systems occurring in the resolution of inverse problems suffer from a severe numerical instability. For example (Bertero and Boccacci 1998), if $m = 10^6$, a small error in the data of the order of 10^{-6} could result in an error in the solution of the order of 100%.

On the other hand, to be thorough, it should also be mentioned that discretization of a direct scattering problem usually results in a well-conditioned matrix to be inverted, as a result of the usual well-posedness of the continuous direct model.

5.6 REGULARIZED SOLUTION OF A LINEAR SYSTEM USING SINGULAR VALUE DECOMPOSITION

The singular value decomposition is very useful also because it allows a direct expression for the Tikhonov regularized solution of the linear system (5.5.1). To achieve this goal, let us consider the Euler equation [see equation (5.3.3)] for the Tikhonov solution of the linear system (5.5.1), which reads as

$$\left([\mathbf{L}]^*[\mathbf{L}] + \alpha[\mathbf{I}]\right)\mathbf{f} = [\mathbf{L}]^*\mathbf{g}. \quad (5.6.1)$$

Since the solution is an element of C^M, it can be expressed as a linear combination of singular vectors as in equation (5.4.16). Moreover, by using equation (5.5.5), it follows that

$$\sum_{i=1}^{M}\langle\mathbf{f},\mathbf{v}_i\rangle\left([\mathbf{L}]^*[\mathbf{L}] + \alpha[\mathbf{I}]\right)\mathbf{v}_i = \sum_{i=1}^{p}\sigma_i\langle\mathbf{g},\mathbf{u}_i\rangle\mathbf{v}_i. \quad (5.6.2)$$

Since

$$\sum_{i=1}^{M}\langle\mathbf{f},\mathbf{v}_i\rangle\left([\mathbf{L}]^*[\mathbf{L}] + \alpha[\mathbf{I}]\right)\mathbf{v}_i = \sum_{i=1}^{M}\langle\mathbf{f},\mathbf{v}_i\rangle\left(\sigma_i^2 + \alpha\right)\mathbf{v}_i, \quad (5.6.3)$$

it follows that

$$\langle\mathbf{f},\mathbf{v}_i\rangle = \begin{cases} \dfrac{\sigma_i}{(\sigma_i^2+\alpha)}\langle\mathbf{g},\mathbf{u}_i\rangle, & i=1,\ldots,p, \\ 0, & i=p+1,\ldots,M. \end{cases} \quad (5.6.4)$$

The Tikhonov regularized solution can then be written as

$$\mathbf{f} = \sum_{i=1}^{p}\frac{\sigma_i}{(\sigma_i^2+\alpha)}\langle\mathbf{g},\mathbf{u}_i\rangle\mathbf{v}_i. \quad (5.6.5)$$

Another commonly used strategy to regularize the solution of ill-conditioned linear systems consists in the so-called truncated singular value decomposition (TSVD). The basic idea is to neglect the smallest singular values of the involved matrix, in order to reduce the condition number. Of course, a tradeoff between stability and accuracy must be found. More precisely, if $\sigma_{p'} > \sigma_p$ ($p' < p$) denotes the smallest considered singular value, the related solution is given by

$$\mathbf{f} = \sum_{i=1}^{p'} \frac{1}{\sigma_i} \langle \mathbf{g}, \mathbf{u}_i \rangle \mathbf{v}_i. \qquad (5.6.6)$$

The corresponding error on the fitting of the known term can easily be deduced as

$$[\mathbf{L}]\sum_{i=1}^{p} \frac{1}{\sigma_i} \langle \mathbf{g}, \mathbf{u}_i \rangle \mathbf{v}_i - [\mathbf{L}]\sum_{i=1}^{p'} \frac{1}{\sigma_i} \langle \mathbf{g}, \mathbf{u}_i \rangle \mathbf{v}_i = \sum_{i=p'+1}^{p} \frac{1}{\sigma_i} \langle \mathbf{g}, \mathbf{u}_i \rangle [\mathbf{L}] \mathbf{v}_i = \sum_{i=p'+1}^{p} \langle \mathbf{g}, \mathbf{u}_i \rangle \mathbf{u}_i. \qquad (5.6.7)$$

It is worth remarking that the choice of p' is often a critical issue, as well as the selection of α for the Tikhonov regularization, and the optimal choice is not at all an easy task (Caorsi et al. 1995). In fact, an estimation of the noise level is often required for selection of the optimal value, and several approaches have been proposed in the literature. One of the most commonly adopted strategies is based on the *Morozov principle* (Colton and Kress 1998), which stipulates that the value of α selected must ensure that the error on the retrieved solution is of the same level as the error on the data. This is a reasonable criterion that is often used in practical applications in which a noise-level estimation is available; for instance, the rule used to choose the value of the regularization parameter in the linear sampling method described below, is based on the Morozov principle.

5.7 QUALITATIVE METHODS FOR OBJECT LOCALIZATION AND SHAPING

Several different methods can be applied to localize an unknown scatterer and define its shape. Among them we can mention the linear sampling method and the level-set method. These methods have been widely studied in the scientific literature, and the reader is referred to the references cited in the following sections. Formulation of the linear sampling method will be outlined in the following section, since this technique is one of the most used, not only as an independent qualitative procedure (with intrinsic computational efficiency) but also in conjunction with quantitative techniques. This important aspect will be discussed in Chapter 8.

A further consideration is necessary. As clearly specified in the previous chapters, this book essentially focuses on short-range imaging techniques. However, it is not possible to set a precise bound for the set of considered techniques based on the distance between the illuminating/receiving elements and the target under test, since, for example, some radar techniques are used in both near-field and far-field ranges. At the same time, some common tomographic methods, which, of course, imply the possibility of sensing the scattered field around the target, in certain frequency ranges, still assume far-field operation conditions, and plane waves are used (at least as approximations) for illumination purposes. The choice of the author is to consider techniques that have been proposed for engineering applications in which the imaging system must be positioned at a relatively short distance, in the *common sense*, to the object or apparatus to be inspected.

5.8 THE LINEAR SAMPLING METHOD

A method for rapid determination of the shape of a scattering object is the so-called linear sampling method (Colton and Kress 1998, Cakoni and Colton 2006, Colton et al. 2003). Under the assumption in Section 3.3 for two-dimensional inverse scattering, we consider the inspection of an inhomogeneous object of support S_o starting from knowledge of the field that it scatters in the far-field region. Moreover, the object is assumed to be illuminated by a set of uniform plane waves, although this is not a strict requirement, since the linear sampling method can be extended to other incident fields and near-field measurements (Colton and Monk 1998). According to equation (2.4.24), we assume again $\hat{\mathbf{p}} = \hat{\mathbf{z}}$ and the propagation vector given by equation (3.5.1).

The scattered electric field admits the asymptotic behavior given by

$$E_z(\mathbf{r}_t) = \frac{e^{-jk_b\rho}}{\sqrt{\rho}} E_{\infty_z}(\hat{\mathbf{r}}_t), \qquad (5.8.1)$$

where $\hat{\mathbf{r}}_t = \mathbf{r}_t / \rho = \hat{\mathbf{x}}\cos\varphi + \hat{\mathbf{y}}\sin\varphi$. Equation (5.8.2) is clearly the two-dimensional counterpart of equation (2.4.22). The scalar field E_{∞_z} is often called the *far-field pattern* in the scientific literature on the linear sampling method (Cakoni and Colton 2006). Since the scattered field E_{scat_z} depends on the incident field, so does the far-field pattern. Consequently, the angle of incidence can be made explicit by writing $E_{\infty_z} = E_{\infty_z}(\hat{\mathbf{r}}_t, \varphi_{\text{inc}})$. The starting relation of the linear sampling method is the following integral equation (Colton et al. 2003)

$$L(g) = \int_0^{2\pi} E_{\infty_z}(\hat{\mathbf{r}}_t, \varphi_{\text{inc}}) g(\mathbf{z}_t, \varphi_{\text{inc}}) d\varphi_{\text{inc}} = \Phi_\infty(\hat{\mathbf{r}}_t, \mathbf{z}_t), \quad \mathbf{z}_t \in S_i, \qquad (5.8.2)$$

where the unknown g is a complex function that, for any $\mathbf{z}_t$, belongs to $L^2(0,2\pi)$, and $\Phi_\infty(\hat{\mathbf{r}}_t, \mathbf{z}_t)$ is given by

$$\Phi_\infty(\hat{\mathbf{r}}_t, \mathbf{z}_t) = e^{jk_b \mathbf{z}_t \cdot \hat{\mathbf{r}}_t}, \qquad (5.8.3)$$

which, in the preceding notation, represents the far-field pattern of the Green function of the background along direction $\hat{\mathbf{r}}_t$ when the impulsive source is located at point $\mathbf{z}_t$. Equation (5.8.2) is a Fredholm linear integral equation of the first kind [see Section 3.1 and equation (3.1.1)].

THE LINEAR SAMPLING METHOD   93

The importance of such an equation is due to the fact that it admits an approximate solution g, whose norm blows up near the boundary of the scatterer and remains large outside; in such a way it is possible to exploit it to visualize the support of the inspected scatterer (Cakoni and Colton 2006).

In real experiments, since the far-field pattern can be measured for only a finite number S of incidence directions and a finite number M of measurement directions (see Section 4.3), equation (5.8.2) needs to be discretized. To this end, a $M \times S$ matrix $[\mathbf{L}]$ is introduced whose (mj)th element l_{mj} is the far-field pattern along the mth observation direction $\hat{\mathbf{r}}_t^m$ when the incident field is a plane wave of incidence angle φ_{inc}^j [i.e., $l_{mj} = E_{\infty z}(\hat{\mathbf{r}}_t^m, \varphi_{inc}^j)$]. For the sake of simplicity, the incidence and the observation directions will be assumed to be uniformly spaced, i.e., $\varphi_{inc}^j = (2\pi/S)(j-1), j = 1,\ldots,S$, and $\varphi^m = (2\pi/M)(m-1)$, $m = 1,\ldots, M$ [see equations (4.3.6) and (4.3.10)].

In a discrete setting, for a fixed $\mathbf{z}_t$, the far-field equation then becomes

$$[\mathbf{L}]\mathbf{g}_{\mathbf{z}_t} = \frac{S}{2\pi} \mathbf{\Phi}_{\infty,\mathbf{z}_t}. \quad (5.8.4)$$

Here, $\mathbf{\Phi}_{\infty,\mathbf{z}_t}$ denotes a vector of M elements, whose entries $\phi_m^{\infty,\mathbf{z}_t}, m = 1,\ldots, M$, are the far-field values of Green's function along the directions $\hat{\mathbf{r}}_t^m, m = 1,\ldots, M$, when the impulsive source is placed at the point $\mathbf{z}_t$, that is, $\phi_m^{\infty,\mathbf{z}_t} = e^{jk_b \hat{\mathbf{r}}_t^m \cdot \mathbf{z}_t}$, $m = 1,\ldots, M$. The term $\mathbf{g}_{\mathbf{z}_t}$ is a vector of size S containing the values of the unknown function at the incidence directions: $g_j^{\mathbf{z}_t} = g(\mathbf{z}_t, \varphi_{inc}^j), j = 1,\ldots, S$.

Moreover, in practice, the far-field pattern $E_{\infty z}$ cannot be known exactly; only noisy measurements of it are available. As a consequence, the datum actually at disposal is the noisy far-field pattern $E_{\infty z}^n$. Accordingly, $[\mathbf{L}]$ is replaced with $[\mathbf{L}^n]$. Moreover, since equation (5.8.2) is ill-posed, its discrete counterpart is ill-conditioned, and a regularization scheme, requiring a regularization parameter, is needed to solve it.

In summary, the linear sampling method can be described as follows:

1. Choose a grid of points $Z \subset \mathfrak{R}^2$, containing the scatterer and for each point $\mathbf{z}_t \in Z$.
    a. Find a one-parameter family of regularized solutions $\{\mathbf{g}_{\mathbf{z}_t,\alpha(\mathbf{z}_t)}\}_{\alpha>0}$ of equation (5.8.4), where $\alpha(\mathbf{z}_t)$ denotes the regularization parameter related to the equation holding for $\mathbf{z}_t$.
    b. Find the optimum regularization parameter $\alpha^*(\mathbf{z}_t)$ and the related solution $\mathbf{g}_{\mathbf{z}_t,\alpha^*(\mathbf{z}_t)}$, according to an optimality criterion.
    c. Store the norm $\|\mathbf{g}_{\mathbf{z}_t,\alpha^*(\mathbf{z}_t)}\|_S$ of the optimum solution.
2. Choose a monotonic function $\Psi: \mathfrak{R}^+ \to \mathfrak{R}$, called the *indicator function*, and, for each point of the grid Z, plot the composite function of the position $\Psi(\|\mathbf{g}_{\mathbf{z}_t,\alpha(\mathbf{z}_t)}\|_S)$.

Following the literature on the linear sampling method, in the sequel the regularized solution of the equation (5.8.4) is determined by using the

Tikhonov method (Tikhonov et al. 1995), which has been described in Section 5.6. Moreover, since the noise affects the matrix of coefficients of the linear systems (5.8.14), the optimal regularization parameter is selected by using the Morozov discrepancy criterion (Tikhonov et al. 1995).

For each $\mathbf{z}_t \in Z$, the Tikhonov regularized solution of the equation (5.8.4) $\mathbf{g}_{\mathbf{z}_t,\alpha(\mathbf{z}_t)}$ of regularization parameter $\alpha(\mathbf{z}_t)$ is such that

$$\mathbf{g}_{\mathbf{z}_t,\alpha(\mathbf{z}_t)} = \operatorname{argmin}\left\{\mathbf{g}_{\mathbf{z}_t,\alpha(\mathbf{z}_t)} \in C^S : \left\|[\mathbf{L}^n]\mathbf{g}_{\mathbf{z}_t,\alpha(\mathbf{z}_t)} - \frac{S}{2\pi}\mathbf{\Phi}_{\infty,\mathbf{z}_t}\right\|_M^2 + \alpha(\mathbf{z}_t)\|\mathbf{g}_{\mathbf{z}_t,\alpha(\mathbf{z}_t)}\|_S^2\right\}$$
(5.8.5)

If the singular value decomposition (Section 5.5) of the $[\mathbf{L}^n]$ matrix is performed, $\mathbf{g}_{\mathbf{z}_t,\alpha(\mathbf{z}_t)}$ can then be expressed as

$$\mathbf{g}_{\mathbf{z}_t,\alpha(\mathbf{z}_t)} = \frac{S}{2\pi}\sum_{i=1}^{p^n}\frac{\sigma_i^n}{(\sigma_i^n)^2 + \alpha(\mathbf{z}_t)}\langle\mathbf{\Phi}_{\infty,\mathbf{z}_t},\mathbf{v}_i^n\rangle_M \mathbf{u}_i^n,$$
(5.8.6)

where $\langle\mathbf{\Phi}_{\infty,\mathbf{z}_t},\mathbf{v}_i^n\rangle_M = \sum_{j=1}^{M}\mathbf{\Phi}_j^{\infty,\mathbf{z}_t}(\mathbf{v}_i^n)_j^*$ is the standard scalar product in C^S, p^n is the rank of the matrix $[\mathbf{L}^n]$, and $\{\mathbf{u}_i^n,\sigma_i^n,\mathbf{v}_i^n\}$ is its singular system. Moreover, since the singular vectors $\{\mathbf{v}_i^n\}_i^S$ are orthonormal, it follows that

$$\|\mathbf{g}_{\mathbf{z}_t,\alpha(\mathbf{z}_t)}\|_S^2 = \frac{S^2}{4\pi^2}\sum_{i=1}^{p^n}\frac{(\sigma_i^n)^2}{((\sigma_i^n)^2 + \alpha(\mathbf{z}_t))^2}|\langle\mathbf{\Phi}_{\infty,\mathbf{z}_t},\mathbf{v}_i^n\rangle_M|^2.$$
(5.8.7)

Selection of the optimal regularization parameter requires an estimate of the noise level. Such an estimate can be given in terms of the positive real number h such that $\|[\mathbf{L}^n] - [\mathbf{L}]\| < h$. Accordingly, the generalized discrepancy principle (Tikhonov et al. 1995) used for determining the optimal regularization parameter requires that $\alpha^*(\mathbf{z}_t)$ be the zero of the generalized discrepancy

$$\rho(\alpha(\mathbf{z}_t)) = \left\|[\mathbf{L}^n]\mathbf{g}_{\mathbf{z}_t,\alpha(\mathbf{z}_t)} - \frac{S}{2\pi}\mathbf{\Phi}_{\infty,\mathbf{z}_t}\right\|_M^2 - h^2\|\mathbf{g}_{\mathbf{z}_t,\alpha(\mathbf{z}_t)}\|_S^2.$$
(5.8.8)

Using (5.8.6) and the properties of the singular value decomposition, we can easily prove that

$$\rho(\alpha(\mathbf{z}_t)) = \frac{S}{2\pi}\sum_{i=1}^{p^n}\frac{(\alpha(\mathbf{z}_t))^2 - h^2(\sigma_i^n)^2}{((\sigma_i^n)^2 + \alpha(\mathbf{z}_t))^2}|\langle\mathbf{\Phi}_{\infty,\mathbf{z}_t},\mathbf{v}_i^n\rangle_M|^2.$$
(5.8.9)

Clearly, the key aspect of the approach is essentially the search for a given set of "amplitudes" (the values of $\mathbf{g}_{\mathbf{z}_t}$) for the incident plane waves so that their combined effect is such that the induced currents on the unknown target are able to generate a symmetric scattering pattern. Since this pattern can be achieved only when the scatterer is a point scatterer or occupies a very small volume, the solution $\mathbf{g}_{z_t}$ corresponds to a set of illumination amplitudes (one for each value of φ_{inc}) such that the incident field is focused on a point or in a small zone of the scatterer. In this case, only this zone essentially contributes to the scattering phenomenon. Moreover, since a regularized solution is searched for, this means that such a distribution of amplitudes for the incident fields is found only approximately, and this *equivalent beamforming* approach concentrates approximately the field in a single point. It should also be noted that the induced current (acting as required) can exist only inside the scatterer, whereas outside it, only incident fields with infinite amplitudes can produce a finite induced current, where the object function is equal to zero [equation (2.6.10)].

It should also be mentioned that, in a discrete setting, since only a finite number of incident directions S can be used to illuminate the target and a fixed number of measurement directions M can be considered, a rule for choosing such parameters is very useful. For the two-dimensional case, it has been estimated (Catapano et al. 2007) that good shapes can be reconstructed if

$$S > 2k_b a, \qquad (5.8.10)$$

where a is the radius of the minimum circle that can include the object cross section.

For illustration purposes, let us consider a numerical example. Figure 5.1 shows the cross section of an infinite homogeneous elliptic cylinder. The semi-major axis of the ellipses is $a = 0.75\lambda$, where λ is the free-space wavelength of the monochromatic incident fields; the semiminor axis is $b = 0.375\lambda$; and the eccentricity is $e = \sqrt{1-(b/a)^2} \approx 0.87$. The cross section center coincides with the center of the investigation area (Section 3.2), which is a square domain of side $d = 2\lambda$. The background is a vacuum ($\varepsilon_b = \varepsilon_0$), whereas the homogeneous cylinder is characterized by $\varepsilon = (1.8 - j0)\varepsilon_0$.

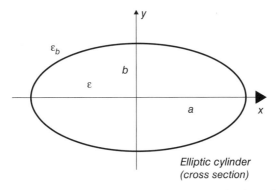

FIGURE 5.1 Cross section of an infinite homogenous lossless dielectric elliptic cylinder. Major axis: $a = 1.5\lambda$. Minor axis: $b = 0.75\lambda$. Propagation medium: vacuum ($\varepsilon_b = \varepsilon_0$).

According to the tomographic configuration described in Section 4.3, the object is successively illuminated by $S = 36$ unit plane waves with incident angles equally spaced. Measurements of the total electric field are performed on a circumference of radius $\rho_M = 100\lambda$. On this circumference, $M = 36$ measurement points are equally distributed; that is, the measurement points have polar coordinates $\mathbf{r}_t^m = \rho_M \cos\varphi_m \hat{\mathbf{x}} + \rho_M \sin\varphi_m \hat{\mathbf{y}}$ such that φ_m, $m = 1,\ldots, M$, is as given by (4.3.10). The total fields at the measurement points are analytically computed by using the eigenfunction expansions in terms of Mathieu functions (see Section 3.5). Then, by subtracting the values of the incident fields, the samples of the scattered fields are calculated.

In order to simulate realistic imaging conditions, the obtained values E_{mi}^s, $i = 1,\ldots, S, m = 1,\ldots, M$, are corrupted by added noise, that is

$$E_{mi}^s \rightarrow E_{mi}^s + N_{mi}, \quad i = 1,\ldots, S, m = 1,\ldots, M, \tag{5.8.11}$$

where the values N_{mi} are the realizations of a Gaussian process with zero mean values and variances corresponding to a fixed signal-to-noise ratio (SNR). The used indicator function is $\Psi(\|\mathbf{g}_{z_t,\alpha(z_t)}\|_S) = -A\log(\|\mathbf{g}_{z_t,\alpha(z_t)}\|_S) + B$, where A (>0) and B are constants such that $\Psi(\|\mathbf{g}_{z_t,\alpha(z_t)}\|_S)$ belongs to the range $[0,1]$.

Figures 5.2a–h show the reconstructed shapes of the elliptic cylinder for input data with noise levels ranging from SNR = 5 to 30 dB. The figures also report the retrieved shapes obtained as the set of points where the indicator function exceeds a prefixed value. In these cases, such a threshold for the normalized indicator function has been selected to be equal to 0.64 by minimizing the reconstruction error in a reference configuration.

The investigation area is a square of dimension $2\lambda \times 2\lambda$, which is partitioned into 127×127 pixels. As can be seen, for higher values of SNR the shape of the object is retrieved quite accurately (no interpolations are used in constructing the final profile), whereas even for low SNR values the object is correctly localized inside the investigation domain, but the shape is retrieved only approximately. In order to evaluate the errors in localizing the cross section, a simple parameter has been used, which is defined as the ratio between the grid points correctly attributed to the scatterer or to the background and the total number of points. Table 5.1 provides the percentage values of such an error parameter. From a computational perspective, each reconstruction required a central processing unit (CPU) time of ~8s on a 3 GHz Pentium PC, with a 1-GB random-access memory (RAM). Finally, as mentioned earlier, it should be noted that the linear sampling method is only one of the qualitative techniques able to provide the shapes of unknown scatterers. Another widely applied approach is based on the so-called level set method, for which the reader is referred to the literature (Litman et al. 1998, Berg and K. Holmstrom 1999, Dorn et al. 2000, Burger 2001, Ramananjaona et al. 2001, Ferraye et. al. 2003, Burger and Osher 2005, Litman 2005, Alexandrov and Santosa 2005, Dorn and Lesselier 2006, Hajihashemii and El-Shenawee 2008 and references cited therein).

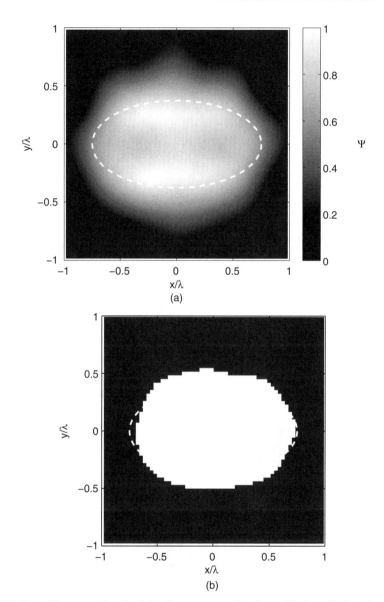

FIGURE 5.2 Shape retrieval of the homogenous lossless elliptic cylinder shown in Figure 5.1 (cross section) using linear sampling method. Plane-wave illumination ($S = 36$, $M = 36$ measurement points on a circumference of radius 100λ). Square investigation domain (side: $d = 2\lambda$; discretization: 127×127 pixels). Complex dielectric permittivity: $\varepsilon = (1.8 - j_0)\varepsilon_0$. Retrieved profiles [(a), (c), (e), (g)] and shape functions [(b), (d), (f), (h)]. Gaussian noise: (a), (b) SNR = 5 dB, (c), (d) SNR = 10 dB, (e), (f) SNR = 20 dB, (g), (h) SNR = 30 dB. The dashed line in each plot indicates the original profile.

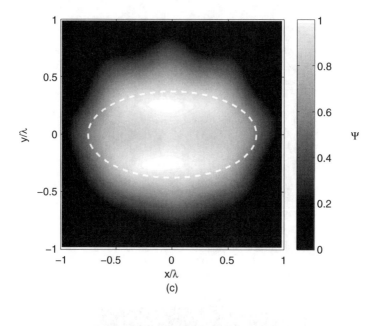

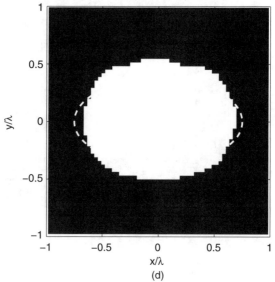

FIGURE 5.2 *Continued*

THE LINEAR SAMPLING METHOD   99

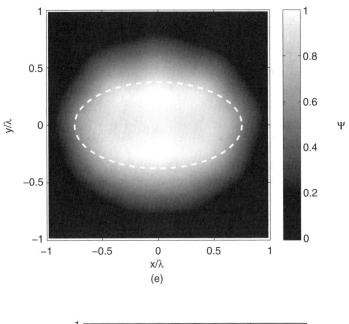

(e)

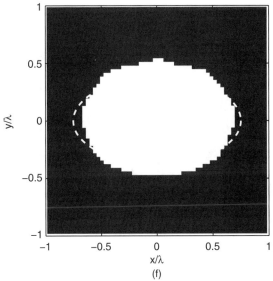

(f)

FIGURE 5.2 *Continued*

100 QUALITATIVE RECONSTRUCTION METHODS

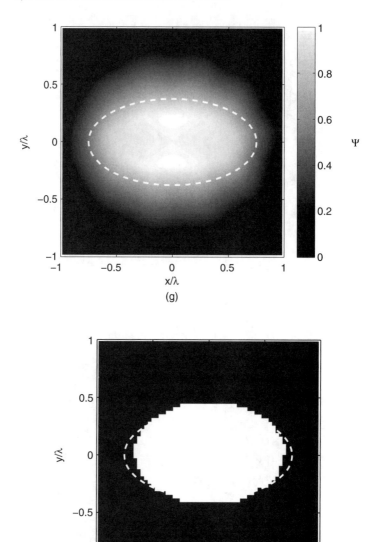

FIGURE 5.2 *Continued*

TABLE 5.1 Percentage Error on Reconstruction of Shape of Elliptic Cylinder in Figure 5.1 for Different Noise Levels on Data Using the Linear Sampling Method

	SNR, dB					
	30	25	20	15	10	5
Error, %	2.4	2.7	3.1	3.5	4.4	7.4

Source: Simulations performed by G. Bozza, University of Genoa, Italy.

5.9 SYNTHETIC FOCUSING TECHNIQUES

Very interesting results in microwave imaging applications have been obtained by using synthetic focusing techniques based on beamforming methods applied using the imaging configurations shown in Figures 4.6 and 4.7. These techniques are not really based on the "inversion" of the data as described in the previous chapters; thus they are mentioned only briefly here, and the reader is referred to the specialized publications cited in the present paragraph.

Synthetic focusing methods are based on the concept of synthetic aperture radar, which is widely used in remote sensing applications. Essentially, the target is illuminated by a single antenna that assumes successively different positions (Fig. 4.6). At each position, the antenna transmits a microwave signal and receives the echo (reflected signal). The echoes are not used directly. On the contrary, they are converted in discrete form and stored. When a sufficient number of measurements have been performed, the measured signal are summed after a proper time (or phase) shifting. If the shifting is correctly performed, only the reflected signal arriving from a reduced zone (ideally, a point scattering) contributes constructively to the sum. On the contrary, the scattered signals due to the other zones of the scatter add destructively, and their contributions can be considered as noise in the measured data. This measurement modality *synthesizes* an array antenna, in which several transmitting/receiving elements work simultaneously (with great improvement with respect to the single element in terms of gain and beamwidth reduction, resulting in very high spatial *resolution*). Synthetic techniques are an alternative to the use of a real array (Fig. 4.7), since they are far less expensive and do not require complex feeding circuits and multiplexing approaches. Mutual coupling among elements is also avoided, although the acquisition time can be much higher. Moreover, a mechanical system for the probe movements is necessary.

An example of this class of methods is constituted by ultrawideband techniques more recently explored mainly, but not exclusively, for medical applications. These techniques seldom attempt to reconstruct the complete profile of the dielectric properties of the region of interest, but instead seek to identify the presence and location of *targets* by their scattering *signature*.

In a basic time-domain formulation, the ultrawideband transmitter radiates bursts of radiofrequency (RF)/microwave energy of extremely short duration (tens of picoseconds to nanoseconds). The energy is scattered by discontinuities present inside the propagation medium. As mentioned, the scattered waves are collected by one or more receiving antennas (Li et al. 2005). Since, by proper combination of the receiving signals, one is able to obtain the response of a very limited zone inside the scatterer, one can construct an image of the *reflection* property of the target by successively exploring, pixel by pixel, the whole target. This can be done simply by changing the shifting values *added* to the received signals at the various measurement positions of the scanning probe. The main difficulty inherent in this simple approach is that, for focusing on a specific space point, the various shifting coefficients need to be defined with knowledge of the propagation velocity inside the propagation medium. However, since the wave velocity inside the target to be inspected (which is unknown) is not known, some a priori information can be used in this case, too (e.g., a reasonable *estimate* of the velocity in a model of the target under test must be assumed).

The basic formulation can be improved in the light of its application to specific cases (e.g., in medical imaging or nondestructive evaluations). First, there is the need for compensating for the reflection due to the external boundaries of the scatterer under test. In some cases, the external shape of the object is known and, consequently, from the distance between the interface and the probes one can deduce which part of the reflected signal is to be eliminated. In other cases, however, this distance is not known. For example, in medical imaging, where there is the need to compensate for the reflection from the air–skin interface, the shape of the body is patient-dependent. Consequently, specific procedures must be developed to deduce the first reflection from the target.

Another important issue concerns the compensation of dispersion effects in the signal propagation, which can introduce notable broadening in the pulse duration. In medical applications, an approach based on a broadband beamformer implementing frequency-dependent amplitudes and phase changes in the various channels has been presented (Li et al. 2005). Such approaches of can be implemented both in frequency and time domains (Bond et al. 2003, Chen et al. 2007, Mohan and Yang 2009). However, since the scientific literature on beamforming and synthetic aperture methods is wide, no further details are given in the present book, and the reader is referred to the cited papers. Some other key references will be provided in Section 10.2.

In general, for the microwave imaging techniques considered so far, including the synthetic focusing techniques, the scattering phenomenon is clearly the fundamental element. However, some other approaches have been developed that try to consider the scattering phenomenon as a disturbing effect to be possibly removed. One of them is based on the use of a chirp radar (Bertero et al. 2000). In this technique a transmitting antenna and a receiving antenna are located on different sides of the object to be inspected (see Fig. 4.5). The transmitting antenna is excited by a chirp pulse given by

$$v(t) = A \sin\left(k T_i t + \frac{1}{2} k t^2 \right), \tag{5.9.1}$$

where T_i and k are parameters to be chosen by the user (Bertero et al. 2000). In constructing the image of the object cross section (cylindrical objects are inspected), only the straight path between the two antennas is considered. This path is distinguished from the others by the fact that the frequency of the beat signal between the input chirp pulse signal and the output signal is determined by the time delay between the two signals. The transmitted wave component on the straight path is extracted by measuring the intensity at the frequency corresponding to the straight line between the two antennas. The key idea is to extract only the signal transmitted on a straight path in order to apply simple back projection algorithms, which are used, for example, in X-ray computerized tomography and, in general, in those imaging techniques for which the assumptions of ray propagation are valid.

It should be noted that the same approach has also been proposed with respect to a backscattering configuration, in which both the transmitting and receiving antennas are located on the same side of the object to be inspected. In this case the receiving signal is essentially a reflected signal.

5.10 QUALITATIVE METHODS FOR IMAGING BASED ON APPROXIMATIONS

According to the discussion on the meaning of qualitative reconstruction methods adopted in this book, this section is devoted to a description of algorithms that are referred to as *qualitative*, since they are based on scattering models involving some approximations, such as those presented in Chapter 4. In particular, the most interesting methods that are described here are the diffraction tomography (Section 5.11) and Born-like methods (Sections 5.12 and 5.13).

It should be noticed that Born approximations of order higher than 1 lead to nonlinear problems, which may require iterative resolution methods. Incidentally, the distinction between linear and nonlinear scattering models is another largely used criterion in inverse scattering literature to classify inversion approaches (Pike and Sabatier 2001).

5.11 DIFFRACTION TOMOGRAPHY

Diffraction tomography (DT) can be seen as an extension to scattered waves of the well-known computerized technique based on Radon transform and valid for X rays. Accordingly, diffraction tomography applies to two-dimensional problems, and the hypothesis of transverse magnetic (TM) illumination will be adopted (Bolomey et al. 1982, Devaney 1983, Pichot et al. 1985).

Moreover, according to the assumptions made in Chapter 4 concerning the imaging configurations, here the scattered field is supposedly collected along a straight probing line, located at distance Y_0 from the origin of the frame and completely outside the investigation domain S_i. Moreover, let φ be the angle between the probing line and the x axis of the reference frame.

The integral equation governing the scattering phenomenon can be written as follows [see equation (3.3.8)]:

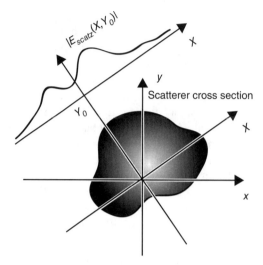

FIGURE 5.3 Two-dimensional imaging configuration for transmission diffraction tomography.

$$E_{\text{scat}_z}(\mathbf{r}_t) = j\omega\mu_b \int_{S_i} J_{\text{eq}_z}(\mathbf{r}'_t) G_{2D}(\mathbf{r}_t/\mathbf{r}'_t)\,d\mathbf{r}'_t. \qquad (5.11.1)$$

It is then very useful to introduce a new reference frame wherein the probing line is parallel to one of the coordinate axes. This goal can be achieved by means of the reference frame of Cartesian coordinates (X,Y,Z) and position vector $\mathbf{R} = \mathbf{R}_t + Z\hat{\mathbf{Z}} = X\hat{\mathbf{X}} + Y\hat{\mathbf{Y}} + Z\hat{\mathbf{Z}}$ (see Fig. 5.3), such that

$$\begin{cases} X = x\cos\varphi + y\sin\varphi \\ Y = -x\sin\varphi + y\cos\varphi \\ Z = z \end{cases} \quad \begin{cases} x = X\cos\varphi - Y\sin\varphi \\ y = X\sin\varphi + Y\cos\varphi \\ z = Z \end{cases} \qquad (5.11.2)$$

In this new frame, the probing line is expressed by the equation $Y = Y_0$, and we obtain

$$E^{\varphi}_{\text{scat}_z}(\mathbf{R}_t) = j\omega\mu_b \int_{S_i} J^{\varphi}_{\text{eq}_z}(\mathbf{R}'_t) G_{2D}(\mathbf{R}_t/\mathbf{R}'_t)\,d\mathbf{R}'_t, \qquad (5.11.3)$$

where $E^{\varphi}_{\text{scat}_z}$ and $J^{\varphi}_{\text{eq}_z}$ denote the scattered field and the equivalent current in the rotated framework, respectively.

By using the plane-wave representation for the free-space Green function [equation (3.3.10)], and since $Y_0 > Y'$, for every $Y' \in S_i$, because the probing line is outside the investigation domain, one obtains

$$E^{\varphi}_{\text{scat}_z}(X,Y_0) = \frac{\omega\mu_b}{2} \int_{-\infty}^{+\infty}\int_{S_i} J^{\varphi}_{\text{eq}_z}(\mathbf{R}'_t) \frac{e^{-j\sqrt{k_b^2 - 4\pi^2\lambda^2}(Y_0 - Y')}}{\sqrt{k_b^2 - 4\pi^2\lambda^2}} e^{-j2\pi\lambda(X-X')}\,d\mathbf{R}'_t\,d\lambda. \qquad (5.11.4)$$

In terms of its Fourier transform $\tilde{E}^{\varphi}_{\text{scat}_z}$ with respect to X, the scattered electric field collected along the probing line of equation $Y = Y_0$ can be written as

$$E^{\varphi}_{\text{scat}_z}(X, Y_0) = \int_{-\infty}^{+\infty} \tilde{E}^{\varphi}_{\text{scat}_z}(\lambda, Y_0) e^{-j2\pi X \lambda} d\lambda, \tag{5.11.5}$$

whence, using (5.11.4), it follows that

$$\int_{-\infty}^{+\infty} \left[\tilde{E}^{\varphi}_{\text{scat}_z}(\lambda, Y_0) - \frac{\omega \mu_b}{2} \int_{S_i} J^{\varphi}_{\text{eq}_z}(\mathbf{R}'_t) \frac{e^{-j\sqrt{k_b^2 - 4\pi^2 \lambda^2}(Y_0 - Y')}}{\sqrt{k_b^2 - 4\pi^2 \lambda^2}} e^{j2\pi \lambda X'} d\mathbf{R}'_t \right] e^{-j2\pi \lambda X} d\lambda = 0, \tag{5.11.6}$$

that is

$$\tilde{E}^{\varphi}_{\text{scat}_z}(\lambda, Y_0) = \frac{\omega \mu_b}{2} \frac{e^{-j\sqrt{k_b^2 - 4\pi^2 \lambda^2} Y_0}}{\sqrt{k_b^2 - 4\pi^2 \lambda^2}}$$
$$\int_{S_i} J^{\varphi}_{\text{eq}_z}(\mathbf{R}'_t) \exp\left\{ j2\pi \left[\lambda X' + \frac{\sqrt{k_b^2 - 4\pi^2 \lambda^2}}{2\pi} Y' \right] \right\} d\mathbf{R}'_t. \tag{5.11.7}$$

By considering the two-dimensional Fourier transform of the equivalent current density

$$\tilde{J}^{\varphi}_{\text{eq}_z}(\mathbf{K}_t) = \int_{-\infty}^{+\infty} \int_{-\infty}^{+\infty} J^{\varphi}_{\text{eq}_z}(\mathbf{R}'_t) e^{j2\pi \mathbf{K}_t \cdot \mathbf{R}'_t} d\mathbf{R}'_t \tag{5.11.8}$$

with $\mathbf{K}_t = K_X \hat{\mathbf{X}} + K_Y \hat{\mathbf{Y}}$, we obtain

$$\tilde{E}^{\varphi}_{\text{scat}_z}(\lambda, Y_0) = \frac{\omega \mu_b}{2} \frac{e^{-j\sqrt{k_b^2 - 4\pi^2 \lambda^2} Y_0}}{\sqrt{k_b^2 - 4\pi^2 \lambda^2}} \tilde{J}^{\varphi}_{\text{eq}_z}(\lambda, \kappa(\lambda)) \tag{5.11.9}$$

with $\kappa(\lambda) = \left(\sqrt{k_b^2 - 4\pi^2 \lambda^2}\right)/2\pi$. Since [see equation (5.11.2)]

$$\tilde{J}_{\text{eq}_z}(\mathbf{k}_t) = \int_{-\infty}^{+\infty}\int_{-\infty}^{+\infty} J_{\text{eq}_z}(\mathbf{r}'_t) e^{j2\pi \mathbf{k}_t \cdot \mathbf{r}'_t} d\mathbf{r}'_t$$
$$= \tilde{J}_{\text{eq}_z}(K_x \cos\varphi - K_y \sin\varphi, K_x \sin\varphi + K_y \cos\varphi)$$
$$= \tilde{J}^{\varphi}_{\text{eq}_z}(K_x, K_y), \tag{5.11.10}$$

with $\mathbf{k}_t = k_x \hat{\mathbf{x}} + k_y \hat{\mathbf{y}}$, from (5.11.9) and (5.11.10) we have

$$\tilde{E}^{\varphi}_{\text{scat}_z}(\lambda, Y_0) = \frac{\omega \mu_b}{2} \frac{e^{-j\sqrt{k_b^2 - 4\pi^2 \lambda^2} Y_0}}{\sqrt{k_b^2 - 4\pi^2 \lambda^2}} \tilde{J}_{\text{eq}_z}(\alpha^{\varphi}(\lambda), \beta^{\varphi}(\lambda)), \tag{5.11.11}$$

where

$$\begin{cases} \alpha^\varphi(\lambda) = \lambda\cos\varphi - \kappa(\lambda)\sin\varphi, \\ \beta^\varphi(\lambda) = \lambda\sin\varphi + \kappa(\lambda)\cos\varphi. \end{cases} \quad (5.11.12)$$

Equation (5.11.11) is the key relation of diffraction tomography and states that the one-dimensional Fourier transform of the z component of the scattered electric field measured along a straight probing line is equal to the two-dimensional Fourier transform of the equivalent current density computed along a line of the transformed plane whose parametric equation is given by (5.11.12), which turns out to be an arc of circumference.

Equation (5.11.11) represents the Fourier diffraction theorem, which is in a certain sense a generalization of the Fourier slice theorem for tomographic approaches based on straight propagation (ray propagation), in which the *projection* collected in a given point of the probing line depends only on the perturbation introduced in the incident wave by the body along the straight path followed by the incident wave (inside the body) to reach that measurement point.

It should be noted that the approach followed here in deriving equation (5.11.11) is the most intuitive. However, an expression for the same Fourier diffraction theorem can be deduced by a direct computation of the Fourier transforms in equation (5.11.1) (Kak and Slaney 1988).

Equation (5.11.11) can be used to obtain $\tilde{J}_{eq_z}$ and hence J_{eq_z}. The problem is that different arcs of circumferences need to be combined in order to fill the frequency plane and obtain the necessary information. This can be done in several ways. The simplest one consists in rotating the illumination and measurement system around the object. It should be noted that this is essentially equivalent to keeping the imaging system fixed and rotating the object around its center.

It is clear that a perfect reconstruction can be obtained only if the two-dimensional Fourier transform of the equivalent current density is bandlimited. For objects with limited cross sections, since $J_{eq_z}(\mathbf{r}_t)$ is bounded [i.e., $J_{eq_z}(\mathbf{r}_t) = 0$ for $\mathbf{r}_t \notin S_i$], it follows that $\tilde{J}_{eq_z}(\mathbf{k}_t)$ is unbounded. Consequently, only a bandlimited version of the original $\tilde{J}_{eq_z}(\mathbf{k}_t)$ can be recovered by using the diffraction tomography, although it has been argued that "in practice, the loss of resolution caused by this band-limiting factor is negligible, being more influenced by considerations such the aperture size of transmitting and receiving elements, etc." (Kak and Slaney 1988 p. 234).

Actually, other practical issues have to be considered. First, the probing line cannot be infinite in extent. This means that a windowed version of the measured values along the probing line is available, namely

$$E^\varphi_{scat_z}(X, Y_0) = 0, |X| > \frac{l}{2}, \quad (5.11.13)$$

where l is the length of the probing line. This is equivalent to multiplying the first side of equation (5.11.5) by a rectangular window function centered at $X = 0$ and of width l. Accordingly, the computed Fourier transform results to be convoluted with a sinc function

$$\tilde{\tilde{E}}^{\varphi}_{\text{scat}_z}(\lambda, Y_0) = \tilde{E}^{\varphi}_{\text{scat}_z}(\lambda, Y_0) * \frac{\sin(2\pi\lambda l^{-1})}{2\pi\lambda l^{-1}}, \qquad (5.11.14)$$

where $*$ denotes the usual convolution operator. According to (5.11.14), the limited probing line results in another approximation on the retrieved $\tilde{J}_{\text{eq}_z}(\mathbf{k}_t)$.

Furthermore, in a discrete setting, the scattered field is measured in a finite set of points. Consequently, the Fourier transforms described above are substituted by discrete Fourier transforms (DFTs) and the $\tilde{J}_{\text{eq}_z}$ is consequently available on a grid in the transformed plane.

At the same time, the discretized version of equation (5.11.11) allows for the use of fast Fourier transforms (FFTs), making the approach extremely fast and effective.

In addition, the previously discussed bandlimitation has another important consequence; specifically, the spatial resolution of diffraction tomography is limited by (Pichot et al. 1985)

$$\delta \geq \frac{\lambda}{2}. \qquad (5.11.15)$$

This limit is particularly significant in microwave imaging, since it indicates that a limited spatial resolution can be obtained at these frequencies by using this technique. Such a resolution is not comparable with the one obtained by using other diagnostic systems working in different bands of the electromagnetic spectrum (e.g., X rays).

It should be noted that the limited spatial resolution of diffraction tomography and its applicability to weakly scattering objects only has been the main reason leading to the exploitation of quantitative approaches (described in Chapter 6), which, in principle, can be applied to strong scatterers because they are not based on approximations different from numerical ones and exhibit spatial resolution theoretically independent of the operating frequency. Furthermore, it is quite difficult to introduce a priori information on the target to be inspected into the diffraction tomography scheme.

It should also be realized that diffraction tomography only reconstructs the distribution of the equivalent current density, which usually provides limited information about the scatterer cross section (see also Section 5.14). However, for weakly scattering objects for which the first-order Born approximation can be applied, the internal electric field is approximated with the incident one and the dielectric properties of the inspected scatterer can be retrieved from the equivalent source. Nevertheless, since diffraction tomography is very simple to implement, it can provide good images of weakly scattering bodies and can also be used as initial guesswork for quantitative reconstruction methods (Franchois et al. 1998).

It should also be noted that, in the scientific literature, a different approach has been developed starting from the Rytov approximation (Devaney 1986), and higher-order versions of diffraction tomography have been proposed as well (Tsihrintzis and Devaney 2000).

Finally, as mentioned at the beginning of this section, diffraction tomography can be essentially seen as an extension of computerized tomography. Concerning this point, it should be mentioned that methods essentially based on geometric ray tracing, which are valid for ray propagation as in the case of X rays, have also been proposed for microwave imaging. Multiple scattering is neglected and the field at a receiving point is assumed to be essentially a perturbed version of the field emitted by a source on the other side of the object, where the perturbation is due to the dielectric properties of the object along a straight line that corresponds to a *ray* connecting the transmitting and receiving antennas. By rotating sources and receiving elements, numerous measured values can be collected for use as input data for simple reconstruction procedures. An example of these approaches is represented by the *algebraic reconstruction technique* (ART), which has been successively modified in order to partially account for the scattering phenomenon, in particular, the scattering effects due to parts of the scatterer *adjacent* to the internal region targeted by the ray propagation. The reader is referred to studies by Maini et al. (1981) and Datta and Bandyopadhyay (1985) and to the references cited therein for further details on these simplified (and fast) methods.

5.12 INVERSION APPROACHES BASED ON BORN-LIKE APPROXIMATIONS

Under the first-order Born approximation, equation (3.3.8) reads as

$$E_z(\mathbf{r}_t) = E_{\text{inc}_z}(\mathbf{r}_t) + j\omega\mu_b \int_{S_i} \tau(\mathbf{r}_t') E_{\text{inc}_z}(\mathbf{r}_t') G_{2D}(\mathbf{r}_t/\mathbf{r}_t') d\mathbf{r}_t', \quad \mathbf{r}_t \in S_m, \quad (5.12.1)$$

which is a linear equation in the only unknown τ, since the incident field is known everywhere. This is an example of a linear inverse problem, generically expressed by a linear functional equation of type $Lf = g$ [equation (3.1.1)]. Consequently, the following considerations are valid for a large class of inverse problems not necessarily related to microwave imaging.

Let us consider now the discretization of equation (5.12.1). Following the same approach used in Section 3.4 and with the same choice of basis and testing functions, after some trivial steps, one obtains an equation similar to (3.4.5), specifically

$$\sum_{n=1}^{N} b_{mn}\tau_n = E_m^s, m = 1, \ldots, M, \quad (5.12.2)$$

in which the coefficients b_{mn} are given by

$$b_{mn} = j\omega\mu_b \int_{S_n} E_{\text{inc}_z}(\mathbf{r}'_t) G_{2D}(\mathbf{r}_m/\mathbf{r}'_t) d\mathbf{r}'_t \tag{5.12.3}$$

[compare with equation (3.4.6)]. Moreover, the known terms E_m^s are still given by equations (3.4.7). The algebraic system of linear equations (5.12.2) can be rewritten in matrix form as

$$[\mathbf{B}]\boldsymbol{\tau} = \mathbf{e}^s, \tag{5.12.4}$$

where $\mathbf{e}^s = [E_1^s, E_2^s, \ldots, E_m^s, \ldots, E_M^s]^T$, $\boldsymbol{\tau} = [\tau_1, \tau_2, \ldots, \tau_n, \ldots, \tau_N]^T$, and b_{mn} is the element of the mth row and nth column of the $N \times M$ matrix $[\mathbf{B}]$.

The linear system (5.12.4) is obtained as the discrete counterpart of a Fredholm integral equation of the first kind, and thus is usually ill-conditioned. For this reason, it needs to be solved with care, and the truncated singular value decomposition (Section 5.6) is often the tool used to obtain a stable solution with respect to noise.

To give the reader an idea of the reconstruction capabilities of the truncated singular value decomposition applied to equation (5.12.4), the reconstruction of the homogenous elliptic cylinder of Figure 5.1 is considered again. In this case, with reference to the tomographic configuration described in Section 4.3, the object is successively illuminated by $S = 8$ line-current sources equally spaced on a circumference of radius $\rho_S = 1.5\lambda$. The incident field generated by these sources is then given by equation (4.3.7), with $\mathbf{r}_t^i = \rho_S \cos\varphi_i \hat{\mathbf{x}} + \rho_S \sin\varphi_i \hat{\mathbf{y}}$, where $\varphi_i = (2\pi/S)(i-1)$, for $i = 1, \ldots, S$. The measurements of the total electric field are performed on a circumference of radius $\rho_M = \rho_S$. On this circumference $M = 55$ measurement points are located on an arc corresponding to an angular sector of $270°$ in the opposite part of illuminating source; that is, for the ith source the measurement points have polar coordinates $\mathbf{r}_t^{mi} = \rho_S \cos\varphi_{mi} \hat{\mathbf{x}} + \rho_S \sin\varphi_{mi} \hat{\mathbf{y}}$, such that

$$\varphi_{mi} = \varphi_i + \frac{\pi}{4} + (m-1)\frac{3\pi}{2(M-1)}, \quad i = 1, \ldots, S, m = 1, \ldots, M. \tag{5.12.5}$$

The total electric field at the measurement points (near-field line-current scattering) is computed analytically (Caorsi et al. 2000). Then, by subtracting the values of the incident fields, the samples of the scattered fields are calculated and corrupted by additive Gaussian noise [equation (5.8.11)]. In particular, for this simulation, we assumed values of the signal-to-ratio ranging from 5 to 50 dB. Furthermore, the investigation area is partitioned into 30×30 square pixels.

Figures 5.4a–d show the reconstructed distributions of the relative dielectric permittivity. As can be seen, due to the weak scattering nature of the target, the truncated singular value decomposition under the first-order Born approximation leads to very good reconstructions except for very noisy data. As the relative dielectric permittivity (with respect to the background) increases, the efficiency of the Born approximation decreases. Figures 5.5a–e show the reconstructed distributions for $\varepsilon'_r = 1.2, 1.4, 1.6, 1.8,$ and 2.0. As expected, as the object function increases, the Born approximation is no longer able

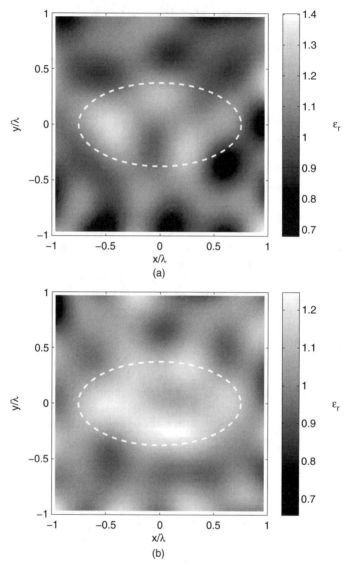

FIGURE 5.4 Reconstruction of the cross section of the homogenous lossless elliptic cylinder shown in Figure 5.1. First-order Born approximation and truncated SVD. Line-current illuminations ($S = 8$ sources, $M = 55$ measurement points on a circumference of radius 1.5 λ). Square investigation domain (side: $d = 2\lambda$; discretization: 30×30 pixels). Complex dielectric permittivity: $\varepsilon = (1.2 - j0)\varepsilon_0$. Gaussian noise: (a) SNR = 5 dB; (b) SNR = 10 dB; (c) SNR = 15 dB; (d) SNR = 25 dB; (e) SNR = 50 dB. The dashed line in each plot indicates the original profile. (Simulations performed by A. Randazzo, University of Genoa, Italy.)

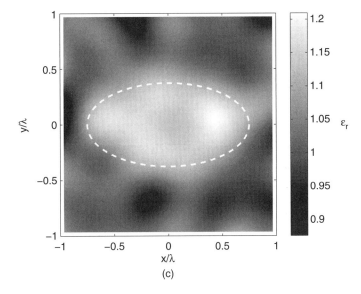

(c)

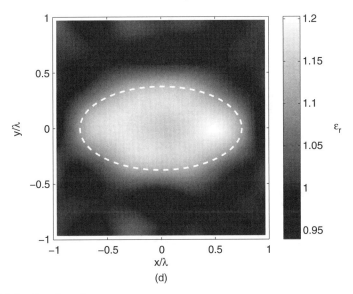

(d)

FIGURE 5.4 *Continued*

112 QUALITATIVE RECONSTRUCTION METHODS

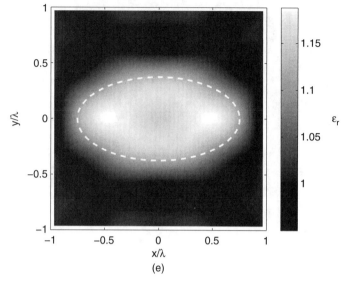

FIGURE 5.4 *Continued*

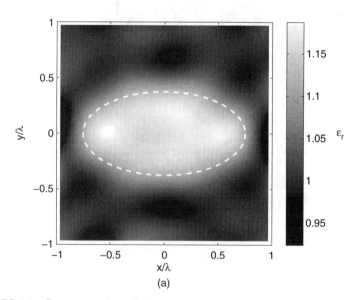

FIGURE 5.5 Reconstruction of the cross section of the homogenous lossless elliptic cylinder shown in Figure 5.1. The reconstruction method and the imaging configuration are the same as in Figure 5.4. Relative dielectric permittivity: (a) $\varepsilon_r' = 1.2$; (b) $\varepsilon_r' = 1.4$; (c) $\varepsilon_r' = 1.6$; (d) $\varepsilon_r' = 1.8$; (e) $\varepsilon_r' = 2.0$. The dashed line in each plot indicates the original profile. (Simulations performed by A. Randazzo, University of Genoa, Italy.)

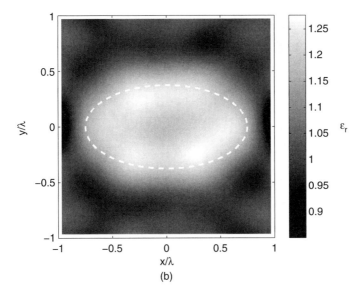

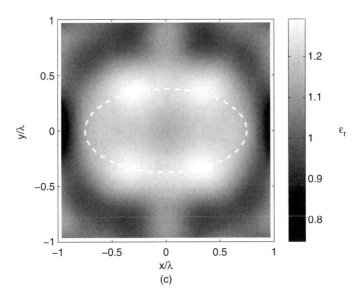

FIGURE 5.5 *Continued*

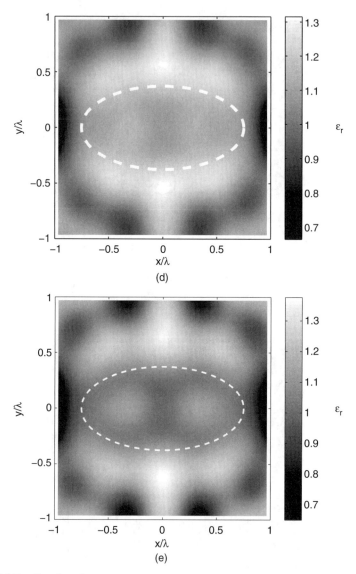

FIGURE 5.5 *Continued*

to approximately locate the object inside the investigation area. For completeness, quantitative data concerning these simulations are reported in Tables 5.2 and 5.3.

Finally, the singular values related to the matrix decomposition are provided in Figure 5.6. It should be noted that a very similar formulation can be derived if the Rytov approximation described in Section 4.8 is applied. The computation is not repeated here.

TABLE 5.2 Relative Errors on Reconstruction of Elliptic Cylinders in Figure 5.4 for Different SNR Values, Calculated by First-Order Born Approximation Using the Truncated SVD Method

	SNR, dB									
	5	10	15	20	25	30	35	40	45	50
Global Errors										
Minimum values	2.5×10^{-4}	6.0×10^{-5}	5.0×10^{-5}	1.1×10^{-4}	1.0×10^{-5}	1.0×10^{-5}	1.0×10^{-5}	3.2×10^{-5}	1.0×10^{-5}	1.5×10^{-5}
Mean values	0.08	0.054	0.036	0.029	0.025	0.025	0.024	0.024	0.024	0.024
Maximum values	0.40	0.342	0.136	0.128	0.130	0.128	0.127	0.125	0.124	0.125
Variance	0.06	0.044	0.027	0.026	0.025	0.025	0.025	0.025	0.025	0.025
Errors on Reconstruction of Object Cross Section										
Minimum values	0.001	0.002	0.001	0.001	0.000	0.005	0.007	0.011	0.010	0.010
Mean values	0.07	0.05	0.05	0.04	0.04	0.04	0.04	0.04	0.04	0.04
Maximum values	0.26	0.15	0.12	0.11	0.09	0.09	0.08	0.08	0.08	0.08
Variance	0.05	0.03	0.02	0.02	0.02	0.02	0.02	0.02	0.02	0.02
Errors on Reconstruction of Background										
Minimum values	2.5×10^{-4}	6.0×10^{-5}	5.0×10^{-5}	1.1×10^{-4}	1.0×10^{-5}	1.0×10^{-5}	1.0×10^{-5}	3.2×10^{-5}	1.0×10^{-5}	1.5×10^{-5}
Mean values	0.08	0.06	0.03	0.03	0.02	0.02	0.02	0.02	0.02	0.02
Maximum values	0.40	0.34	0.14	0.13	0.13	0.13	0.13	0.12	0.12	0.12
Variance	0.06	0.05	0.03	0.03	0.03	0.03	0.03	0.03	0.03	0.03

TABLE 5.3 Relative Errors on Reconstruction of Elliptic Cylinders in Figure 5.5 for Different Relative Dielectric Permittivity Values, Calculated by First-Order Born Approximation Using the Truncated SVD Method

	\multicolumn{11}{c}{Relative Dielectric Permittivity}										
	1.0	1.2	1.4	1.6	1.8	2.0	2.2	2.4	2.6	2.8	3.0
					Global Errors						
Minimum values	6.0×10^{-5}	1.0×10^{-5}	2.1×10^{-4}	9.9×10^{-5}	1.0×10^{-5}	5.0×10^{-4}	1.0×10^{-5}	5.3×10^{-5}	1.0×10^{-4}	8.9×10^{-5}	8.0×10^{-5}
Mean values	0.01	0.03	0.06	0.12	0.18	0.21	0.23	0.23	0.25	0.27	0.30
Maximum values	0.06	0.12	0.24	0.29	0.41	0.51	0.61	0.71	0.75	0.75	0.74
Variance	0.01	0.02	0.06	0.10	0.13	0.16	0.19	0.21	0.22	0.22	0.23
				Errors on Reconstruction of Object Cross Section							
Minimum values	—	—	0.009	0.20	0.32	0.43	0.47	0.47	0.45	0.47	0.50
Mean values	—	0.04	0.12	0.24	0.37	0.48	0.55	0.59	0.62	0.64	0.66
Maximum values	—	0.09	0.16	0.29	0.41	0.51	0.61	0.71	0.75	0.75	0.74
Variance	—	0.02	0.02	0.02	0.02	0.02	0.03	0.06	0.07	0.07	0.06
				Errors on Reconstruction of Background							
Minimum values	6.0×10^{-5}	1.0×10^{-5}	2.1×10^{-4}	9.9×10^{-5}	1.0×10^{-5}	5.0×10^{-4}	1.0×10^{-5}	5.3×10^{-5}	1.0×10^{-4}	8.9×10^{-5}	8.0×10^{-5}
Mean values	0.009	0.022	0.048	0.083	0.120	0.137	0.133	0.132	0.143	0.168	0.198
Maximum values	0.055	0.122	0.244	0.292	0.337	0.377	0.349	0.366	0.393	0.416	0.537
Variance	0.007	0.025	0.056	0.077	0.090	0.094	0.092	0.089	0.096	0.107	0.128

Source: Simulations performed by A. Randazzo, University of Genoa, Italy.

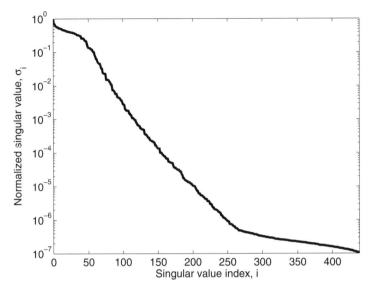

FIGURE 5.6 Singular values of the matrix used to obtain the reconstructions in Figure 5.5. (Simulations performed by A. Randazzo, University of Genoa, Italy.)

5.13 THE BORN ITERATIVE METHOD

The Born approximation discussed in Section 4.6 is also applied to develop iterative methods. The first method to be mentioned is the so-called Born iterative method, in which the initial object function is iteratively updated until a stopping criterion is satisfied. At each iteration the first-order Born approximation is applied.

With reference to equation (5.12.1), let us consider τ^0 as a known initial guess distribution. If no a priori information is available, we can assume $\tau^0(\mathbf{r}) = 0$ (the *object* coincides with the background). After the initialization step ($k = 0$) the approach iteratively evolves. At the kth step ($k \geq 1$), the equations to be solved are as follows:

$$E_z^{k-1}(\mathbf{r}_t) = E_{\text{inc}_z}(\mathbf{r}_t) + j\omega\mu_b \int_{S_i} \tau^{k-1}(\mathbf{r}_t') E_z^{k-1}(\mathbf{r}_t') G_{2D}(\mathbf{r}_t/\mathbf{r}_t') d\mathbf{r}_t', \quad \mathbf{r}_t \in S_i, \quad (5.13.1)$$

$$E_z(\mathbf{r}_t) = E_{\text{inc}_z}(\mathbf{r}_t) + j\omega\mu_b \int_{S_i} \tau^k(\mathbf{r}_t') E_z^{k-1}(\mathbf{r}_t') G_{2D}(\mathbf{r}_t/\mathbf{r}_t') d\mathbf{r}_t', \quad \mathbf{r}_t \in S_m. \quad (5.13.2)$$

Equations (5.13.1) and (5.13.2) are not used together. In particular, at the kth iteration, the electric field E_z^{k-1} inside the investigation domain S_i is first obtained by solving the state equation (5.13.1), where τ^{k-1} is a known quantity. Afterward, the scattering potential τ^k is determined by using the data equation (5.13.2). In this equation the left-hand side is the known term of the problem (it is obtained from measurements performed in the observation domain S_m), and hence it does not depend on the iteration number k. The iterative approach finishes when a stopping criterion is satisfied.

It should be noted that the whole computation involves the solution of only linear integral equations, which can be numerically performed as mentioned in Section 3.4. It should also be noted that the same Green function is used at any iteration. This Green function is related to the background medium.

A variational form of this procedure has also been proposed in which the unknown distribution to be retrieved is not directly the object function, but the difference between the object functions at two different iterations (Zaiping et al. 2000).

5.14 RECONSTRUCTION OF EQUIVALENT CURRENT DENSITY

Besides inverse scattering problems, *inverse source* problems can be considered in microwave imaging. Such problems are aimed at reconstructing the equivalent current density, $\mathbf{J}_{eq}$, and, as indicated by equation (3.2.3), they are linear. The issue is the information content of $\mathbf{J}_{eq}$. It has been shown that some methods are based on the reconstruction of $\mathbf{J}_{eq}$, including diffraction tomography (Section 5.11). However, in those cases, one needs some a priori information on the object, as in cases in which the Born approximation can be used (i.e., since the incident field is known, from $\mathbf{J}_{eq}$ one can immediately deduce the dielectric parameters of the unknown target, in particular, for dielectrics, the object function τ [equations (2.6.10) and (2.7.3)]. In the general case, the situation is different. In fact, the equivalent current density can be written as the sum of radiating and nonradiating components, that is

$$\mathbf{J}_{eq}(\mathbf{r}) = \mathbf{J}_R(\mathbf{r}) + \mathbf{J}_{NR}(\mathbf{r}), \quad \mathbf{r} \in V_o, \tag{5.14.1}$$

where the nonradiating source is such that the field that it produces is everywhere equal to zero except inside its support (Devaney and Wolf 1973). Since $\mathbf{J}_{NR}$ does not radiate outside V_o, it cannot be reconstructed starting from measurements performed outside V_o (i.e., in the observation domain). On the contrary, $\mathbf{J}_{NR}$ contributes to the field inside V_o. Consequently, the total field inside V_o can be written as

$$\mathbf{E}(\mathbf{r}) = \mathbf{E}_{inc}(\mathbf{r}) + \mathbf{E}_{scat}^{R}(\mathbf{r}) + \mathbf{E}_{scat}^{NR}(\mathbf{r}), \tag{5.14.2}$$

where

$$\mathbf{E}_{scat}^{R}(\mathbf{r}) = j\omega\mu_b \int_{V_o} \mathbf{J}_R(\mathbf{r}') \cdot \overline{\mathbf{G}}(\mathbf{r}/\mathbf{r}')d\mathbf{r}' \tag{5.14.3}$$

and

$$\mathbf{E}_{scat}^{NR}(\mathbf{r}) = j\omega\mu_b \int_{V_o} \mathbf{J}_{NR}(\mathbf{r}') \cdot \overline{\mathbf{G}}(\mathbf{r}/\mathbf{r}')d\mathbf{r}'. \tag{5.14.4}$$

Consequently, if only the radiating component of $\mathbf{J}_{eq}$ is available, the term $\mathbf{E}_{scat}^{NR}$ cannot, in general, be computed. Accordingly, the dielectric properties of the object to be inspected cannot be deduced from equation (2.6.10).

Moreover, the information on $\mathbf{J}_{eq}$ can be important when a priori information on the object is available, as in the case in which the Born approximation can be used. It has also been observed that nonradiating sources contribute to the higher spatial frequencies of the object to be inspected. Accordingly, neglecting them can result in a smoothing effect on the reconstruction, which can be acceptable in certain cases.

However, a possible approach that factors in both components of the equivalent current density is to first reconstruct its visible part with a standard method [applied to the solution of equation (5.14.3)] and, successively, retrieving the invisible part by enforcing the constraint that the unknown total electric field $\mathbf{E}^i$ produced by the ith source and the unknown object function τ be *consistent* (for any i) with the known incident field, by means of an optimization approach similar to that described in Section 6.7. These points are discussed further in the literature (Gamliel et al. 1989, Wallacher and Louis 2006, Caorsi and Gragnani 1999).

Alternatively, by considering a multiillumination approach (see Section 4.2), one obtains

$$\mathbf{J}_{eq}^i(\mathbf{r}) = \tau(\mathbf{r})\mathbf{E}^i(\mathbf{r}), \quad i = 1, \ldots, S. \qquad (5.14.5)$$

We can observe that τ is independent of i, and, unlike trying to reconstruct both τ and $\mathbf{E}^i$, $\mathbf{E}^i$ can be viewed as "an unwanted factor or multiplicative noise term which contains a certain range of spatial frequencies determined by the distribution of energy of the radiation field and its effective wavelength within the object" (Morris et al. 1995).

REFERENCES

Alexandrov, O. and F. Santosa, "A topology preserving level set method for shape optimization," *J. Comput. Phys.* **204**, 121–130 (2005).

Berg, J. M. and K. Holmström, "On parameter estimation using level sets," *SIAM J. Control Optim.* **37**, 1372–1393 (1999).

Bertero, M. and P. Boccacci, *Introduction to Inverse Problems in Imaging*, Institute of Physics, Bristol, UK, 1998.

Bertero, M., M. Miyakawa, P. Boccacci, F. Conte, K. Orikasa, and M. Furutani, "Image restoration in chirp pulse microwave CT (CP-MCT)," *IEEE Trans. Biomed. Eng.* **47**, 690–699 (2000).

Bolomey, J.-C., A. Izadnegahdar, L. Jofre, Ch. Pichot, G. Peronnet, and M. Solaimani, "Microwave diffraction tomography for biomedical applications," *IEEE Trans. Microwave Theory Tech.* **30**, 1998–2000 (1982).

Bond, E. J., X. Li, S. C. Hagness, and B. D. Van Veen, "Microwave imaging via space-time beamforming for early detection of breast cancer," *IEEE Trans. Anten. Propag.* **51**, 1690–1705 (2003).

Burger, M., "A level set method for inverse problems," *Inverse Problems* **17**, 1327–1355 (2001).

Burger, M. and S. Osher, "A survey on level set methods for inverse problems and optimal design," *Eur. J. Appl. Math.* **16**, 263–301 (2005).

Cakoni, F. and D. Colton, *Qualitative Methods in Inverse Scattering Theory*, Springer, Berlin, 2006.

Caorsi, S., S. Ciaramella, G. L. Gragnani, and M. Pastorino, "On the use of regularization techniques in numerical inverse-scattering solutions for microwave imaging applications," *IEEE Trans. Microwave Theory Tech.* **43**, 632–640 (1995).

Caorsi, S. and G. L. Gragnani, "Inverse-scattering method for dielectric objects based on the reconstruction of the nonmeasurable equivalent current density," *Radio Sci.* **34**, 1–8 (1999).

Caorsi, S., M. Pastorino, and M. Raffetto, "Electromagnetic scattering by a multilayer elliptic cylinder under line-source illumination," *Microwave Opt. Technol. Lett.* **24**, 322–329 (2000).

Catapano, I., L. Crocco, and T. Isernia, "On simple methods for shape reconstruction of unknown scatterers," *IEEE Anten. Propag.* **55**, 1431–1435 (2007).

Chen, Y., E. Gunawan, K. S. Low, S. C. Wang, C. B. Soh, and L. L. Thi, "Time of arrival data fusion method for two-dimensional ultra wideband breast cancer detection," *IEEE Trans. Anten. Propag.* **55**, 2852–2865 (2007).

Colton, D., H. Haddar, and M. Piana, "The linear sampling method in inverse electromagnetic scattering theory," *Inverse Problems* **19**, S105–S137 (2003).

Colton, D. and R. Kress, *Inverse Acoustic and Electromagnetic Scattering Theory*, Springer, Berlin, 1998.

Colton, D. and P. Monk, "A linear sampling method for the detection of leukemia using microwaves," *SIAM J. Appl. Math.* **58**, 926–941 (1998).

Datta, A. N. and B. Bandyopadhyay, "An improved SIRT-style reconstruction algorithm for microwave tomography," *IEEE Trans. Biomed. Eng.* **32**, 719–723 (1985).

Devaney, A. J., "A computer simulation study of diffraction tomography," *IEEE Trans. Biomed. Eng.* **30**, 377–386 (1983).

Devaney, A. J., "Reconstructive tomography with diffracting wavefields," *Inverse Problems* **2**, 161–183 (1986).

Devaney, A. J. and E. Wolf, "Radiating and nonradiating classical current distributions and the fields they generate," *Phys. Rev. D* **8**, 1044–1047 (1973).

Dorn, O. and D. Lesselier, "Level set methods for inverse scattering," *Inverse Problems* **22**, R67–R131 (2006).

Dorn, O., E. Miller, and C. Rappaport, "A shape reconstruction method for electromagnetic tomography using adjoint fields and level sets," *Inverse Problems* **16**, 1119–1156 (2000).

Ferraye, R., J. Y. Dauvignac, and C. Pichot, "An inverse scattering method based on contour deformations by means of a level set method using frequency hopping technique," *IEEE Trans. Anten. Propag.* **51**, 1100–1113 (2003).

Franchois, A., A. Joisel, C. Pichot, and J.-C. Bolomey, "Quantitative microwave imaging with a 2.45-GHz planar microwave camera," *IEEE Trans. Med. Imag.* **17**, 550–561 (1998).

Gamliel, A., K. Kim, A. I. Nachman, and E. Wolf, "A new method for specifying nonradiating monochromatic scalar sources and their fields," *J. Opt. Soc. Am.* **6**, 1388–1393 (1989).

Hajihashemii, M. R. and M. El-Shenawee, "Shape reconstruction using the level-set method for microwave applications," *IEEE Anten. Wireless Propag. Lett.* **7**, 92–96 (2008).

Kak, A. C. and M. Slaney, *Principles of Computerized Tomographic Imaging*. IEEE Press, New York, 1988.

Li, X., E. J. Bond, B. D. Van Veen, and S. C. Hagness, "An overview of ultra-wideband microwave imaging via space-time beamforming for early-stage breast-cancer detection," *IEEE Anten. Propag. Mag.* **47**, 19–29 (2005).

Litman, A., "Reconstruction by level sets of *n*-ary scattering obstacles," *Inverse Problems* **21**, S131–S152 (2005).

Litman, A., D. Lesselier, and D. Santosa, "Reconstruction of a two-dimensional binary obstacle by controlled evolution of a level-set," *Inverse Problems* **14**, 685–706 (1998).

Maini, R., M. F. Iskander, C. H. Durney, and M. Berggren, "On the sensitivity and the recolution of microwave imaging using ART," *Proc. IEEE* **69**, 1517–1519 (1981).

Mohan, A. S. and F. Yang, "Improved beamforming for microwave imaging to detect breast cancer," *Proc. 2009 Antennas Propagation Society Int. Symp.* 2009.

Morris, J. B., F. C. Lin, D. A. Pommet, R. V. McGahan, and M. A. Fiddy, "A homomorphic filtering method for imaging strongly scattering penetrable objects," *IEEE Anten. Propag.* **43**, 1029–1035 (1995).

Oden, T. J. and L. F. Demkowicz, *Applied Functional Analysis*, CRC Press, Boca Raton, FL, 1996.

Pichot, C., L. Jofre, G. Peronnet, and J.-C. Bolomey, "Active microwave imaging of inhomogeneous bodies," *IEEE Trans. Anten. Propag.* **33**, 416–425 (1985).

Pike, E. R. and P. C. Sabatier, eds., *Scattering and Inverse Scattering in Pure and Applied Science*, Academic Press, London, 2001.

Ramananjaona, C., M. Lambert, D. Lesselier, and J. P. Zolesio, "Shape reconstruction of buried obstacles by controlled evolution of a level set: From a min-max formulation to numerical experimentation," *Inverse Problems* **17**, 1087–1111 (2001).

Rudin, W., *Real and Complex Analysis*, McGraw-Hill, New York, 1970.

Sarkar, T., D. Weiner, and V. Jain, "Some mathematical considerations in dealing with the inverse problem," *IEEE Trans. Anten. Propag.* **29**, 373–379 (1981).

Tikhonov, A. N., A. V. Goncharsky, V. V. Stepanov, and A. G. Yagola, *Numerical Methods for the Solution of Ill-Posed Problems*, Springer, Berlin, 1995.

Tsihrintzis, G. A. and A. J. Devaney, "Higher order (nonlinear) diffraction tomography: Inversion of the Rytov series," *IEEE Trans. Inform. Theory* **46**, 1748–1761 (2000).

Wallacher, E. and A. K. Louis, "Complete set of radiating and nonradiating parts of a source and their fields with applications in inverse scattering limited-angle problems," *Int. J. Biomed. Imag.* **2006**, 1–13 (2006).

Zaiping, N., Y. Feng, Z. Yanwen, and Z. Yerong, "Variational Born iteration method and its applications to hybrid inversion," *IEEE Trans. Geosci. Remote Sens.* **38**, 1709–1715 (2000).

CHAPTER SIX

Quantitative Deterministic Reconstruction Methods

6.1 INTRODUCTION

This chapter is devoted to a description of quantitative deterministic reconstruction methods, that is, to procedures aimed at retrieving the values of the electromagnetic parameters of an investigation region by means of deterministic algorithms.

Such approaches attempt to solve the equations describing the scattering phenomena that were introduced in Chapters 2 and 3 by using algorithms not exploiting randomness for determining the problem solution (see Chapter 7 for stochastic approaches).

Quantitative methods are based on "exact" models, which are theoretically valid for any scatterers, even those with high contrast with respect to the background and to a large extent in terms of wavelength. In principle, they are able to inspect strong scatterers since they take into account the nonlinear nature of the inverse scattering problem, without the approximations used by qualitative methods.

As stated previously, the key relationships for a general inverse scattering problem formulation are equations (3.2.1) and (3.2.2) within the EFIE framework, or equations (3.2.3) and (3.2.4) if the contrast source approach is considered. In both cases, the inverse problem becomes nonlinear and involves two sets of unknowns. Namely, if the EFIE formulation is considered, the electric field inside the investigation domain as well as the dielectric properties of the inspected scatterers are unknown. Within the contrast source framework, the unknowns are instead the contrast function and the contrast source.

Microwave Imaging, By Matteo Pastorino
Copyright © 2010 John Wiley & Sons, Inc.

Essentially, it is possible to handle these two sets of unknowns in two different ways. In the first case, both the unknowns are searched for simultaneously or at least in an alternate way, usually by applying iterative approaches (at a given iteration one searches for the object function, supposing the total field to be known, and, at the next iteration, the total field is retrieved, assuming the object function to be known). In the second case, only the object function is considered as an unknown to be retrieved and, at each iteration, the internal total electric field is computed by solving a direct scattering problem (see Chapter 2 for a definition of this problem).

Besides nonlinearity, quantitative methods need to address the instability issues associated with the nature of the scattering equations. In fact, the pathologies typical of inverse problems and identified in Chapter 5 for the linear case have to be mitigated by using some specific regularization tools. In this context, several approaches are available in the scientific literature, some of which are discussed in the following sections. One key tool in tackling nonlinear inverse scattering problems is iterative linearization, which allows one to deal with linear problems, which, although still affected by ill-posedness pathologies, can be addressed by means of the powerful and mature regularization theory of linear equations.

The following sections introduce two different, but essentially equivalent, approaches to the resolution of inverse scattering problems. The first one attempts to directly solve the nonlinear systems of equations relating the unknown dielectric properties to the measured fields; the other one recasts such a problem to the minimization of a suitable functional.

The optimization approach is of significant interest for several reasons and is also crucial for the development of stochastic approaches. One of the most important advantages of using optimization formulations consists in quite great flexibility in taking into account a priori information available on the inspected scene. A priori information is in fact fundamental in limiting the set of admissible solutions and also in improving the efficiency and effectiveness of the inversion procedures.

For the motivations, mentioned above, when compared with qualitative methods, quantitative methods are usually characterized by a quite heavy computational burden, but have a more ambitious goal since they attempt to provide the most complete information on the inspected targets as far as the electromagnetic properties are concerned.

According to the general setting of this book, this chapter is intended to provide a description of some quantitative procedures for the inversion of scattered data, along with the optimization approach mentioned above. The methods addressed are essentially Newton-like and gradient-based techniques and are described mainly from an operating perspective, rather than on theoretical grounds, in order to provide the readers with information for practical insights into their implementations and necessary computational resources.

6.2 INEXACT NEWTON METHODS

Newton methods are iterative approaches for the resolution of nonlinear problems. Accordingly, they can be used for solving inverse scattering problems formulated within their natural nonlinear framework. The key idea consists in building a sequence of approximations of the problem unknowns computed by solving the iteratively linearized involved equations, obtained by using the Fréchet derivative.

One of the first studies on the application of Newton methods to inverse scattering dates back to 1981 (Roger 1981), and now the related literature is really wide (Dembo et al. 1982, Engl et al. 1996, Kaltenbacher 1997, Hanke et al. 1995, Rieder 1999, Semenov et al. 2000, Franchois and Tijhuis 2003, Rubk et al. 2007, Mojabi and LoVetri 2009).

A critical issue concerns the resolution of the linearized equations, which turn out to suffer from instability problems, due to the pathological features of inverse problems identified in previous chapters. Accordingly, linearized equations cannot be solved exactly; rather, regularized solutions are sought. In principle, any regularization method can be adopted to this end, resulting in a different "inexact" Newton algorithm. In this context, the word "inexact" is due to the regularizing resolution procedure of the linearized equations.

Several inexact Newton methods have been proposed in the literature (Dembo et al. 1982), according to the regularization method adopted for solving the linearized equations. Among the several different methods that can be used to accomplish this task, there are the Tikhonov method, the truncated singular value decomposition, the truncated Landweber method, and the conjugate gradient. An inexact Newton algorithm for microwave imaging based on the truncated Landweber algorithm is discussed in detail below.

As shown in Section 6.3, the truncated Landweber method is a regularization algorithm (Engl et al. 1996), based on the Landweber method, which is an iterative algorithm for the resolution of linear problems. Since the first iterates of the methods filter out the components of the input data corrupted by noise, which are usually related to the highest-frequency Fourier components, a few iterations are able to provide a regularized solution of the linear problem.

Moreover, since the convergence of the Landweber method is usually quite slow, its regularizing capabilities are very high and this advantage is exploited in solving the nonlinear inverse problem of interest (Hanke et al. 1995, Hanke 1991).

The resulting inexact Newton method is characterized by two nested loops, which can be summarized as follows (Kaltenbacher 1997):

1. *Outer steps*—linearization by means of the Fréchet derivative;
2. *Inner steps*—resolution, in a regularized fashion, of the linear equation by means of the truncated Landweber method.

In the approach described, above, the number of inner (i.e., Landweber) iterations is fixed, in order to have ensure that only one parameter controls the so-called *semiconvergence* (Bertero and Boccacci 1998), that is, the number of outer (i.e., Newton) iterations.

Some comments are in order regarding the choice of starting solution in the resolution procedure, which are valid not only for the deterministic methods detailed in the remainder of this chapter.

In fact, if a global optimization method is used (e.g., genetic algorithms or other methods discussed in Chapter 7), the choice of the initial trial solution is not very critical, since such methods are able to escape from local minima. On the contrary, if the methods described in the present chapter are used, which are essentially local search techniques, a suitable initialization of the process is of fundamental importance in order to avoid entrapment of the procedure in false solutions.

In practice, there are several options. In some cases the object to be inspected is partially known. For example, this is the case when a model of the inspected target is available and the actual object is known to slightly differ from it. Of course, in this case the reference model is then a good choice for starting the procedure. This is essentially a simple way to insert a priori information into the solution process. However, as mentioned previously, accounting for a priori information is not very easy when deterministic methods are used, and this is one of the simplest approaches for accomplishing this task. Another possibility for initialization of the iterative process is based on the use of a solution provided by a linearized model. For example, diffraction tomography, discussed in Section 5.11, can be effectively used to provide a starting solution to initialize the nonlinear resolution procedure.

Other qualitative methods can be used as well, but it is rather difficult to quantify the *information* that an approximate (e.g., linear) model can in general provide in the inspection of a strong scatterer. A smoothly shaped version of the original profile is very often obtained, which can at least provide an approximate support, but no guarantee can be theoretically provided in general. It is worth noting that, in practice, when no information is available on the inspected scene, the initial solution is often chosen to be an empty configuration.

These considerations suggest the introduction of another quite common strategy used in many microwave imaging techniques, namely, *frequency hopping*, which consists in inverting a set of data collected at a certain frequency and in using the result obtained as a starting solution for the inversion of data measured at a higher operating frequency (Bozza et al. 2006, Ferraye et al. 2003). This process has two main advantages: (1) it allows taking into account more information on the scene because it is based on multifrequency data; and (2) since the starting solution improves when the operating frequency increases, frequency hopping usually is also a tool for successfully inspecting strong scatterers. On the other hand, processing time is longer, and the inspection of time dispersive scatterers must be done carefully. In fact, if the dielectric properties of the scatterers under test are different when the operating frequency changes, this fact must be to accounted for in the inversion procedure, for example, by assuming a dispersion model, as suggested in Section 2.3. Of course, if a dispersion model is adopted, the unknown to retrieve will be the model parameters of interest.

6.3 THE TRUNCATED LANDWEBER METHOD

The Landweber algorithm is an iterative method for solving linear problems. Although it can be introduced for general problems in Hilbert spaces, for simplicity, we shall address the case of finite-dimensional spaces, namely, the Landweber method for resolution of linear systems of equations.

The general problem we consider here is the same as in Section 5.5; specifically, we intend to determine the generalized solution of the linear system (5.5.1)

$$[\mathbf{L}]\mathbf{f} = \mathbf{g}, \tag{6.3.1}$$

where $[\mathbf{L}]$ is a $N \times M$ matrix. Since the generalized solution satisfies the normal form

$$[\mathbf{L}]^*(\mathbf{g} - [\mathbf{L}]\mathbf{f}) = 0, \tag{6.3.2}$$

we start from this relation to introduce the Landweber algorithm. The key idea is to apply a fixed-point approach to equation (6.3.2), by rewriting it in the following equivalent form

$$\mathbf{f} = \mathbf{f} + \beta[\mathbf{L}]^*(\mathbf{g} - [\mathbf{L}]\mathbf{f}), \tag{6.3.3}$$

where β is any nonzero number. Equation (6.3.3) suggests considering the sequence defined by the recursive relations

$$\begin{aligned}\mathbf{f}_0 &= 0, \\ \mathbf{f}_{n+1} &= \mathbf{f}_n + \beta[\mathbf{L}]^*(\mathbf{g} - [\mathbf{L}]\mathbf{f}_n), \quad n \geq 0.\end{aligned} \tag{6.3.4}$$

By induction it can be easily proved that

$$\mathbf{f}_n = \beta \sum_{m=0}^{n-1} \left([\mathbf{I}] - \beta[\mathbf{L}]^*[\mathbf{L}]\right)^m [\mathbf{L}]^* \mathbf{g}, \quad n \geq 1. \tag{6.3.5}$$

Accordingly, by using the singular value decompositions (5.4.14) and (5.4.15), we obtain

$$\mathbf{f}_n = \beta \sum_{m=0}^{n-1} \left([\mathbf{I}] - \beta[\mathbf{V}][\mathbf{\Sigma}]^T [\mathbf{\Sigma}][\mathbf{V}]^*\right)^m [\mathbf{V}][\mathbf{\Sigma}]^T [\mathbf{U}]^* \mathbf{g}, \quad n \geq 1. \tag{6.3.6}$$

Moreover, since

$$([\mathbf{I}] - \beta [\mathbf{V}][\mathbf{\Sigma}]^T [\mathbf{\Sigma}][\mathbf{V}]^*)^m = \sum_{p=0}^{m} \binom{m}{p} [\mathbf{I}]^{m-p} (-\beta)^m ([\mathbf{V}][\mathbf{\Sigma}]^T [\mathbf{\Sigma}][\mathbf{V}]^*)^m$$

$$= [\mathbf{V}] \left[\sum_{p=0}^{m} \binom{m}{p} (-\beta)^m ([\mathbf{\Sigma}]^T [\mathbf{\Sigma}])^m \right] [\mathbf{V}]^*$$

$$= [\mathbf{V}] ([\mathbf{I}] - \beta [\mathbf{\Sigma}]^T [\mathbf{\Sigma}])^m [\mathbf{V}]^*, \qquad (6.3.7)$$

it follows that

$$\mathbf{f}_n = [\mathbf{V}][\mathbf{\Lambda}_n][\mathbf{\Sigma}]^T [\mathbf{U}]^* \mathbf{g}, \quad n \geq 1, \qquad (6.3.8)$$

where

$$[\mathbf{\Lambda}_n] = \beta \sum_{m=0}^{n-1} ([\mathbf{I}] - \beta [\mathbf{\Sigma}]^T [\mathbf{\Sigma}])^m \qquad (6.3.9)$$

is a diagonal matrix such that its ith diagonal element γ_{nii} is equal to β for $i > p$, where p is the rank of $[\mathbf{L}]$. As a consequence, equation (6.3.8) can be written as

$$\mathbf{f}_n = \sum_{i=1}^{p} \gamma_{nii} \sigma_i \langle \mathbf{g}, \mathbf{u}_i \rangle \mathbf{v}_i, \quad n \geq 1. \qquad (6.3.10)$$

Using well-known properties of geometric series, because it holds for $i \leq p$, that

$$\gamma_{nii} = \beta \left(\sum_{m=0}^{n-1} ([\mathbf{I}] - \beta [\mathbf{\Sigma}]^T [\mathbf{\Sigma}])^m \right)_{ii} = \frac{1 - (1 - \beta \sigma_i^2)^n}{\sigma_i^2}, \qquad (6.3.11)$$

the nth iterate of the Landweber method can be expressed in terms of the singular system of $[\mathbf{L}]$ as follows:

$$\mathbf{f}_n = \sum_{i=1}^{p} \frac{1 - (1 - \beta \sigma_i^2)^n}{\sigma_i} \langle \mathbf{g}, \mathbf{u}_i \rangle \mathbf{v}_i, \quad n \geq 1. \qquad (6.3.12)$$

If

$$|1 - \beta \sigma_i^2| < 1, i = 1, \ldots, p, \qquad (6.3.13)$$

then

$$\lim_{n \to \infty} \mathbf{f}_n = \sum_{i=1}^{p} \lim_{n \to \infty} \frac{1 - (1 - \beta \sigma_i^2)^n}{\sigma_i} \langle \mathbf{g}, \mathbf{u}_i \rangle \mathbf{v}_i = \sum_{i=1}^{p} \frac{1}{\sigma_i} \langle \mathbf{g}, \mathbf{u}_i \rangle \mathbf{v}_i. \qquad (6.3.14)$$

Equation (6.3.14) states that the Landweber sequence tends to the generalized solution of the linear problem (6.3.1), provided that condition (6.3.13) is satisfied. It is useful to observe that condition (6.3.13) is fulfilled if and only if

$$0 < \beta < \frac{2}{\sigma_1^2} = \frac{2}{\|[\mathbf{L}]\|^2}. \qquad (6.3.15)$$

Let us now discuss the behavior of the *truncated* Landweber method. If the number of performed iterates is I, then the solution that the truncated method provides can be written [see (6.3.12)] as

$$\mathbf{f}_I = \sum_{i=1}^{p} \frac{1-\left(1-\beta\sigma_i^2\right)^I}{\sigma_i} \langle \mathbf{g}, \mathbf{u}_i \rangle \mathbf{v}_i. \qquad (6.3.16)$$

A comparison between equations (6.3.16) and (5.5.10) shows that applying the truncated Landweber method is equivalent to perturb the singular values in computation of the pseuodoinverse, according to the rule

$$\sigma_i \to \sigma_i \theta_I(\sigma_i), \qquad (6.3.17)$$

where

$$\theta_I(\sigma_i) = \frac{1}{1-\left(1-\beta\sigma_i^2\right)^I}. \qquad (6.3.18)$$

In particular, for the smallest singular values, responsible for the ill-conditioning, we can write

$$\left(1-\beta\sigma_i^2\right)^I \approx 1 - I\beta\sigma_i^2. \qquad (6.3.19)$$

As a consequence, the smallest singular values are transformed according to the following law

$$\sigma_i \to \sigma_i \theta_I(\sigma_i) = \sigma_i \frac{1}{I\beta\sigma_i^2} = \frac{1}{I\beta\sigma_i}, \qquad (6.3.20)$$

so that the contribution of related components to the solution is minimized and the stability is improved.

This reasoning, although quite qualitative, gives insight into the behavior of the truncated Landweber algorithm, which will be used for the simulations reported in Section 6.6. However, for a more exhaustive discussion of the method in a general framework, the reader is referred to the mathematical literature (Bertero and Boccacci 1998, Engl et al. 1996).

6.4 INEXACT NEWTON METHOD FOR ELECTRIC FIELD INTEGRAL EQUATION FORMULATION

With reference to the inspection of targets under the tomographic hypotheses described in Section 4.3, the equations to be solved are (4.3.1) and (4.3.2) for a multiple illumination. In order to describe in detail the application of the inexact Newton method, the inverse scattering problem, given by the above-mentioned integral equations, can be written in an abstract way as

$$\mathbf{T}(\mathbf{x}) = \mathbf{y}, \tag{6.4.1}$$

where $\mathbf{x}$ is an array containing the unknown scattering potential and the unknown electric field inside the investigation domain as

$$\mathbf{x} = [\tau, E_z^1, E_z^2, \ldots, E_z^S]^T, \tag{6.4.2}$$

where S is the number of sources, and

$$\mathbf{T}(\tau, E_z^1, E_z^2, \ldots, E_z^S) = \begin{pmatrix} j\omega\mu_b \int_{S_i} \tau(\mathbf{r}_t') E_z^1(\mathbf{r}_t') G_{2D}(\mathbf{r}_t/\mathbf{r}_t') d\mathbf{r}_t' \\ \vdots \\ j\omega\mu_b \int_{S_i} \tau(\mathbf{r}_t') E_z^S(\mathbf{r}_t') G_{2D}(\mathbf{r}_t/\mathbf{r}_t') d\mathbf{r}_t' \\ E_z^1(\mathbf{r}_t) - j\omega\mu_b \int_{S_i} \tau(\mathbf{r}_t') E_z^1(\mathbf{r}_t') G_{2D}(\mathbf{r}_t/\mathbf{r}_t') d\mathbf{r}_t' \\ \vdots \\ E_z^S(\mathbf{r}_t) - j\omega\mu_b \int_{S_i} \tau(\mathbf{r}_t') E_z^S(\mathbf{r}_t') G_{2D}(\mathbf{r}_t/\mathbf{r}_t') d\mathbf{r}_t' \end{pmatrix}. \tag{6.4.3}$$

In equation (6.4.1), $\mathbf{y}$ is the known term including the scattered field measured in the observation domain S_m and the incident field inside S_i, which are both known quantities:

$$\mathbf{y} = [E_{scat_z}^1, \ldots, E_{scat_z}^S, E_{inc_z}^1, \ldots, E_{inc_z}^S]^T. \tag{6.4.4}$$

We remark that in equation (6.4.2) E_z^i is the total field inside the investigation domain, whereas in (6.4.4) $E_{scat_z}^i$ and $E_{scat_z}^i$ denote the scattered field in the observation domain and the incident field inside the investigation domain due to the ith source ($i = 1, \ldots, S$).

In order to solve the inverse scattering problem, the following steps are considered:

1. Choose a guess solution $\mathbf{x}^0 = (\tau^0, E_z^{1,0}, E_z^{2,0}, \ldots, E_z^{S,0})^T$ and set $j = 0$ (counter for the outer iterations).
2. Linearize equation (6.4.1) by means of the Fréchet derivative $\mathbf{T}'_{\mathbf{x}^j}$ of the operator $\mathbf{T}$ at point $\mathbf{x}^j = (\tau^j, E_z^{1j}, E_z^{2j}, \ldots, E_z^{Sj})^T$, obtaining the linear equation

$$\mathbf{T}'_{\mathbf{x}^j}(\mathbf{h}^j) = \mathbf{y} - \mathbf{T}(\mathbf{x}^j). \tag{6.4.5}$$

3. Find a regularized solution $\tilde{\mathbf{h}}^j$ for the ill-posed equation (6.4.5) by means of the truncated Landweber algorithm (as discussed in Section 6.3).
4. Update the solution by setting $\mathbf{x}^{j+1} = \mathbf{x}^j + \tilde{\mathbf{h}}^j$ and $j \leftarrow j + 1$.
5. If a stopping rule is satisfied, terminate; otherwise, go to step 2).

As can be seen, the algorithm is composed of two nested loops: in the outer one, (6.4.1) is linearized, while in the inner one, a regularized solution of the equation is found by means of the truncated Landweber method.

It is quite evident that computation of the Fréchet derivative of the operator $\mathbf{T}$ plays a fundamental role in the resolution algorithm. Since $\mathbf{T}$ is bilinear with respect to the object function and the electric field, we can then express the Fréchet derivative of $\mathbf{T}$ at point $\mathbf{x}^j$ as

$$\mathbf{T}_{\mathbf{x}^j} = \begin{pmatrix} T_\tau^1 & T_{E_z}^1 & 0 & \cdots & 0 \\ T_\tau^2 & 0 & T_{E_z}^2 & \ddots & 0 \\ \vdots & \vdots & & \ddots & \vdots \\ T_\tau^S & 0 & \cdots & 0 & T_{E_z}^S \\ -T_\tau^1 & I-T_{E_z} & 0 & \cdots & 0 \\ -T_\tau^2 & 0 & I-T_{E_z} & \ddots & 0 \\ \vdots & \vdots & & \ddots & \vdots \\ -T_\tau^S & 0 & \cdots & 0 & I-T_{E_z} \end{pmatrix}, \quad (6.4.6)$$

where the linear operators T_τ^j, $j = 1,\ldots,S$, $T_{E_z}^j$, $j = 1,\ldots,S$, and T_{E_z} are such that

$$T_\tau^j(h) = j\omega\mu_b \int_{S_i} E_z^j(\mathbf{r}_t') h(\mathbf{r}_t') G_{2D}(\mathbf{r}_t/\mathbf{r}_t') d\mathbf{r}_t', \quad (6.4.7)$$

$$T_{E_z}^j(h) = j\omega\mu_b \int_{S_i} \tau(\mathbf{r}_t') h(\mathbf{r}_t') G_{2D}(\mathbf{r}_t/\mathbf{r}_t') d\mathbf{r}_t', \quad \mathbf{r}_t \in S_m^j \quad (6.4.8)$$

$$T_{E_z}(h) = j\omega\mu_b \int_{S_i} \tau(\mathbf{r}_t') h(\mathbf{r}_t') G_{2D}(\mathbf{r}_t/\mathbf{r}_t') d\mathbf{r}_t', \quad \mathbf{r}_t \in S_i \quad (6.4.9)$$

and I is the identity operator.

From an operating perspective, it is very useful to observe that the method presented above can be implemented quite efficiently since the discrete representation of the Fréchet derivative leads to sparse and structured matrices, for which fast and efficient tools, such as the fast Fourier transform, can be used.

The approach described above has been tested against the same configuration as in Figure 5.1. The results presented here were obtained by using a square investigation domain subdivided into 30×30 square subdomains, and by using 40 outer iterations and, for each of them, 75 Landweber iterations.

Moreover, $S = 8$ sources and $M = 241$ measurement points are used for each source. The sources are located on a circumference of radius 1.5λ at the points

$$\mathbf{r}_t^s = 1.5\lambda \left[\hat{\mathbf{x}} \cos\left[\frac{2\pi}{S}(s-1)\right] + \hat{\mathbf{y}} \sin\left[\frac{2\pi}{S}(s-1)\right]\right], \quad s = 1,\ldots,S. \quad (6.4.10)$$

The measurement points are placed on a circumference of the same radius at positions dependent on the source:

$$\mathbf{r}_t^m = 1.5\lambda \left[\hat{\mathbf{x}} \cos\left[\frac{2\pi}{S}(s-1) + \frac{2\pi}{M-1}(m-1) + \frac{\pi}{3}\right] \right. \\ \left. + \hat{\mathbf{y}} \sin\left[\frac{2\pi}{S}(s-1) + \frac{2\pi}{M-1}(m-1) + \frac{\pi}{3}\right]\right], \quad s = 1,\ldots,S, m = 1,\ldots,M.$$

(6.4.11)

The input data are corrupted by noise [as described in equation (5.8.11)], so that their signal-to-noise ratio is equal to 25 dB. The results obtained are shown in Figures 6.1 and 6.2 and Table 6.1. The starting solution is always the background, except for Fig. 6.1d, for which an estimate of the object with $\varepsilon_r = 1.8$ has been used.

Finally, it is worth mentioning that the formulation presented here is capable of inspecting strong scatterers since no approximation has been

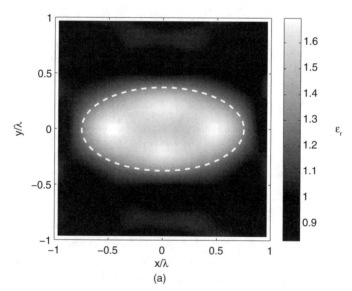

FIGURE 6.1 Quantitative reconstruction of the lossless elliptic cylinder shown in Figure 5.1 (cross section) using the inexact Newton method with EFIE formulation, $S = 8$, $M = 241$. Square investigation domain (side: $d = 2\lambda$; discretization: 30×30 pixels); SNR = 25 dB. Distribution of the retrieved relative dielectric permittivity: (a) $\varepsilon = (1.6 - j0)\varepsilon_0$; (b) $\varepsilon = (1.8 - j0)\varepsilon_0$; (c) $\varepsilon = (2.0 - j0)\varepsilon_0$; (d) $\varepsilon = (2.8 - j0)\varepsilon_0$. (Simulations performed by G. Bozza, University of Genoa, Italy.)

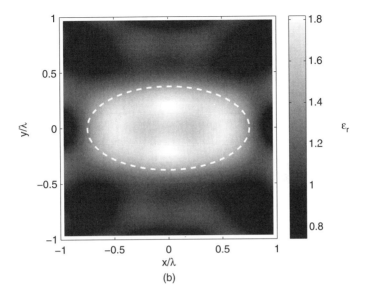

(b)

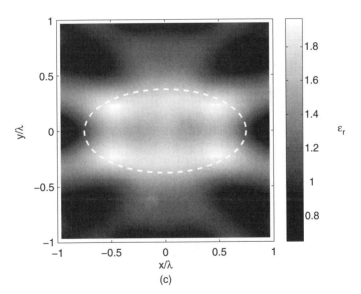

(c)

FIGURE 6.1 *Continued*

134 QUANTITATIVE DETERMINISTIC RECONSTRUCTION METHODS

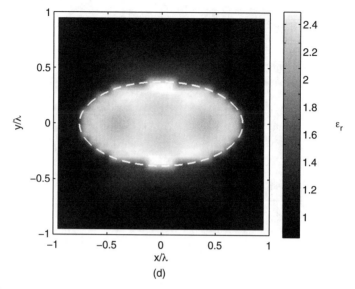

(d)

FIGURE 6.1 *Continued*

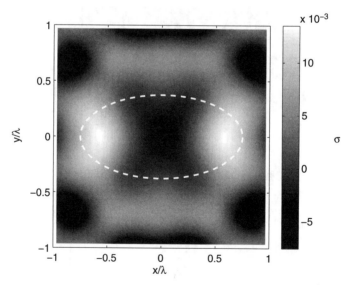

FIGURE 6.2 Quantitative reconstruction of the lossless elliptic cylinder shown in Figure 5.1 (cross section) using the inexact Newton method with EFIE formulation and $S = 8$, $M = 241$. Square investigation domain (side: $d = 2\lambda$; discretization: 30×30 pixels); SNR = 25 dB. Distribution of the electric conductivity: $\varepsilon = (2.8 - j0)\varepsilon_0$. (Simulations performed by G. Bozza, University of Genoa, Italy.)

TABLE 6.1 Relative Errors on Reconstruction of Elliptic Cylinders in Figures 6.1 and 6.2 for Different Relative Dielectric Permittivity Values, Calculated by Electric Field Integral Equation Formulation Using the Inexact Newton Method

	Relative Dielectric Permittivity								
	1.2	1.4	1.6	1.8	2.0	2.2	2.4	2.6	2.8
Global Errors									
Minimum values	9.0×10^{-4}	1.9×10^{-3}	3.5×10^{-3}	6.7×10^{-3}	6.5×10^{-3}	5.2×10^{-3}	0.011	0.011	4.5×10^{-3}
Mean values	0.02	0.04	0.07	0.12	0.20	0.30	0.44	0.65	0.17
Maximum values	0.11	0.18	0.26	0.39	0.76	0.96	1.38	2.55	0.42
Variance	4.3×10^{-4}	1.4×10^{-3}	2.7×10^{-3}	5.5×10^{-3}	1.4×10^{-2}	3.6×10^{-2}	0.14	0.49	8.8×10^{-3}
Errors on Reconstruction of Object Cross Section									
Minimum values	6.8×10^{-4}	1.9×10^{-3}	3.5×10^{-3}	0.043	0.19	0.47	0.61	0.52	0.15
Mean values	0.03	0.05	0.08	0.14	0.30	0.59	1.06	1.76	0.22
Maximum values	0.11	0.18	0.22	0.38	0.76	0.96	1.38	2.55	0.34
Variance	0.032	2.3×10^{-3}	3.5×10^{-3}	4.1×10^{-3}	0.013	9.6×10^{-3}	0.046	0.41	1.7×10^{-3}
Errors on Reconstruction of Background									
Minimum values	3.4×10^{-4}	2.2×10^{-3}	3.8×10^{-3}	6.7×10^{-3}	6.5×10^{-3}	5.2×10^{-3}	0.011	0.011	4.5×10^{-3}
Mean values	0.11	0.038	0.065	0.12	0.17	0.22	0.27	0.33	0.16
Maximum values	0.10	0.18	0.25	0.39	0.54	0.47	0.60	0.87	0.42
Variance	0.024	1.2×10^{-3}	2.5×10^{-3}	5.7×10^{-3}	0.011	0.015	0.025	0.028	0.01

Source: Simulations performed by G. Bozza, University of Genoa, Italy.

considered in the scattering model. However, the same algorithm can be applied to other scattering models, provided that the unknown and known terms and the operator expressing the relationship among them are properly defined.

For example, when the second-order Born approximation is used, the unknown term coincides with τ, whereas the operator $\mathbf{T}$ and the known term become [see equation (4.6.6)]

$$\mathbf{T}(\tau) = \begin{pmatrix} j\omega\mu_b \int_{S_i} \tau(\mathbf{r}')\left[E^1_{\text{inc}_z}(\mathbf{r}') + j\omega\mu_b \int_{S_i} \tau(\mathbf{r}'') E^1_{\text{inc}_z}(\mathbf{r}'') G_{2D}(\mathbf{r}'/\mathbf{r}'') d\mathbf{r}'' \right] G_{2D}(\mathbf{r}_t/\mathbf{r}') d\mathbf{r}' \\ \vdots \\ j\omega\mu_b \int_{S_i} \tau(\mathbf{r}')\left[E^S_{\text{inc}_z}(\mathbf{r}') + j\omega\mu_b \int_{S_i} \tau(\mathbf{r}'') E^S_{\text{inc}_z}(\mathbf{r}'') G_{2D}(\mathbf{r}'/\mathbf{r}'') d\mathbf{r}'' \right] G_{2D}(\mathbf{r}_t/\mathbf{r}') d\mathbf{r}' \end{pmatrix}$$

(6.4.12)

$$\mathbf{y} = \left[E^1_{\text{scat}_z}, \ldots, E^S_{\text{scat}_z} \right]^T.$$ (6.4.13)

Further details on application of the inexact Newton method to the second-order Born approximation–based model can be found in the article by Estatico et al. (2005). In both cases (exact and second-order Born formulations), the problem can be discretized as discussed in Section 3.4.

6.5 INEXACT NEWTON METHOD FOR CONTRAST SOURCE FORMULATION

As discussed in Sections 3.2 and 3.3, the contrast source formulation of inverse scattering problems is often used in microwave imaging. In this section, the application to the contrast source framework of the inexact Newton procedure introduced above will be discussed and the advantages with respect to the EFIE formulation will be outlined. The starting point is represented by the equation of the contrast source model for a multiillumination configuration [equations (4.3.3) and (4.3.4)], which can be immediately rewritten as

$$E^i_{\text{scat}_z}(\mathbf{r}_t) = j\omega\mu_b \int_{S_i} J^i_z(\mathbf{r}') G_{2D}(\mathbf{r}_t/\mathbf{r}') d\mathbf{r}' \quad \mathbf{r}_t \in S^i_m, \quad (6.5.1)$$

$$J^i_z(\mathbf{r}_t) = \tau(\mathbf{r}_t) E^i_{\text{inc}_z}(\mathbf{r}_t) + j\omega\mu_b \tau(\mathbf{r}_t) \int_{S_i} J^i_z(\mathbf{r}_t) G_{2D}(\mathbf{r}_t/\mathbf{r}') d\mathbf{r}'_t \quad \mathbf{r}_t \in S_i, \quad (6.5.2)$$

where J^i_z is the contrast source due to the ith transmitting antenna. If S again denotes the number of incident fields, the problem unknowns are J^i_z, $i = 1, \ldots, S$, and the scattering potential τ. All the unknown functions are defined on the investigation domain S_i.

In order to develop the inexact Newton method, it is important to note that equation (6.5.1) is linear and that it does not involve the object function, which is directly related to the electromagnetic properties of the scatterers. However, the ill-posedness of the inverse problem prevents us from using only such an equation to find the contrast source and, for this reason, equation (6.5.2) is needed.

With reference to the more abstract statement of the inverse problem provided by equation (6.4.1), it follows that

$$\mathbf{x} = [\tau, J_z^1, J_z^2, \ldots, J_z^S]^T \qquad (6.5.3)$$

$$\mathbf{y} = [0, E_{\text{scat}_z}^1, 0, E_{\text{scat}_z}^2, \ldots, 0, E_{\text{inc}_z}^S]^T. \qquad (6.5.4)$$

and

$$\mathbf{T}(\mathbf{x}) = \begin{bmatrix} T_1(\tau, J_z^1) \\ T_2(\tau, J_z^2) \\ \vdots \\ T_S(\tau, J_z^S) \end{bmatrix}, \qquad (6.5.5)$$

where

$$T_i(\tau, J_z^i) = \begin{bmatrix} T_i^1(\tau, J_z^i) \\ T_i^2(\tau, J_z^i) \end{bmatrix}, \qquad (6.5.6)$$

and T_i^1 and T_i^2 are defined as follows:

$$T_i^1(\tau, J_z^i) = J_z^i(\mathbf{r}_t) - \tau(\mathbf{r}_t) E_{\text{inc}_z}^i(\mathbf{r}) \\ - j\omega\mu_b \tau(\mathbf{r}_t) \int_{S_i} G_{2D}(\mathbf{r}_t/\mathbf{r}_t') J_z^i(\mathbf{r}_t') d\mathbf{r}_t', \quad \mathbf{r}_t \in S_i \qquad (6.5.7)$$

$$T_i^2(\tau, J_z^i) = j\omega\mu_b \int_{S_i} G_{2D}(\mathbf{r}_t/\mathbf{r}_t') J_z^i(\mathbf{r}_t') d\mathbf{r}_t', \quad \mathbf{r}_t \in S_m^i. \qquad (6.5.8)$$

By following the definition, we can express the Fréchet derivative $\mathbf{T}_\mathbf{x}'$ of the operator $\mathbf{T}$ at point $\mathbf{x}$ as follows:

$$\mathbf{T}_\mathbf{x}' = \begin{bmatrix} \frac{\partial T_1^1}{\partial \tau}\bigg|_\mathbf{x} & \frac{\partial T_1^1}{\partial J_z^1}\bigg|_\mathbf{x} & \cdots & \frac{\partial T_1^1}{\partial J_z^S}\bigg|_\mathbf{x} \\ \frac{\partial T_1^2}{\partial \tau}\bigg|_\mathbf{x} & \frac{\partial T_1^2}{\partial J_z^1}\bigg|_\mathbf{x} & \cdots & \frac{\partial T_1^2}{\partial J_z^S}\bigg|_\mathbf{x} \\ \frac{\partial T_2^1}{\partial \tau}\bigg|_\mathbf{x} & \frac{\partial T_2^1}{\partial J_z^1}\bigg|_\mathbf{x} & \cdots & \frac{\partial T_2^1}{\partial J_z^S}\bigg|_\mathbf{x} \\ \frac{\partial T_2^2}{\partial \tau}\bigg|_\mathbf{x} & \frac{\partial T_2^2}{\partial J_z^1}\bigg|_\mathbf{x} & \cdots & \frac{\partial T_2^2}{\partial J_z^S}\bigg|_\mathbf{x} \\ \vdots & \vdots & \vdots & \vdots \\ \frac{\partial T_S^1}{\partial \tau}\bigg|_\mathbf{x} & \frac{\partial T_S^1}{\partial J_z^1}\bigg|_\mathbf{x} & \cdots & \frac{\partial T_S^1}{\partial J_z^S}\bigg|_\mathbf{x} \\ \frac{\partial T_S^2}{\partial \tau}\bigg|_\mathbf{x} & \frac{\partial T_S^2}{\partial J_z^1}\bigg|_\mathbf{x} & \cdots & \frac{\partial T_S^2}{\partial J_z^S}\bigg|_\mathbf{x} \end{bmatrix}. \qquad (6.5.9)$$

By observing that, for every $i = 1, \ldots, S, i \neq t$

$$\left.\frac{\partial T_i^p}{\partial J_z^t}\right|_{\mathbf{x}} = 0, \quad p = 1, 2 \tag{6.5.10}$$

and that, for every $i = 1, \ldots, S$

$$\left.\frac{\partial T_i^2}{\partial \tau}\right|_{\mathbf{x}} = 0, \tag{6.5.11}$$

it follows that

$$\mathbf{T}'_{\mathbf{x}} = \begin{bmatrix} \left.\frac{\partial T_1^1}{\partial \tau}\right|_{\mathbf{x}} & \left.\frac{\partial T_1^1}{\partial J_z^1}\right|_{\mathbf{x}} & 0 & 0 & \cdots & 0 \\ 0 & \left.\frac{\partial T_1^2}{\partial J_z^1}\right|_{\mathbf{x}} & 0 & 0 & \cdots & 0 \\ \left.\frac{\partial T_2^1}{\partial \tau}\right|_{\mathbf{x}} & 0 & \left.\frac{\partial T_2^1}{\partial J_z^2}\right|_{\mathbf{x}} & 0 & \cdots & 0 \\ 0 & 0 & \left.\frac{\partial T_2^2}{\partial J_z^2}\right|_{\mathbf{x}} & 0 & \cdots & 0 \\ \vdots & \vdots & \vdots & \vdots & \vdots & \vdots \\ \left.\frac{\partial T_S^1}{\partial \tau}\right|_{\mathbf{x}} & 0 & 0 & 0 & \cdots & \left.\frac{\partial T_S^1}{\partial J_z^S}\right|_{\mathbf{x}} \\ 0 & 0 & 0 & 0 & \cdots & \left.\frac{\partial T_S^2}{\partial J_z^S}\right|_{\mathbf{x}} \end{bmatrix} \tag{6.5.12}$$

It can be easily shown that, for every $i = 1, \ldots, S$

$$\left.\frac{\partial T_i^1}{\partial J_z^i}\right|_{\mathbf{x}} = I - \tau G_{S_i}, \tag{6.5.13}$$

$$\left.\frac{\partial T_i^2}{\partial J_z^i}\right|_{\mathbf{x}} = G_{S_m^s}, \tag{6.5.14}$$

$$\left.\frac{\partial T_i^1}{\partial \tau}\right|_{\mathbf{x}} = -E_{\text{inc}_z}^i - G_{S_i}(J_z^i), \tag{6.5.15}$$

where

$$G_{S_i}(J) = j\omega\mu_b \int_{S_i} G_{2D}(\mathbf{r}_t|\mathbf{r}'_t) J(\mathbf{r}'_t) d\mathbf{r}'_t, \quad \mathbf{r}_t \in S_i, \tag{6.5.16}$$

$$G_{S_m^i}(J) = j\omega\mu_b \int_{S_i} G_{2D}(\mathbf{r}_t|\mathbf{r}'_t) J(\mathbf{r}'_t) d\mathbf{r}'_t, \quad \mathbf{r}_t \in S_m^i. \tag{6.5.17}$$

By exploiting such relations, we can express $\mathbf{T}'_{\mathbf{x}}$ as follows:

$$\mathbf{T}'_x = \begin{bmatrix} -E^1_{\text{inc}_z} - G_{S_i}(J^1_z) & I - \tau G_{S_i} & 0 & \cdots & 0 \\ 0 & G_{S^1_m} & 0 & \cdots & 0 \\ -E^2_{\text{inc}_z} - G_{S_i}(J^2_z) & 0 & I - \tau G_{S_i} & \cdots & 0 \\ 0 & 0 & G_{S^2_m} & \cdots & 0 \\ \vdots & \vdots & \vdots & \vdots & \vdots \\ -E^S_{\text{inc}_z} - G_{S_i}(J^S_z) & 0 & 0 & \cdots & I - \tau G_{S_i} \\ 0 & 0 & 0 & \cdots & G_{S^S_m} \end{bmatrix}. \quad (6.5.18)$$

As in the case of the EFIE formulation, the Fréchet derivative is represented by a highly sparse and structured matrix in a discrete setting. In particular, it is important to note that the number of independent blocks appearing in (6.5.18) is reduced, and also that the superdiagonal blocks $I - \tau G_{S_i}$ are all are equal to each other. Moreover, if the measurement domains are the same for each source, that is, if it is supposed that $S^i_m = S_m$, for every $i = 1, \ldots, S$, then the operators $G_{S^i_m}$, $i = 1, \ldots, S$, are equal to each other and the number of independent elements in the Fréchet derivative further decreases.

If the Fréchet derivative (6.5.18) is compared with the one resulting from the EFIE formulation [see equation (6.4.6)], the contrast source formulation turns out to require the memorization of a more sparse matrix. This is a consequence of the structure of equation (6.5.1), involving only the contrast source J^i_z, but not the scattering potential τ.

Consequently, the truncated Landweber–inexact Newton method is in fact more efficient in memory requirements if applied to the inverse scattering problem within the contrast source framework (Bozza and Pastorino 2009). This can be understood by observing that, in a discrete setting, the operators defined in equations (6.5.13)–(6.5.15) are represented by matrices of sizes $N \times N$, $M \times N$, and $N \times N$, respectively, where N is the number of basis functions used to represent the contrast source and the contrast function and M is the number of receivers for each transmitter. The total number of nonzero elements in the matrix representing the Fréchet derivative operator is then $M_{\text{CS}} = SN(M + 2N)$, whereas that in the Fréchet derivative obtained in the EFIE formulation is $M_{\text{EFIE}} = 2SN(M + N)$. As a consequence, the relative memory saving can be quantified to be equal to

$$\frac{M_{\text{EFIE}} - M_{\text{CS}}}{M_{\text{EFIE}}} = \frac{1}{2} \frac{M}{M + N}. \quad (6.5.19)$$

The same elliptic cylinder considered as test throughout this book (e.g., see Fig. 5.1) has been reconstructed by using the inexact Newton method (described above) applied to the contrast source model.

As in the EFIE framework, the investigation domain has been chosen as a square region subdivided into 30×30 subdomains, and 40 outer and 75 inner iterations have been performed, using $S = 8$ sources and $M = 241$ measurement points [see also equations (6.4.10) and (6.4.11), respectively]. The initialization is also the same. The results are shown in Figure 6.3 for a signal-to-noise ratio corresponding to the input data, which are still equal to 25 dB.

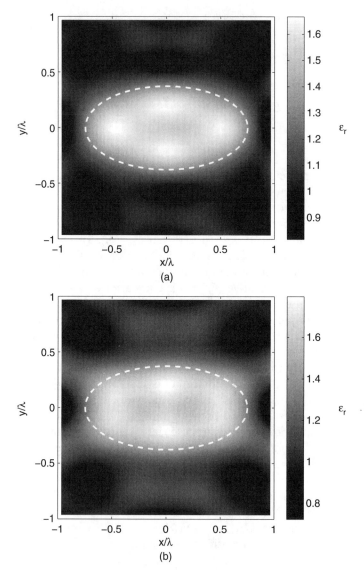

FIGURE 6.3 Quantitative reconstruction of the lossless elliptic cylinder shown in Figure 5.1 (cross section) using the inexact Newton method, with contrast source formulation, and $S = 8$, $M = 241$. Square investigation domain (side: $d = 2\lambda$; discretization: 30×30 pixels); SNR = 25 dB. Distribution of the retrieved relative dielectric permittivity: (a) $\varepsilon = (1.6 - j0)\varepsilon_0$; (b) $\varepsilon = (1.8 - j0)\varepsilon_0$; (c) $\varepsilon = (2.0 - j0)\varepsilon_0$; (d) $\varepsilon = (2.8 - j0)\varepsilon_0$.

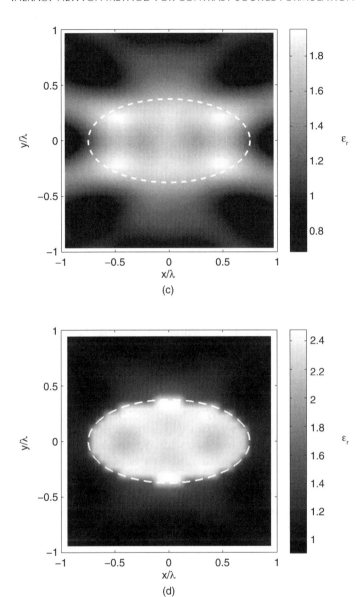

FIGURE 6.3 *Continued*

6.6 THE DISTORTED BORN ITERATIVE METHOD

Another very important quantitative approach is the *distorted Born iterative method* (Chew and Wang 1990). In this case, with reference again to Figure 4.10, the approach is similar to the one described in Section 5.13, but, at each iteration, Green's function is also updated. In particular, after the initialization phase, in which the initial distribution τ^0 is chosen, the following equation is solved (kth step):

$$E_z(\mathbf{r}_t) = E^k_{inc_z}(\mathbf{r}_t) + j\omega\mu_b \int_{S_i} \delta\tau^k(\mathbf{r}'_t) E^k_{inc_z}(\mathbf{r}_t) G^k_I(\mathbf{r}_t/\mathbf{r}'_t) d\mathbf{r}'_t, \quad \mathbf{r}_t \in S_m. \quad (6.6.1)$$

Once $\delta\tau^k$ has been determined, the scattering potential is updated as follows:

$$\tau^{k+1}(\mathbf{r}_t) = \tau^k(\mathbf{r}_t) + \delta\tau^k(\mathbf{r}_t). \quad (6.6.2)$$

In equation (6.6.1) G^k_I is the inhomogeneous Green function related to a background medium characterized by τ^k. Moreover, $E^k_{inc_z}$ denotes the incident field in the inhomogeneous medium, that is, the field produced by the sources in the background at the kth iteration. As a consequence, Green's function and the incident field need to be updated at each iteration, too. The required computations can be performed as described in detail in Chapter 4, in particular in Section 4.10, with respect to the inhomogeneous Green function. Accordingly, if $\tau^0 = 0$, then $G^0_I(\mathbf{r}_t/\mathbf{r}'_t) = G_{2D}(\mathbf{r}_t/\mathbf{r}'_t)$ and $E^0_{inc_z}(\mathbf{r}_t) = E_{inc_z}(\mathbf{r}_t)$.

It is very important to observe that the distorted Born iterative method has been proved to be equivalent to the Newton–Kantorovich method (Remis and van den Berg 2000). This reveals a close link between the present approach and Newton methods, which could be very useful in the choice of the more suitable inversion procedure.

In order to demonstrate the imaging capabilities of the distorted Born iterative method method, the reconstructions of some real scatterers are provided. The results concern a multiillumination, multiview approach (Section 4.2). At each iteration, equation (6.6.1) is discretized as described in Section 3.4 and the resulting (ill-posed) linear equation is regularized by using the Landweber method (described in Section 6.3).

Figures 6.4a–c show the results obtained by the method described above, by inverting the input data measured at the Institute Frésnel, Marseille, France, which are widely used as a benchmark for inversion procedures at microwave frequencies (Belkebir and Saillard 2001).

The first example concerns the reconstruction of a two-layer circular cylinder of radii 0.015 and 0.04 m. The inner layer has a relative dielectric permittivity equal to 3.0, whereas the external layer has a relative permittivity equal to 1.45. The two layers are not concentric; the outermost is centered at the origin, whereas the innermost one is centered at the point (−0.005, 0).

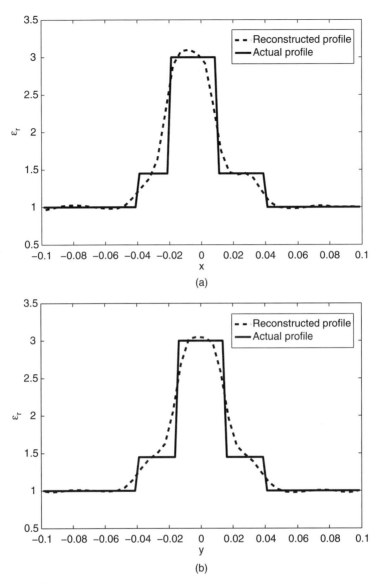

FIGURE 6.4 Reconstruction of a two nonconcentric circular cylinder (experimental data provided by the Institute Frésnel, Marseille, France). Radii: 0.015 and 0.04 m; relative dielectric permittivities of 3.0 and 1.45. Original and reconstructed profiles along (a) the x axis ($y = 0$) and (b) the y axis ($x = 0$) and (c) reconstructed distribution of the relative dielectric permittivity, using the distorted Born iterative method with Landweber iterations. (Simulations performed by A. Randazzo, University of Genoa, Italy.)

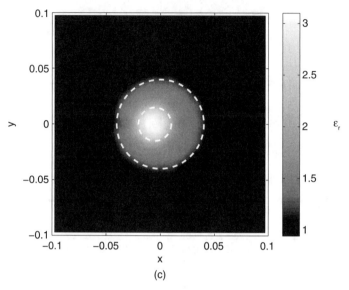

FIGURE 6.4 *Continued*

The results obtained are reported in Figure 6.4, which provides the original and reconstructed dielectric profiles along the x and y axes and the final reconstructed image (distribution of the relative dielectric permittivity).

In the second example two adjacent circular cylinders are reconstructed. They have different diameters (0.03 and 0.08 m) and are made of different materials [$\varepsilon_r = 3.0$ (smaller cylinder) and $\varepsilon_r = 1.45$ (larger cylinder)]. Figure 6.5 provides the original and reconstructed dielectric profiles along the x and y axes and the final image for this scatterer.

Finally, in the third example a more complex target is inspected. It represents a *combination* of the previous two scatterers; again, it consists of two adjacent circular cylinders, but one of them is a two-layer cylinder. The homogeneous cylinder has a radius of 0.015 m, has a relative dielectric permittivity equal to 3.0, and is centered at point (−0.055, 0). The two layers of the other cylinder have radii equal to 0.015 and 0.04 m, with corresponding relative dielectric permittivities equal to 3.0 and 1.45, respectively, and are centered at (0, 0) and (−0.005, 0). The final image is provided in Figure 6.6. In all the cases, the method is able to obtain very good reconstructions even for quite strong scatterers and real (noisy) input data.

More exhaustive discussions of the distorted Born iterative method can be found in the literature. For example, for lossy propagation media, the approach considered by Isernia et al. (2004) can provide very good reconstructions.

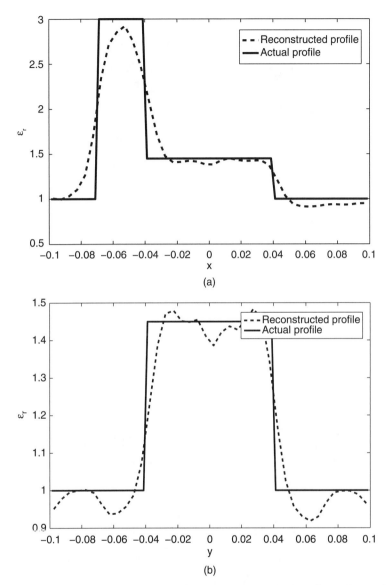

FIGURE 6.5 Reconstruction of two adjacent circular cylinders (experimental data provided by the Institute Frésnel, Marseille, France). Diameters: 0.03 and 0.08 m; relative dielectric permitivities $\varepsilon_\gamma = 3.0$ (smaller cylinder) and $\varepsilon_\gamma = 1.45$ (larger cylinder). Original and reconstructed profiles along (a) the x axis ($y = 0$) and (b) the y axis ($x = 0$) and (c) reconstructed distribution of the relative dielectric permittivity, using the distorted Born iterative method with Landweber iterations. (Simulations performed by A. Randazzo, University of Genoa, Italy.)

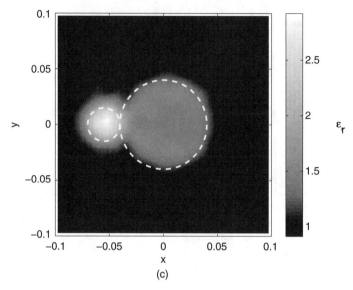

(c)

FIGURE 6.5 Continued

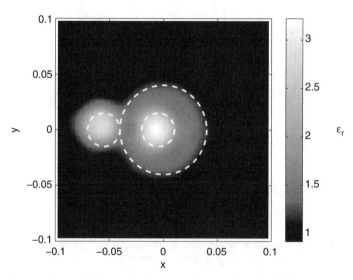

FIGURE 6.6 Reconstruction of two adjacent circular cylinders, one of them being a two-layer nonconcentric cylinder (for dimensions and dielectric parameters, see Figs. 6.4 and 6.5) (experimental data provided by the Institute Frésnel, Marseille, France). Reconstructed distribution of the relative dielectric permittivity; distorted Born iterative method with Landweber iterations. (Simulations performed by A. Randazzo, University of Genoa, Italy.)

6.7 INVERSE SCATTERING AS AN OPTIMIZATION PROBLEM

As mentioned in Section 6.1, the quantitative inverse scattering problem can also be tackled by recasting it as an optimization problem. At the beginning of this chapter, we discussed the reasons why this formulation is useful and pointed out that it is flexible mainly because it can include a priori information in the scattering model. We note here that such a formulation is, of course, independent of the specific method used to minimize the functional obtained, and so the optimization approach is not restricted to deterministic methods.

In this section, we shall address a quite general formulation of the quantitative inverse scattering problem, whereas Section 6.8 describes a minimization strategy based on the conjugate gradient. The inverse problem expressed by equations (3.3.11) and (3.3.12) can be recast as an optimization problem by defining a functional to be minimized, which usually has the form

$$F(\tau, E_z) = \alpha \|R_{\text{data}}(\tau, E_z)\|^2 + \beta \|R_{\text{state}}(\tau, E_z)\|^2 + \gamma P(\tau, E_z), \quad (6.7.1)$$

where, according to (3.3.11) and (3.3.12), the residual terms $R_{\text{data}}(\tau, E_z)$ and $R_{\text{state}}(\tau, E_z)$ are given by

$$R_{\text{data}}(\tau, E_z) = E_z(\mathbf{r}_t) - E_{\text{inc}_z}(\mathbf{r}_t) - j\omega\mu_b \int_{S_i} \tau(\mathbf{r}'_t) E_z(\mathbf{r}'_t) G_{2D}(\mathbf{r}_t/\mathbf{r}'_t) d\mathbf{r}'_t, \quad \mathbf{r}_t \in S_m, \quad (6.7.2)$$

$$R_{\text{state}}(\tau, E_z) = E_z(\mathbf{r}_t) - E_{\text{inc}_z}(\mathbf{r}_t) - j\omega\mu_b \int_{S_i} \tau(\mathbf{r}'_t) E_z(\mathbf{r}'_t) G_{2D}(\mathbf{r}_t/\mathbf{r}'_t) d\mathbf{r}'_t, \quad \mathbf{r}_t \in S_i, \quad (6.7.3)$$

and $P(E_z, \tau)$ is a term that can include a priori information on the configuration to be inspected. It is worth noting that such a term can factor in several kinds of information about the inspected scene, such as constraints on the dielectric parameters or their gradients, and can also play the role of a regularization term (see Section 5.3). Clearly, the quantities that are computed for $\mathbf{r}_t \in S_m$ in (6.7.1) represent known quantities (special symbols to denote *measured* elements are not adopted).

Moreover, the choice of the constants α, β, and γ can be performed in different ways. From a practical point of view, they can be heuristically determined for a certain class of scatterers. This is a sort of *calibration* in the sense that a known object can be inspected and the optimal values for these constants are used for a class of similar scatterers. The following is another common choice for α and β:

$$\alpha = \frac{1}{\int_{S_m} |E_{\text{inc}_z}(\mathbf{r}_t)|^2 d\mathbf{r}_t}, \quad (6.7.4)$$

$$\beta = \frac{1}{\int_{S_i} |E_{\text{inc}_z}(\mathbf{r}_t)|^2 d\mathbf{r}_t}. \quad (6.7.5)$$

148 QUANTITATIVE DETERMINISTIC RECONSTRUCTION METHODS

If the multiillumination approach described in Section 4.2 is considered, equation (6.7.1) can be simply generalized as

$$F(\tau, E_z) \rightarrow F(\tau, E_z^1, E_z^2, \ldots, E_z^S), \tag{6.7.6}$$

where the terms $R_{\text{data}}(\tau, E_z^1, E_z^2, \ldots, E_z^S)$ and $R_{\text{state}}(\tau, E_z^1, E_z^2, \ldots, E_z^S)$ are direct generalizations to S sources of equations (6.7.2) and (6.7.3) when equations (4.3.1) and (4.3.2) are used. Moreover, it must be noted that other forms of the functional in equation (6.7.1) can be chosen. For example, a multiplicative combination of terms can be used (De Zaeytijd et al. 2007)

$$F(\tau, E_z) = \|R_{\text{data}}(\tau, E_z)\|^2 [1 + \gamma P(\tau, E_z)], \tag{6.7.7}$$

where, in this case

$$P(\tau) = \int_{S_i} |\nabla \tau(\mathbf{r}_t)|^2 d\mathbf{r}_t. \tag{6.7.8}$$

A similar approach can be used for the inverse problem when formulated in terms of the contrast source (see Section 3.2). In some cases, the normalization constants can possibly vary over the steps of the iterative minimization (van den Berg and Abubakar 2001). Another possibility is to apply a log transformation to the functional in (6.7.1) (in particular, considering the data term only). For some targets, this approach provided better final imaging than did the standard least-squares criterion (Meaney et al. 2001) and seems to be effetive in dealing with some "local minima" causes (Meaney et al. 2009). In general, the choice of structure of the functional to be minimized, including the choice of the normalization constants, can have a significant impact on the optimization approach, and current research activity is conducted to better evaluate the implications of the various assumptions. Finally, it should be noted that the above mentioned functional can be discretized by using the approach described in Section 3.4. The obtained expressions usually include multiple summations over the number of measurement points, the number of expansion basis functions, the number of illuminating sources, and (in the multifrequency case) the number of considered frequencies.

6.8 GRADIENT-BASED METHODS

There are several different ways to minimize the functional in equation (6.7.1). One of the most effective deterministic techniques is the conjugate-gradient method. The conjugate gradient is a powerful tool in nonlinear optimization based on the iterative update of a starting solution along some conjugate directions. The choice of conjugate directions can be performed according to several criteria, as the description below suggests.

In this section, we shall describe the conjugate gradient method for solving the general optimization problem

$$\mathbf{x} = \arg\min_{\mathbf{x} \in C^N}\{F(\mathbf{x})\}, \quad (6.8.1)$$

where F is a real-valued function dependent on $\mathbf{x} \in C^N$. Of course, due to the generality of the cost function F, the method can be in principle applied to the general functional (6.7.1).

The conjugate-gradient method for the minimization of F is based on the selection of a guess solution $\mathbf{x}_0$ and on the iterative update of the current solution. The underlying idea is to update the current solution by solving a scalar minimization problem; thus, the solution $\mathbf{x}_n$ obtained at the nth iteration is updated as

$$\mathbf{x}_{n+1} = \mathbf{x}_n + \alpha_n \mathbf{w}_n, \quad (6.8.2)$$

where $\mathbf{w}_n$ is the conjugate direction and α_n is the solution of the scalar optimization problem

$$\alpha_n = \arg\min_{\alpha}\{F(\mathbf{x}_n + \alpha \mathbf{w}_n)\}. \quad (6.8.3)$$

Of course, $\mathbf{w}_n$ also needs to be iteratively updated, according to the law

$$\mathbf{w}_n = -\nabla F(\mathbf{x}_n) + \beta_n \mathbf{w}_{n-1}, \quad (6.8.4)$$

where $\nabla F(\mathbf{x}_n)$ is the gradient of F evaluated at point $\mathbf{x}_n$. Several rules are available for update of the coefficient β_n. For example, the Fletcher–Reeves approach consists in choosing

$$\beta_n = \begin{cases} 0, & n = 0, \\ \dfrac{\langle \nabla F(\mathbf{x}_n), \nabla F(\mathbf{x}_n) \rangle}{\langle \nabla F(\mathbf{x}_{n-1}), \nabla F(\mathbf{x}_{n-1}) \rangle}, & n \geq 1. \end{cases} \quad (6.8.5)$$

Another possibility is the Polak–Ribière approach, which is expressed as follows:

$$\beta_n = \begin{cases} 0, & n = 0, \\ \dfrac{\langle \nabla F(\mathbf{x}_n), \nabla F(\mathbf{x}_n) - \nabla F(\mathbf{x}_{n-1}) \rangle}{\langle \nabla F(\mathbf{x}_{n-1}), \nabla F(\mathbf{x}_{n-1}) \rangle}, & n \geq 1. \end{cases} \quad (6.8.6)$$

In the literature, the gradient method has been widely used for microwave imaging problems (e.g., Harada et al. 1995, Lobel et al. 1997, Franchois and Pichot 1997, Rekanos et al. 1999, Maniatis et al. 2000) and is commonly applied in conjunction with contrast source formulation (van den Berg and Kleinman 1997). In that approach, the structure of the functional to be minimized allows some analytical computations in implementation of the conjugate gradient method, which results in faster reconstructions (Abubakar et al. 2002).

Several improved versions of the method have been introduced in the more recent literature of numerical methods. Although it is clearly beyond the scope of the present book to provide further details on this matter, one of these improved procedures, which has been prove to be very fast and efficient, is the *biconjugate stabilized gradient* method (van der Vorst 1992), which has been outlined in Section 3.4 and implemented to obtain the results reported in that section (Fig. 3.3).

REFERENCES

Abubakar, A., P. M. van den Berg, and J. J. Mallorqui, "Imaging of biomedical data using a multiplicative regularized contrast source inversion method," *IEEE Trans. Microwave Theory Tech.* **50**, 1761–1771 (2002).

Belkebir, K. and M. Saillard, "Special section: Testing inversion algorithms against experimental data," *Inverse Problems* **17**, 1565–1571 (2001).

Bertero, M. and P. Boccacci, *Introduction to Inverse Problems in Imaging*, Institute of Physics, Bristol, UK, 1998.

Bozza, G., C. Estatico, M. Pastorino, and A. Randazzo, "An inexact Newton method for microwave reconstruction of strong scatterers," *IEEE Anten. Wireless Propag. Lett.* **5**, 61–64 (2006).

Bozza, G. and M. Pastorino, "An Inexact Newton-based approach to microwave imaging within the contrast–source formulation," *IEEE Trans. Anten. Propag.* **57**, 1122–1132 (2009).

Chew, W. C. and Y. M. Wang, "Reconstruction of two-dimensional Permittivity distribution using the distorted Born iterative method," *IEEE Trans. Med. Imag.* **9**, 218–225 (1990).

Dembo, R. S., S. C. Eisenstat, and T. Steihaug, "Inexact Newton methods," *SIAM J. Num. Anal.* **19**, 400–408 (1982).

De Zaeytijd, J., A. Franchois, C. Eyraud, and J.-M. Geffrin, "Full-wave three-dimensional microwave imaging with a regularized Guass-Newton method—theory and experiment," *IEEE Trans. Anten. Propag.* **55**, 3279–3292 (2007).

Engl, H. W., M. Hanke, and A. Neubauer, *Regularization of Inverse Problems*. Kluwer Academic Publishers, Dordrecht, 1996.

Estatico, C., M. Pastorino, and A. Randazzo, "An inexact Newton method for short-range microwave imaging within the second-order Born approximation," *IEEE Trans. Geosci. Remote Sens.* **43**, 2593–2605 (2005).

Ferraye, R., J.-Y. Dauvignac, and C. Pichot, "An inverse scattering method based on contour deformations by means of a level set method using frequency hopping technique," *IEEE Trans. Anten. Propag.* **51**, 1100–1113 (2003).

Franchois, A. and A. G. Tijhuis, "A quasi-Newton reconstruction algorithm for a complex microwave imaging scanner environment," *Radio Sci.* **38**, 1–13 (2003).

Franchois, A. and C. Pichot, "Microwave imaging—complex permittivity reconstruction with a Levenberg-Marquardt method," *IEEE Trans. Anten. Propag.* **45**, 203–215 (1997).

Hanke, H. M., A. Neubauer, and O. Scherzer, "A convergence analysis of the Landweber iterations for nonlinear ill-posed problems," *Num. Math.* **72**, 21–37 (1995).

Hanke, M., "Accelerated Landweber iterations for the solution of illposed equations," *Num. Math.* **60**, 341–373 (1991).

Harada, H., D. J. N. Wall, T. Takenaka, and M. Tanaka, "Conjugate gradient method applied to inverse scattering problems," *IEEE Trans. Anten. Propag.* **43**, 784–792 (1995).

Isernia, T., L. Crocco, and M. D'Urso, "New tools and series for forward and inverse scattering problems in lossy media," *IEEE Geosci. Remote Sens. Lett.* **1**, 327–331 (2004).

Kaltenbacher, B., "Some Newton type methods for the regularization of nonlinear ill-posed problems," *Inverse Problems* **13**, 729–753 (1997).

Lobel, P., C. Pichot, L. Blanc-Féraud, and M. Barlaud, "Conjugate gradient algorithm with edge-preserving regularization for image reconstruction from Ipswich data for mystery objects," *IEEE Anten. Propag. Mag.* **39**, 12–14 (1997).

Maniatis, T. A., K. S. Nikita, and N. K. Uzunoglu, "Two-dimensional dielectric profile reconstruction based on spectral-domain moment method and nonlinear optimization," *IEEE Trans. Microwave Theory Tech.* **48**, 1831–1840 (2000).

Meaney, P., T. Grzegorczyk, S. I. Jeon, and K. Paulsen, "Log transformation with Gauss-Newton microwave image reconstruction reduces incidence of local minima convergence," *Proc. 2009 IEEE Antennas Propagation Society Int. Symp., Charleston, SC, USA,* 2009.

Meaney, P. M., K. D. Paulsen, B. W. Pogue, and M. I. Miga, "Microwave image reconstruction utilizing log-magnitude and unwrapped phase to improve high contrast object recovery," *IEEE Trans. Med. Imag.* **20**, 104–116 (2001).

Mojabi, P. and J. LoVetri, "Microwave biomedical imaging using the multiplicative regularized Gauss-Newton inversion," *IEEE Anten. Wireless Propag. Lett.* **8**, 645–648 (2009).

Rekanos, I. T., S. M. Panas, and T. D. Tsiboukis, "Microwave imaging using the finite-element method and a sensitivity analysis approach," *IEEE Trans. Med. Imag.* **18**, 1108–1114 (1999).

Remis, R. F. and P. M. van den Berg, "On the equivalence of the Newton-Kantorovich and distorted Born methods," *Inverse Problems* **16**, L1–L4 (2000).

Rieder, A., "On the regularization of nonlinear ill-posed problems via inexact Newton iterations," *Inverse Problems* **15**, 309–327 (1999).

Roger, A., "Newton-Kantorovich algorithm applied to an electromagnetic inverse problem," *IEEE Trans. Anten. Propag.* **29**, 232–238 (1981).

Rubk, T., P. M. Meaney, P. Meincke, and K. D. Paulsen, "Nonlinear microwave imaging for breast-cancer screening using Gauss–Newton's method and the CGLS inversion algorithm," *IEEE Trans. Anten. Propag.* **55**, 2320–2331 (2007).

Semenov, S. Y. et al., "Three-dimensional microwave tomography: Experimental imaging of phantoms and biological objects," *IEEE Trans. Microwave Theory Tech.* **48**, 1071–1074 (2000).

van den Berg, P. M. and A. Abubakar, "Contrast source inversion method: State of art," *Progress Electromagn. Res.* **34**, 189–218 (2001).

van den Berg, P. M. and R. E. Kleinman, "A contrast source inversion method," *Inverse Problems* **13**, 1607–1620 (1997).

van der Vorst, H. A., "Bi-CGSTAB: A fast and smoothly converging variant of Bi-CG for the solution of nonsymmetric linear systems," *SIAM J. Sci. Stat. Comput.* **13**, 631–644 (1992).

CHAPTER SEVEN

Quantitative Stochastic Reconstruction Methods

7.1 INTRODUCTION

The current generation of computers makes it possible to utilize nonlinear global optimization procedures, which are in principle able to reach the global minimum of a given cost function. Stochastic optimization methods, which are now very common in electromagnetic imaging as well as in a great number of different inverse problems, are of particular interest.

The simulated annealing technique was one of the first stochastic minimization methods applied to electromagnetic imaging. It is an iterative method that, at each iteration, updates the current trial solution by exploiting probabilistic concepts. More recently, however, several inversion procedures founded on population-based algorithms have been proposed, since these global optimization techniques (especially the genetic algorithm) are now commonly applied in a plethora of engineering fields. Population-based algorithms can be used for imaging purposes with different implementation schemes (sometimes they assume different names depending on the choice of mechanism for generating the new solutions at the various iterative steps). Essentially, these methods start from sets of trial solutions (the *populations*) and iteratively evolve by applying some specific operators that are usually inspired on some biological or natural mechanisms (e.g., selection, crossover, and mutation in the genetic algorithm). During the minimization process, these algorithms tend to keep the information on the whole search space, avoiding (in principle) entrapment of the solution in local minima. Moreover, because the operators are applied not on a single trial solution, but on a

---

Microwave Imaging, By Matteo Pastorino
Copyright © 2010 John Wiley & Sons, Inc.

population of solutions, these approaches are usually intrinsically suitable for parallel computing (parallel computers are now rather common and not too expensive). Consequently, one of the main limitations of population-based methods (the high computational load; in particular, the high CPU time) tends to become decreasingly significant, allowing the application of procedures that would have been considered impractical even a few years ago. Furthermore, stochastic optimization approaches allow simple combinations with other methods, such as fast deterministic methods (see Chapter 5), in order to develop more efficient hybrid methods like the ones described in Chapter 8.

It should also be noted that population-based algorithms seem to be particularly suitable for the inclusion of a priori information into the model. From an engineering standpoint, a priori information is very important, since it can be used for accelerating the reconstruction procedure (essentially, by reducing the search space). Actually, as will be discussed further in this chapter, a significant of a priori information on the target under investigation may be available. An example is represented by the medical field (see Section 10.2), where the *external shape* of the human body is known, as well as the boundaries between adjacent organs, which can be deduced from other classical imaging modalities; moreover, the dielectric properties of the various biological tissues belong to only a fixed and known set. Another example is the area of nondestructive testing (see Section 10.1), where the *object* to be detected by the noninvasive analysis is often only a defect in a completely known configuration. In many cases, the a priori information cannot be taken into account in a simple way. In terms of this aspect, stochastic optimization methods are usually very effective.

In the following text, simulated annealing is discussed first. Then, some population-based methods, which have already been proposed for microwave imaging applications, will be considered. Specifically, there are the genetic algorithm, the differential evolution method, the particle swarm optimization method, and the ant colony optimization method.

7.2 SIMULATED ANNEALING

Simulated annealing (Kirkpatrick et al. 1983, Johnson et al. 1989, Aarts and Korst 1990) is an iterative technique in which a trial solution is generated at each iteration, and it is accepted or rejected on a probabilistic basis according to the Metropolis criterion (Kirkpatrick et al. 1983). The solution at the kth iteration is denoted by ξ^k. Let us consider a cost function $F(\xi)$ to be minimized. $F(\xi)$ can be, for example, the discrete counterpart of the functional in equation (6.7.1). In that case ξ^k coincide with the estimate of the discretized version of the array of unknowns $\mathbf{x}$ [equation (6.4.2)] at the kth iteration. Simulated annealing is basically characterized by the following steps:

Initialization. A trial solution ξ^0 is chosen in the initialization phase ($k = 0$).

Iterations. At step k, a new solution is generated starting from the solution at the previous step $k - 1$ by randomly modifying ξ^{k-1}. Once ξ^k has been obtained, the value of $F^k = F(\xi^k)$ is computed. Then, if $\Delta F = F^k - F^{k-1} < 0$, the solution ξ^k is accepted, since the cost function decreases. On the contrary, if $\Delta F = F^k - F^{k-1} \geq 0$, ξ^k is accepted with a probability

$$p = e^{-\Delta F/T} \quad \text{(Boltzmann probability factor)} \quad (7.2.1)$$

and rejected with probability $q = 1 - p$, where T is a parameter (viz., *temperature*) controlling the iteration process. This selection strategy is called the *Metropolis criterion*. The iterative process is stopped when the stopping criterion is satisfied or the maximum number of solutions have been generated.

It is evident from the Metropolis criterion that, during the minimization process, the functional can sometimes assume a value greater that the one at the previous iteration. Therefore, the solution potentially has the ability to escape from local minima of the functional to be minimized. To this end, it is crucial to choose the suitable probability p. The control parameter T is usually updated during the iterative process by following a fixed scheduling rule (usually of logarithmic type).

To be applied to a specific problem, simulated annealing requires (as mentioned previously) the definition of a suitable cost function, the choice of an initial solution, a strategy to update the solution at any iteration, the choice of proper scheduling for the temperature parameter, and a stopping criterion. Since simulated annealing is able to escape from local minima, it is, in principle, able to reach the global minimum. However, it is computationally very heavy, like most of the stochastic optimization procedures. Nevertheless, this technique has yielded very good results in several applications, including array pattern synthesis and design of microwave components.

To illustrate how simulated annealing can be used in imaging applications, let us follow one of the first proposed applications. Garnero et al. (1991) applied simulated annealing to solve a two-dimensional inverse scattering problem for dielectric cylinders. The formulation of the problem, based on the electric field integral equation, is essentially the one reported in Chapter 3.

The integral equation (data equation) is discretized by using the method of moments with pulse basis functions, and the solution is reduced to an algebraic system of equations (exactly as described in Section 3.4). The unknowns are the values of the complex dielectric permittivity in a grid of cells (the dielectric permittivity is assumed to be constant in each cell). Since the internal total electric field is another problem unknown, Garnero et al. (1991) computed this field at each iteration by solving a forward scattering problem, where the current distribution of dielectric parameters is used as the "object" at the current iteration. As already discussed in Chapter 6, although this approach is computationally expensive, it allows one keep to as unknowns only the values of the dielectric permittivities in the various cells.

Let $\varepsilon_n = \varepsilon(x_n, y_n)$, $n = 1, \ldots, N$ be the value of the dielectric permittivity in the nth cell. The solution at the kth iteration is constructed as follows:

$$\xi^k = \left[\varepsilon_1^k, \varepsilon_2^k, \ldots, \varepsilon_N^k\right]. \tag{7.2.2}$$

Garnero et al. (1991) assumed the cost function to be a data equation similar to equation (6.7.2) for two-dimensional problems. The initial solution ξ^0 has been constructed by randomly choosing values in a prescribed range $\xi_{\min} \leq \xi^0 \leq \xi_{\max}$. It should be noted that this approach represents the first possible way to insert a priori information into the model, since one can search for solutions that assume values in different ranges in various different regions of the investigation domain or even that assume only specific discrete values belonging to a prescribed set.

The new solution ξ^{k+1} is obtained by adding a random value $\Delta\varepsilon$ to one of the elements of ξ^k. The Metropolis criterion is then applied as a serial raster-scan process, in which the various elements of ξ^k are sequentially considered. Obviously, other strategies can be followed.

Concerning the temperature parameter, in order to accelerate the convergence, T is reduced when a fixed number of iterations have been completed, without waiting for the equilibrium. Finally, the minimization process stops when the number of accepted changes in a given time interval is below a fixed threshold.

A similar approach has been considered (Caorsi et al. 1991), but simulated annealing has been applied by starting with a description of the inverse scattering formulation in terms of probabilistic concepts. Following this approach, an a priori definition of the cost function is not required, but the functional to be minimized is obtained as a consequence of the assumptions on the model. The problem is still the microwave inspection of infinite dielectric cylinders of arbitrary cross sections. An interesting point is, however, represented by choice of the temperature scheduling, which, through equation (7.2.1), determines the acceptation probability p. Obviously, the value of probability p can be established a priori or calculated in an adaptive way during development of the relaxation method. For instance, it could be possible to start from a value of $p = 1$ and decrease it as the number of iterations increases. In this way, at the beginning of the iterative process, the algorithm randomly chooses the values of ξ^k because of both the value of p and the high starting temperature, so that configurations leading to a sensible increase in global *energy* (the value of the functional) could also be accepted. With development of the algorithm, the probability that the choice of ξ^k is such that the global energy increases is low, and this fact is reasonable if we suppose that, as the algorithm proceeds, it becomes increasingly probable that the reconstruction obtained more closely approaches the real solution. As for the scheduling used to modify the temperature at every step of the annealing process, we can say that the decrease in temperature should be adapted to the energy variations in the different configurations. This means that, to maintain the random aspect of the stochastic relaxation, the temperature decrease should be proportional to the energy decrease. Another important point to keep in mind is initialization of the temperature. Actually, if the starting value is too high, the method could not converge for an excess of randomness in the solution, while a low value could lead to a deterministic development of the method, which could in turn produce a nonoptimum solution. It has been shown that the optimum scheduling should follow a logarithmic law; specifically, a temperature decrease based on a logarithmic law is necessary to ensure convergence (Geman and Geman 1984, Hajek 1988). However, in practical implementations, this scheduling cannot be used, since it takes too long. So, a faster scheduling can be used. An example is (Sekihara et al. 1992)

$$T^k = T^0 \frac{\ln 2}{\ln(1+\mu^k)}\Theta, \qquad (7.2.3)$$

where T^k is the temperature at the kth iteration, T^0 is the starting temperature, μ is a parameter chosen to ensure the maximum possible range of temperature variation during annealing (its value is computer-specific), and Θ is a parameter introduced to control the temperature scheduling according to the results obtained from a test of the convergence of the solution, applied at every step of the simulated annealing algorithm.

This algorithm should converge when the temperature is very small or when there is no more activity in the solution, which means no changes in any of the cells. Since these conditions are difficult to reach in a finite number of steps, additional constraints can be added in order to control the convergence of the method and to stop it after a reasonable amount of iterations, thus accepting a suboptimum solution (Caorsi et al. 1991).

The iterative process terminates when a stopping criterion is satisfied or when a stationary condition is obtained. In particular, a minimum number of updates at each raster scan is defined for each of the involved sets of unknowns, and, at each generic raster-scan iteration k, the stationarity of the cost function is evaluated, by comparing the current energy value and the values at steps $k-1, k-2, \ldots, k-w$ (where w is the predefined dimension of the iteration window in which the comparison takes place). The stationary condition is satisfied if

$$|F^k - F^{k-j}| < \delta, \qquad (7.2.4)$$

for each j such that $1 \le j \le w$ and δ is a given constant (Caorsi et al. 1994). The parameter Θ is then used for a fine-tuning the temperature scheduling, according to the following empirical criteria:

1. If the number of updates is smaller than the fixed minimum number, if the energy is stationary, and $F^k > F_{th}$, there is a high probability that a local minimum has been reached, from which is impossible to climb up at the current temperature value. Then, in order to raise the temperature, the value of Θ is slightly increased.
2. If the number of updates is large but the value of the cost function is stationary, the value of Θ is slightly decreased, in order to speed up the process. In any other case, the value of Θ is left unchanged.

It is worth noting that excessively restrictive conditions on the criteria 1 and 2 listed above may sensibly reduce the speed of the method.

7.3 THE GENETIC ALGORITHM

The genetic algorithm is an iterative optimization method, in which a set of trial solutions, called *population*, is continuously modified by using operators inspired by the principles of genetic and natural selection (Goldberg 1989, Holland 1992, Haupt 1995, Johnson and Ramat-Samii 1997, Weile and Michielssen 1997). Figure 7.1 is a the flowchart of a basic genetic algorithm.

Let us denote the population at the kth iteration by $\wp^k = \{\xi_p^k\}_{p=1,\ldots,P^k}$, where $\xi_p^k, p = 1,\ldots,P^k$, is the pth element of the population $\wp^k$, whose number of elements is P^k. The elements of the population are usually coded; at each element there is a corresponding string called the *chromosome*, which consists of *genes*:

$$\xi_p^k \rightarrow \Psi_p^k = [\psi_{p1}^k, \psi_{p2}^k, \ldots, \psi_{pn}^k, \ldots, \psi_{pN}^k]. \qquad (7.3.1)$$

The genetic algorithm iteratively modifies the population by applying three types of operators: selection, crossover, and mutation. The principal steps and operators of the algorithm are briefly described as fowllows (Rahmat-Samii and Michielssen 1999):

Initialization. The approach starts with an initialization phase ($k = 0$), in which the $\wp^0$ population is constructed. To this end, chromosomes are randomly generated or some a priori information is used. The concept is similar to that discussed in Section 7.2 for the simulated annealing algorithm and is further discussed below.

Selection. At each iteration, good chromosomes are selected on the basis of their objective functions. In particular, a value $F_p^k = F(\xi_p^k)$ is associated with each chromosome Ψ_p^k. The selection phase produces a new population, indicated by $\wp_S^k$, which is a "function" of the original population $\wp^k$ and can be written as $\wp_S^k = S(\wp^k)$.

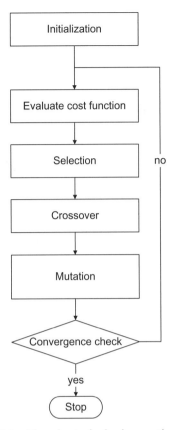

FIGURE 7.1 Flowchart of a basic genetic algorithm.

Crossover and Mutation. The selection phase is followed by the application of two genetic operators: crossover and mutation. The crossover operator combines two selected chromosomes and can be indicated as $\Psi_c^k = C(\Psi_a^k, \Psi_b^k)$, where Ψ_a^k and Ψ_b^k are two randomly selected chromosomes of the $\wp_S^k$ population. Application of the crossover operation results in a new population $\wp_C^k = C(\wp_S^k)$.

The mutation operation acts on the elements of the $\wp_C^k$ population by perturbing the genes of the chromosomes of the population. The result of the mutation operation is a new population $\wp_M^k = M(\wp_C^k)$. At this point, the population of chromosomes obtained is used to start the process at the $(k + 1)$th iteration. In other words, the starting population at the $(k + 1)$th iteration can be viewed as a composed function $\wp^{k+1} = M\{C[S(\wp_S^k)]\}$ (Weile and Michielssen 1997).

Elitism. Often the so-called elitism operator is used. If Ψ_q^k is the best chromosome at the kth iteration, that is, if $\xi_q^k = \operatorname{argmin}\{F_p^k\}$, this chromosome is directly included in the new population $\wp^{k+1}$. In this way, the best solution is directly propagated into the next iteration.

The iterative process terminates when a given stopping criterion is fulfilled or when a prescribed maximum number of iterations is reached.

Concerning the possible selection strategies, the most widely used are the population decimation, the proportionate selection, and the tournament selection. In the population decimation (ranking selection), a fixed (arbitrary) maximum fitness is chosen as a cutoff. If F_{th}^k denotes such a threshold values, the Q individuals with fitness such that $F_q^k < F_{th}^k$, $q = 1, \ldots, Q$, are used to generate new individuals through pairing and reproduction. This deterministic procedure is very simple, but a given discarded individual can possess unique important characteristics (some *genetic information*) that could become important in successive stages of the optimization process. These characteristics are lost by using this selection scheme.

In the proportional selection, the population P_S^k is randomly constructed by selecting the individuals with a probability that is proportional to values of their cost functions. The selection probability for the qth element at the kth iteration is given by

$$p_{S_q}^k = \frac{\Phi(F_q^k)}{\sum_p \Phi(F_p^k)}. \tag{7.3.2}$$

The function Φ is used to map the objective function to a non-negative interval such that, in minimization problems, low cost values correspond to high probabilities. Consequently, individuals with small cost have a high probability to be selected. This selection procedure is often called a *roulette wheel*, since an angular sector of a wheel (proportional to the fitness function) can be associated with each individual. Then the wheel is "spun" and the individual pointed at the end of the rotation is selected. Although the proportional selection is probably the most popular, it can result in a premature convergence if the chromosomes in a given population have very similar or very different fitness levels (Rahmat-Samii and Michielssen 1999). A possible solution is to use a fitness *scaling* to ensure that populations with many similar individuals do not tend to stagnate or that the best elements are not overrepresented in $\wp_S^k$. The scaling approach requires that scaling values be calculated and checked at each iteration [e.g., in maximizing a function, since the proportional selections work only with positive values of the cost function, it must be ensured that negative scaled values are not obtained (Rahmat-Samii and Michielssen 1999)].

For these reasons, tournament selection is often preferred. In this selection strategy, a subset of $R < P^k$ individuals is randomly selected from the original population (R is called *tournament size*). These elements "compete," and the element with the best fitness wins the tournament and is selected for the subsequent stages of the process. The choice of R determines the *selective pressure*, and the tournament selection acts as an implicit fitness scaling. Sometimes, one assumes $R = 2$. It seems that the tournament selection behaves better than the proportional selection in the initial phases of the optimization. Moreover, by using the tournament selection, positive values of the fitness function are not required and the strategy can be simply used for example in minimization problems (Rahmat-Samii and Michielssen 1999).

Other strategies can be used, because of great flexibility offered by the genetic algorithm. For example, if one uses the tournament selection with $R = 2$, after the choice of the two chromosomes, it is possible to determine the new element of the population by using the Metropolis criterion [equation (7.2.1)]. In this way the genetic algorithm is simply hybridized with the simulated annealing, but it should be noted that the temperature becomes another parameter to control during the optimization process.

None of the previous selection strategies guarantees that the fittest individual of a given population will be reproduced in the new population. Consequently, usually the elitism strategy mentioned above is used.

Owing to the flexibility of the method, a plethora of different versions of the genetic algorithm can be implemented by simply changing the structures of chromosomes, operators, and so on (Haupt and Haupt 2004).

As far as the coding of chromosomes is concerned, in the simplest implementation, the chromosomes Ψ_p^k are represented by strings of 0 and 1. Moreover, the crossover operator is defined as

$$\Psi_c^k = C(\Psi_a^k, \Psi_b^k) = [\psi_{a_1}^k, \psi_{a_2}^k, \ldots, \psi_{a_r}^k, \psi_{b_{r+1}}^k, \psi_{b_{r+2}}^k, \ldots, \psi_{b_N}^k], \qquad (7.3.3)$$

$$\Psi_d^k = C(\Psi_a^k, \Psi_b^k) = [\psi_{b_1}^k, \psi_{b_2}^k, \ldots, \psi_{b_r}^k, \psi_{a_{r+1}}^k, \psi_{a_{r+2}}^k, \ldots, \psi_{a_N}^k], \qquad (7.3.4)$$

where r is a random integer number uniformly distributed that determines the *crossover point*. The crossover operator is applied with a fixed probability p_c, which represents one of the parameters of the method that can be chosen by the user.

Analogously, the mutation operation, in the binary coding, can be simply obtained by randomly changing one of the bits (genes) of the chromosome of the population $\wp_c^k$ (from 1 to 0 or from 0 to 1). All the chromosomes of the population are considered for mutation, but the mutation operation, too, is applied with a certain probability p_m (usually much lower than p_c).

The basic formulation of the genetic algorithm can be evidently improved by introducing extended versions of the genetic operators, as well as different selection and reproduction strategies. In-depth discussions about these improved versions can be found in the books listed in the reference section. However, since the encoding and decoding operations required by the binary-coded version of the genetic algorithm are computationally expensive, real-coded versions are now usually preferred. In these cases, the unknown parameters are not represented by a string of bits, but any real parameter is itself a gene of the chromosome (i.e., $\xi_p^k \equiv \Psi_p^k$ for any p and k). Obviously, new definitions for the genetic operators are needed. In particular, concerning crossover and mutations, the simplest extensions of the previous definitions are

$$\Psi_c^k = C(\Psi_a^k, \Psi_b^k) = q\Psi_a^k + (1-q)\Psi_b^k, \qquad (7.3.5)$$

$$\Psi_d^k = C(\Psi_a^k, \Psi_b^k) = (1-q)\Psi_a^k + q\Psi_b^k, \tag{7.3.6}$$

where q is a random value between 0 and 1, which is selected with a uniform probability. In the same way, the mutation can be simply obtained by adding random numbers to certain genes. In the real-coded version of the genetic algorithm, the mutation operator can be applied with a fixed probability p_m.

Genetic algorithms are very helpful in solving highly constrained problems. Moreover, they are particularly suitable for applications in which the unknowns to be retrieved can create problems for deterministic methods and local search methods. From this perspective, they can be successfully applied to inverse problems in which there is significant information on the unknown target, but this information is difficult to factor in using local methods.

An example of application of the real-coded genetic algorithm is as follows. In particular, the reference configuration introduced in Chapter 5 is considered again (Fig. 5.1). The signal-to-noise ratio (SNR) is 25 dB. The cost functional F is the discrete version of equation (6.7.1) with only the data term (6.7.2). In order to reduce the number of unknowns, spline basis functions have been used to discretize the scattering equations, following the approach in the study by Baussard et al. (2004). In this case, the number of unknowns (which are the coefficients of the splines) is 81. According to guidelines reported in the literature, the following parameters of the genetic algorithm have been chosen: $P^k = P = 100$ for any k, probability of crossover $p_c = 0.8$, probability of mutation $p_m = 0.2$, and threshold on the cost function $F_{th} = 0.1$. The maximum number of iterations k_{max} is 500. Figure 7.2 shows the reconstructed profile of the relative dielectric permittivity. As can be seen, the approach is able to correctly identify the scatterer. Figures 7.3 and 7.4 present the cost function values and the corresponding reconstruction errors (e) for the best individual of the population versus the iteration number, respectively.

A second example concerns a more complex situation, in which two separate lossless scatterers are present in the investigation domain. In particular, two circular cylinders of radius $r = 0.25\lambda_0$ are located at $(0.5\lambda_0, 0.5\lambda_0)$ and $(-0.5\lambda_0, 0)$. The dielectric characteristics of the two objects are $\varepsilon_1 = 1.6\varepsilon_0$ and $\varepsilon_1 = 1.4\varepsilon_0$, respectively.

The measured data are acquired by using a multiview configuration. In particular, $S = 8$ views are considered. At each view, the incident electric field is generated by a line-current source located at

$$\mathbf{r}_t^s = 1.5\lambda_0 \cos\left(2\pi \frac{s-1}{S}\right)\hat{\mathbf{x}} + 1.5\lambda_0 \sin\left(2\pi \frac{s-1}{S}\right)\hat{\mathbf{y}}, \quad s = 1 \cdots S. \tag{7.3.7}$$

and the scattered electric field is measured in $M = 55$ points, located at

$$\mathbf{r}_t^{m_s} = 1.5\lambda_0 \cos\left(2\pi \frac{(s-1)}{S} + \frac{3\pi}{2}\frac{(m-1)}{M-1} + \frac{\pi}{4}\right)\mathbf{x}$$
$$+ 1.5\lambda_0 \sin\left(2\pi \frac{(s-1)}{S} + \frac{3\pi}{2}\frac{(m-1)}{M-1} + \frac{\pi}{4}\right)\mathbf{y}, \quad s = 1 \cdots S, m = 1 \cdots M. \tag{7.3.8}$$

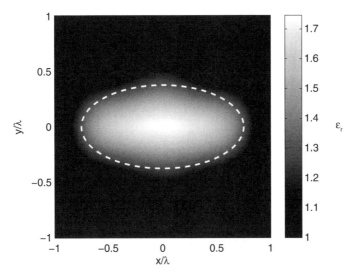

FIGURE 7.2 Reconstructed distribution of the relative dielectric permittivity of an elliptic cylinder ($\varepsilon_r = 1.6$); genetic algorithm.

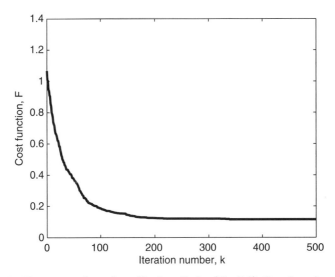

FIGURE 7.3 Reconstruction of an elliptic cylinder (Fig. 7.2). Cost function of the best individual of the population versus the iteration number; genetic algorithm.

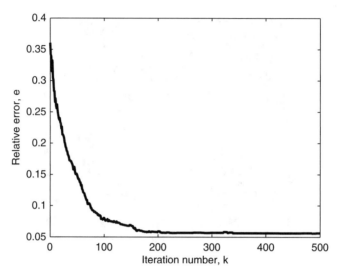

FIGURE 7.4 Reconstruction of an elliptic cylinder (Fig. 7.2). Relative reconstruction error of the best individual of the population versus the iteration number; genetic algorithm.

The electric field data (input data) are computed by using a direct solver based on the method of moments, which discretize the scene by means of $N = 1681$ pulse basis functions. Moreover, the computed values are corrupted with an additive Gaussian noise with zero mean value and a variance corresponding to a SNR of 25 dB.

As in the previous case, spline basis functions are used to discretize the inverse problem and the same parameters of the genetic algorithm are chosen. Figure 7.5 shows the reconstructed profile of the relative dielectric permittivity and proves that both the scatterers are correctly localized. Figures 7.6 and 7.7 present, for the best individual case, the cost function values and the related reconstruction errors.

The genetic algorithm has been proposed for solving several inverse problems related to microwave imaging. The key differences among the various approaches reported in the literature concern not only the adopted version of the genetic algorithm (mentioned previously) but also the inverse scattering formulation and the procedure used, at each iteration, for solving the direct scattering problem and computing the cost function.

From a historical perspective, one of the first applications of the genetic algorithm to inspect dielectric bodies was reported by Kent and Gunel (1997), who reconstructed two-layer dielectric cylinders. In particular, they assumed the outer radii of the cylinders to be known, and formed the chromosomes by binary-coding the values of the dielectric permittivities of the materials constituting the two layers as well as the value of the internal radius. They used 7 bits for any unknown parameter. This represents an example in which some a priori information on the scatterers under test (the external shape) is directly

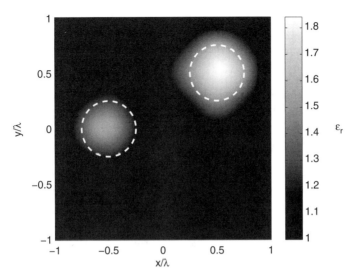

FIGURE 7.5 Reconstructed distribution of the relative dielectric permittivity of two circular cylinders, calculated using the genetic algorithm.

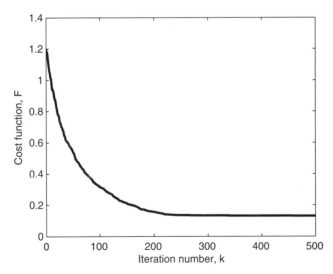

FIGURE 7.6 Reconstruction of two circular cylinders (Fig. 7.5). Cost function of the best individual of the population versus the iteration number, calculated using the genetic algorithm.

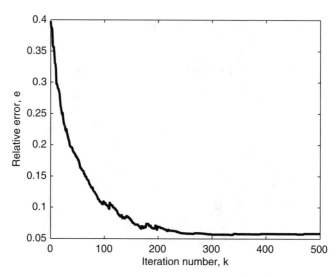

FIGURE 7.7 Relative reconstruction error of the best individual of the population versus the iteration number (two circular cylinders), calculated using the genetic algorithm.

included into the model. The simulations that they reported assumed a plane-wave transverse magnetic illumination at a frequency $f = 3\,\text{GHz}$, and the scattered data were collected on a linear probing line whose distance from the center of the cylinder was equal to 10λ with measurements performed on 64 points equally spaced at $\lambda/4$.

For this very simple application, the cost function [equation (6.7.1)] was constituted only by the data term [equation (6.7.2)] and the forward computation, at each iteration of the genetic algorithm, was performed analytically. Moreover, the cylinders under test had relatively small dielectric permittivities ($\varepsilon_r \leq 1.5$), which have been searched in the range $1.0 \leq \varepsilon_r \leq 2.0$. It should be noted that the ability of the approach in reducing the search space has been exploited. The accuracy of the reconstructions has been of course very good since a reduced parameterization and noiseless input data have been used. The binary-coded genetic algorithms has been successively applied in several other cases. A significant example is the retrieval of the external shape of PEC cylinders. To inspect PEC cylinders, according to the formulation described in Section 3.3, the shape of the object is defined by a *shape function* $\rho(\varphi)$ (in cylindrical coordinates), which defines a closed curve (the profile of the infinite cylinder is a closed line in the transversal plane). A common representation concerns the expansion in trigonometric series as

$$\rho(\varphi) = \sum_{n=0}^{N/2} A_n \cos(n\varphi) + \sum_{n=1}^{N/2} B_n \sin(n\varphi), \quad (7.3.9)$$

where the coefficients A_n and B_n are now the unknowns to be retrieved. Chiu and Liu (1996) applied a binary genetic algorithm to recover these parameters. In the construction of the function, a regularization term (see Section 6.7) given by $P = \alpha|\rho'(\varphi)|^2$ has been added to the data term, and the regularization parameter α (heuristically determined) essentially depends on the problem dimensions [$0.0001 < \alpha < 10$ (Chiu and Liu 1996)]. The imaging configuration contained 32 measurement points equally spaced on a circumference of radius 10λ, whereas the scatterer cross section was illuminated by a set of eight plane waves with different incident angles. The binary chromosomes have been constructed by using 8 bits for each parameter, and the nine coefficients of equation (7.3.9) have been considered for reconstructing simple scatterers with contours represented by smooth functions, suitable for trigonometric representations. The crossover and mutation probabilities were $p_c = 0.8$ and $p_m = 0.04$, respectively, and the population consisted of $P = 300$ elements. Quite good results have been obtained for cylindrical objects with maximum transversal dimensions slightly exceeding one wavelength starting from simulated input data numerically corrupted by Gaussian noise with zero mean values and standard deviations (normalized to the square amplitude of the simulated scattered values) $\sigma_n \leq 0.4$.

This work has been successively extended to inspect different dielectric and conducting cylinders in order to evaluate the capabilities and limitations of the genetic algorithm. Some examples are mentioned. Homogeneous dielectric cylinders with finite electrical conductivities have been reconstructed (Chiu and Chen 2000), whereas other PEC cylinders partially immersed in a homogeneous half-space (Fig. 4.8) (Yang and Chiu 2002) or located inside a multilayer material (Lin et al. 2004) have been inspected. The studies above cited have provided some further indications concerning the application of binary-coded genetic algorithms to microwave imaging. In particular, good results have been obtained by using a population of about 300–600 individuals, in which any parameter has been coded by means of 8–16 bits, whereas the crossover probability and mutation probability have been chosen in the ranges 0.5–0.7 and 0.005–0.05, respectively.

Qualitative *inverse* scattering formulations have also been used in conjunction with the genetic algorithm in order to inspect inhomogeneous dielectric cylinders of arbitrary shapes under transverse magnetic (TM) illumination conditions. The first-order Born approximation [see Section 4.6 and equation (4.6.7)] has been applied (Caorsi and Pastorino 2000), and the second-order Born approximation [see Section 4.6, where the solving equation is the two-dimensional counterpart of equation (4.6.6)] has also been considered (Caorsi et al. 2001). However, it is worth mentioning that the very high computational load associated with stochastic optimization procedures leaves the reconstruction of pixelated images of inhomogeneous targets still a very challenging problem.

Despite the interesting results obtained by using binary-coded genetic algorithms, real-coded versions of the method have been applied more frequently.

For example, the external contours of arbitrarily shaped PEC cylinders have been retrieved (Qing and Zhong 1998, Qing and Lee 1999; Qing et al. 2001). These contours have been represented by equation (7.3.9), but the unknown coefficients A_n and B_n directly formed the chromosomes [nine real elements used by Qing et al. (2001)]. The suggested parameters for the genetic algorithm application were crossover and mutation probabilities: $p_c = 1.0$ and $p_m = 0.04$, respectively, whereas the population was composed of $P = 300$ elements. The cost function was constituted by the data term only.

The reconstruction capabilities of the real-coded genetic algorithm can be drawn by the example presented in Figure 7.8, which shows the reconstruction, starting from noiseless data, of a single PEC cylinder whose contour is represented by the equation

$$\rho(\varphi) = 0.3 + 0.05\sin(2\varphi). \quad (7.3.10)$$

In particular, Figure 7.8a shows the original and estimated contours. It should be noted that initial contour is very different from the original one. Nevertheless, the final reconstruction is almost perfect. The genetic algorithm, as well as other evolutionary algorithms, does not require that the initial solution be close to the exact one. The behavior of the cost function, for the same simulation, is shown in Figure 7.8b.

Another application for which the genetic algorithm has been proposed is the determination of the widths and positions of strips located in a given *investigation* region (Takenaka et al. 1998). For a two-dimensional geometry, the metallic strip may be located along a certain line. This line can be discretized in small segments, and each segment may or may not belong to a strip. This problem is essentially a binary problem, in which a *shape function* is associated with the line where the strips are present. In particular, for each segment of the line, the shape function is equal to 1 if the segment is occupied by the metal strip, and 0 in the other case. By using a surface scattering formulation, the measured scattered electric field can be expressed in terms of this shape function. This approach has been considered in the study by Takenaka et al. (1998), where the scattering of a series of separated objects was computed by using an additional theorem (Chew 1990). However, the related inverse problem has been recast as an optimization problem by using a single cost function term, and the binary nature of the problem has been exploited. In particular, as the shape function is the only problem unknown, the inverse problem has been solved by using a standard binary genetic algorithm. It should be observed that this formulation implies a straightforward extension to two-dimensional configurations. In particular, the reconstruction of a two-dimensional configuration of small cylinders of square cross sections has also been reported (Takenaka et al. 1998).

Some examples of the different implementations of the genetic algorithms can be mentioned. Nonuniform probability densities for the genetic operators of a binary-coded genetic algorithm have been used (Chien and Chiu 2003) for the reconstruction of dissipative dielectric cylinders, and "start and restart"

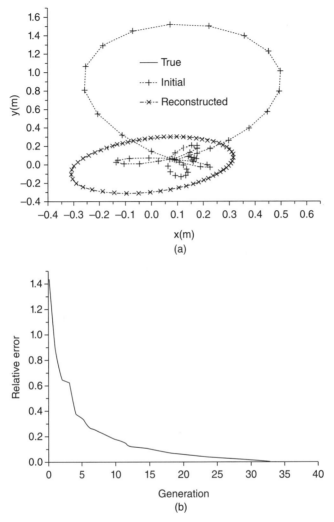

FIGURE 7.8 Reconstruction of a single PEC cylinder with the shape function of equation (7.3.10), using the genetic algorithm: (a) plots of the contours of the cross section at different iterations; (b) cost function (*error function*) versus the iteration number. [Reproduced from A. Qing, C. K. Lee, and L. Jen, "Electromagnetic inverse scattering of two-dimensional perfectly conducting objects by real-coded genetic algorithm," *IEEE Trans. Geosci. Remote Sens.* **39**(3), 665–674 (March 2001), © 2001 IEEE.]

procedures have been applied as well. In the latter case, the genetic algorithm is applied with a very small population. The approach reaches the near-optimal region of the solution space by a repeated computation. This version of the stochastic procedure, called the *micro genetic algorithm* (micro-GA) (Krishnakumar 1989), has been applied for reconstruction of the contours of

PEC cylinders (Xiao and Yabe 1998; Huang and Mohan 2004). As an example, the approach in the study by Xiao and Yabe (1998) was a binary-coded version that uses a population composed by five elements only.

Some other ideas on implementation of the genetic algorithms have been exploited in implementing codes for microwave imaging applications. For example, application of a parabolic form of the crossover operator has been proposed (Bort et al. 2005) and, since the same trial solutions can be repeatedly considered during the optimization approach, the use of a so-called tabu list has been proposed in conjunction with standard genetic algorithms for the detection of metallic objects with cavities (Zhou et al. 2003).

7.4 THE DIFFERENTIAL EVOLUTION ALGORITHM

The differential evolution algorithm (Storm and Price 1997; Chakraborty 2008) is a more recently proposed stochastic optimization method. A flowchart of this algorithm is shown in Figure 7.9.

The overall structure of the differential evolution algorithm is similar to those of the GA. The differential evolution algorithm iteratively modifies a

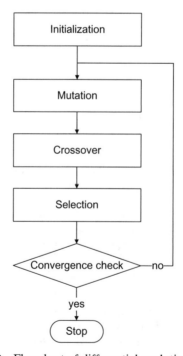

FIGURE 7.9 Flowchart of differential evolution algorithm.

population $\wp^k$ of P^k elements, $\xi_p^k, p = 1,\ldots, P^k$, where k is the iteration number. Usually, the number of elements in the population does not change during the evolution of the algorithm and the trial solutions are real-coded: $\xi_p^k \to \Psi_p^k = [\psi_{p_1}^k, \psi_{p_2}^k, \ldots, \psi_{p_n}^k, \ldots, \psi_{p_N}^k]$ [equation (7.3.1)], with $\psi_{p_n}^k \in \Re$. As is the genetic algorithm, the differential evolution algorithm is based on *genetic operators*. The main difference, however, relies on the scheme used for generation of the offspring. In particular, all the trial solutions are used once and in a deterministic way as parents to producing offspring. This property ensures a good exploration of the search space. A schematic representation of the offspring generation process is shown in Figure 7.10, and the steps in the algorithm are briefly described as follows:

Initialization. In the initialization phase, the starting population of the differential evolution algorithm can be constructed by randomly creating P^0 arrays of N real values:

$$\psi_{p_n}^0 = \psi_{n,L} + (\psi_{n,U} - \psi_{n,L})\,\text{rand}(0,1), \, n = 1,\ldots, N, \, p = 1,\ldots, P^0, \quad (7.4.1)$$

where $\Psi_{n,L}$ and $\Psi_{n,U}$ are the lower and upper bounds of the nth component of the elements of the population, respectively. In the same equation, rand(0,1) is a function that returns a random number uniformly distributed in the range [0,1]. Clearly, it is possible to introduce some a priori information in the initialization phase by properly modifying equation (7.4.1). Moreover, it is evident that the comments made in Section 7.3 about the possibility of including a priori information into the model are still valid here.

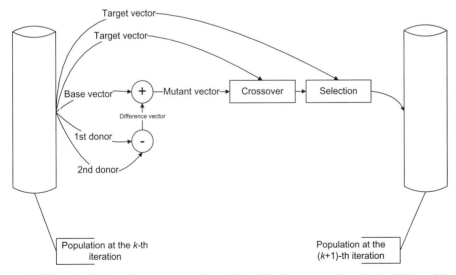

FIGURE 7.10 Schematic representation of the offspring generation in the differential evolution algorithm.

Mutation. In the differential evolution algorithm, the first applied operator is the mutation, which generates a set of P^k new elements. In particular, a new trial solution V_p^k is generated by perturbing an element of the population (called a *base vector*) by using a difference-based mutation. In the literature, several different versions of the mutation operators have been proposed, which differ mainly in the strategy of selection of the base vectors and in the number of difference vectors used (Chakraborty 2008). The various types of strategies are usually identified by the notation $DE/x/y/z$ (Storn and Price 1997), where x denotes the base vector selection strategy, y is the number of difference vectors, and z indicates the type of crossover (the last point will be discussed later). As an example, in the $DE/rand/1/*$ variant, the base vector is chosen randomly in the current population and only one difference vector is used to mutate it:

$$V_p^k = \Psi_{p_0}^k + c_f\left(\Psi_{p_1}^k - \Psi_{p_2}^k\right), \tag{7.4.2}$$

Here, c_f is a weighting factor that controls the *amplification* of the difference array, and p_0, p_1, and p_2 are randomly selected indices, which identify the base vector and the two elements used to generate the difference array, respectively. The reader is referred to the text by Chakraborty (2008) for an overview of the various mutation operators developed in the framework of differential evolution algorithms. It is worth noting that the mutant vector is generated by using only randomly chosen elements of the population, which allows one to account for the population diversity and limit the *destructive* effect of the mutation operator.

Crossover. Crossover works on the arrays generated in the mutation stage and on the elements Ψ_p^k of the current population (which, in the crossover stage, are called *target vectors*) in order to generate a set of *trial vectors* U_p^k. Analogously to the mutation operator, several variants of crossover have been proposed in the literature (Storn and Price 1997). A crossover operator commonly used is the uniform crossover, defined as

$$u_{p_n}^k = \begin{cases} v_{p_n}^k & \text{if rand}(0,1) \leq c_r \text{ or } n = j \\ \psi_{p_n}^k & \text{otherwise} \end{cases} \quad n = 1,\ldots,N, p = 1,\ldots,P^k, \tag{7.4.3}$$

where j is a random integer selected in $[0, N]$, $u_{p_n}^k$ and $v_{p_n}^k$ are the nth components of the arrays U_p^k and V_p^k, respectively, and c_r is the crossover rate.

Selection. In the differential evolution algorithm, the pth element of the population at the $(k+1)$th iteration is generated from the trial and target vectors by using a simple one-to-one survivor selection:

$$\Psi_p^{k+1} = \begin{cases} U_p^k & \text{if } F(U_p^k) \leq F(\Psi_p^k) \\ \Psi_p^k & \text{if } F(U_p^k) > F(\Psi_p^k) \end{cases}. \tag{7.4.4}$$

As in the genetic algorithm, the procedure stops when a given stopping criterion is fulfilled or when a prescribed maximum number of iterations, k_{max}, is reached.

The choice of the values of the control parameters, c_f and c_r, is essentially heuristic and related to the particular problem in which the differential evolution is used. However, some general suggestions are provided in the literature (Chakraborty 2008, Price et al. 2005). In particular, c_f is usually chosen in the range [0.5, 1.0], and the value of c_r in the range [0.8, 1.0]. If the parameters c_r and c_f are suitably chosen, a good balance between good convergence speed and low possibility of a premature convergence to local minima can be obtained.

It is worth noting that the differential evolution algorithm contains a kind of implicit elitism. In fact, since the fitness of the trial vector U_p^k is compared only with the one of the target vector Ψ_p^k from which it is generated, the corresponding element in the next generation has a fitness that can only be better than or at least equal to the one of Ψ_p^k.

In order to reduce the computational load, application of the differential evolution to imaging problems requires a reduced parameterization. Michalski (2000, 2001) obtained this by approximating the cylindrical scatterers with canonical objects with circular and elliptic cross sections. The proposed application concerns the inspection of tunnels and pipes in a cross-borehole configuration (Michalski 2001), in which the effects of the interface between the upper and lower media were neglected assuming deeply buried objects. In the inverse problem, the problem unknowns are represented by the cylinder center and the radius (circular cross section) or the semimajor axis, the eccentricity, and the tilt angle (elliptic cross section). The differential evolution algorithm has been applied to a cost function based on the data equation only [equation (6.7.2)]. Excellent reconstructions have been obtained with the following control parameters: $P = 25$, $k_{max} = 40$ (maximum number of iterations of the differential evolution), $c_r = 0.9$ and $c_f = 0.7$.

Massa et al. (2004) applied the differential evolution algorithm by combining two of the various possible implementation strategies for this evolutionary approach. In particular, the *DE/1/best/bin* version (Storn and Price 1997) is used until the cost function has reached a predefined value; successively, the *DE/1/rand/bin* strategy (Storn and Price 1997) is applied. It has been found that the *DE/1/best/bin* strategy is quite able to rapidly locate the *attraction basin* of a minimum, but, since it uses the best individual of the population to perform the mutation, it can sometimes be trapped in a local minimum. This drawback is overcome by switching, after a predefined threshold, to the *DE/1/rand/bin* strategy, which is able to explore more efficiently the search space, without modifying the previous best solution if it is inside the correct attraction basin. Concerning the choice of the control parameters, c_f has been chosen in the range [0.5, 1.0], whereas good reconstructions have been obtained with $c_r = 0.8$.

The differential evolution algorithm has been further applied to inspect single and multiple PEC cylinders of arbitrary shapes (Qing 2003, 2006). The approach has been successfully used to obtain good reconstructions of the profiles of PEC cylinders (with both synthetic and real data) by using the following values of control parameters: $c_r = 0.9$ and $c_f = 0.7$. For this specific application, the differential evolution method has been compared with a standard real-coded genetic algorithm under the same operating conditions. In this case, the differential evolution algorithm has been found to outperform the genetic algorithm, due to the need for a smaller population. It should be mentioned that in Qing (2006) used the *dynamic differential evolution strategy*, in which an additional competition is introduced between the trial vector and the current optimal individual. The current optimal element is replaced if the new element corresponds to a better solution and the updated element is immediately included in the new population.

In an example application of the differential evolution, the reference configuration is the same as in Section 7.3 (Fig. 7.2). The parameters of the differential evolution algorithm are $P = 100$, $k_{max} = 500$, $F_{th} = 0.1$, $c_r = 0.8$, and c_f is randomly chosen in the range [0.5, 1.0]. In this example, the *DE/rand/1/bin* strategy (Storn and Price 1997) is applied. The same cost function [equation (6.7.2)] and the same spline basis functions (Baussard et al. 2004) are also used. Figure 7.11 shows the reconstructed distribution of the relative dielectric permittivity of the investigation area obtained by the differential evolution algorithm. Figures 7.12 and 7.13 plot the behavior of the cost function and the

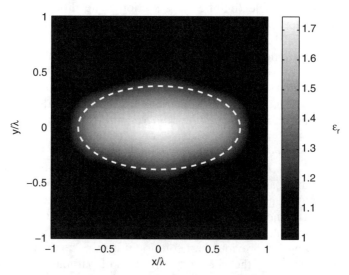

FIGURE 7.11 Reconstructed distribution of relative dielectric permittivity of an elliptic cylinder, calculated using the differential evolution algorithm.

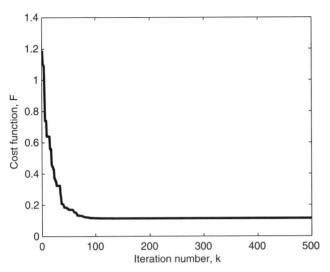

FIGURE 7.12 Cost function of the best individual of the population versus the iteration number calculated for an elliptic cylinder using the differential evolution algorithm.

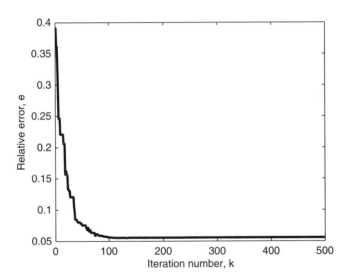

FIGURE 7.13 Relative reconstruction error of the best individual of the population versus the iteration number for an elliptic cylinder, calculated using the differential evolution algorithm.

176 QUANTITATIVE STOCHASTIC RECONSTRUCTION METHODS

relative reconstruction errors (related to the best individual of the population) versus the iteration number, respectively. As can be seen from these figures, the approach correctly reconstructs the shape and dielectric properties of the scatterer.

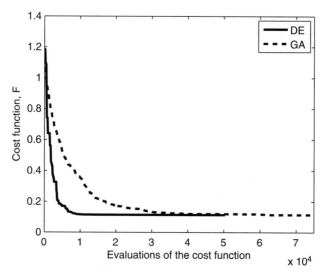

FIGURE 7.14 Comparison between the differential evolution algorithm and the genetic algorithm for calculation of cost function value versus number of cost function evaluations for an elliptic cylinder.

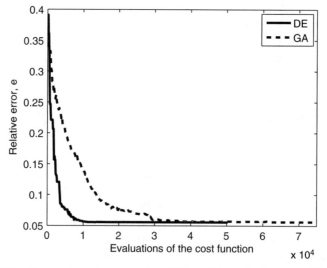

FIGURE 7.15 Comparison between the differential evolution algorithm and the genetic algorithm for estimating relative reconstruction error versus number of cost function evaluations for an elliptic cylinder.

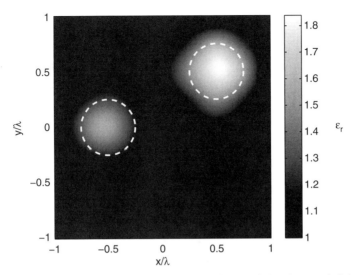

FIGURE 7.16 Reconstructed distribution of the relative dielectric permittivity of two circular cylinders, calculated using the differential evolution algorithm.

Moreover, as can be seen from Figure 7.12, the differential evolution algorithm converges more rapidly than does the genetic algorithm. This behavior can be better evaluated from Figures 7.14 and 7.15, which provide a direct comparison in terms of the cost function values and of the relative errors for the best individual of the population. For the present case, the differential evolution algorithm requires fewer cost function evaluations than does the standard genetic algorithm to reach the final solution.

As a second example, the configuration adopted in Section 7.3 for the reconstruction shown in Figure 7.5 is considered. The parameters of the differential evolution are $P = 100$, $k_{max} = 500$, $F_{th} = 0.1$, and $c_r = 0.8$, and c_f is randomly chosen in the range [0.5, 1.0]. The *DE/rand/1/bin* strategy (Storn and Price 1997) is employed. Figure 7.16 shows the distribution of the reconstructed relative dielectric permittivity obtained by applying the differential evolution method. Figures 7.17 and 7.18 give the plots of the cost function and of the corresponding reconstruction errors. As can be seen, in this case, too, the differential evolution algorithm converges faster than does the genetic algorithm.

7.5 PARTICLE SWARM OPTIMIZATION

Particle swarm optimization is another global optimization algorithm, which is inspired by the social behavior of bird flocking or fish schooling (Kennedy

QUANTITATIVE STOCHASTIC RECONSTRUCTION METHODS

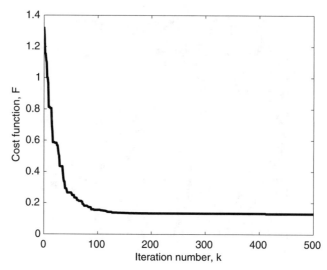

FIGURE 7.17 Cost function of the best individual of the population versus the iteration number for two circular cylinders, estimated using the differential evolution algorithm.

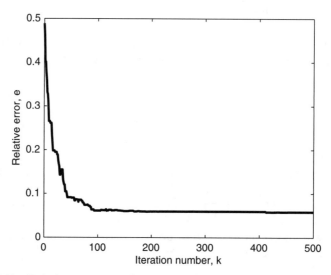

FIGURE 7.18 Relative reconstruction error of the best individual of the population versus the iteration number for two circular cylinders, calculated using the differential evolution algorithm.

and Eberhart 1995, 1999; Robinson and Rahmat-Samii 2004). Similarly to other evolutionary computation techniques, it is based on a population (*swarm*) of trial solutions (*particles*), which is iteratively updated by following a predefined set of rules. Unlike other approaches, however, particle swarm optimization does not rely on *evolution* operators, but modifications are introduced in order to *follow* the current best solutions. One of the main advantages of particle swarm optimization, with respect to other techniques such as the genetic algorithm is that it is simple to implement and requires adjustment of only a few parameters. Moreover, a priori information can be easily inserted into the updating scheme. Finally, particle swarm optimization generally requires a rather small population size, which results in a reduced computational cost of the overall minimization process (Clerk and Kennedy 2002, Robinson and Rahmat-Samii 2004).

In order to describe how particle swarm optimization works, let us consider a population $\wp^k$ of P elements, ξ_p^k, $p = 1, \ldots, P$ (as usual, k is the iteration number). Let us suppose that the elements are again real-coded, $\xi_p^k \to \Psi_p^k = [\psi_{p1}^k, \psi_{p2}^k, \ldots, \psi_{pn}^k, \ldots, \psi_{pN}^k]$ [equation (7.3.1)], with $\psi_{pn}^k \in \Re$. In the particle swarm optimization framework, ξ_p^k is usually referred to as the position of the pth particle of the swarm. Moreover, a velocity vector $V_p^k = [v_{p1}^k, v_{p2}^k, \ldots, v_{pn}^k, \ldots, v_{pN}^k]$ is associated with each trial solution of the population.

Particle swarm optimization is characterized by the following steps:

Initialization. When no a priori information is available, the particle population is initialized with randomly chosen values.

Swarm Ranking. At each iteration, the cost function values of the particles are evaluated, and two "best" solutions are identified. The first one, $P_{\text{best}} = [p_{\text{best}_1}, p_{\text{best}_2}, \ldots, p_{\text{best}_n}, \ldots, p_{\text{best}_N}]$, is the best solution achieved so far by a given particle (i.e., by the pth element of the population). The second one, $G_{\text{best}} = [g_{\text{best}_1}, g_{\text{best}_2}, \ldots, g_{\text{best}_n}, \ldots, g_{\text{best}_N}]$, is the "best" value obtained so far by any particle of the population (i.e., it is a global "best").

Solution Update. At each iteration, both the elements ξ_p^k (positions of the particles) and the velocity vectors V_p^k are updated. In particular, the following rules are used to construct the $\wp^{k+1}$ population:

$$v_{pn}^{k+1} = w v_{pn}^k + c_1 r_1 (p_{\text{best}_n} - \psi_{pn}^k) + c_2 r_2 (g_{\text{best}_n} - \psi_{pn}^k), \quad n = 1, \ldots, N, \, p = 1, \ldots, P, \tag{7.5.1}$$

$$\psi_{pn}^{k+1} = \psi_{pn}^k + v_{pn}^{k+1}, \quad n = 1, \ldots, N, \, p = 1, \ldots, P, \tag{7.5.2}$$

Here, r_1 and r_2 are random numbers uniformly distributed in the range [0, 1], c_1 and c_2 are positive constants called *acceleration coefficients*

```
Initialization
While (convergence check not passed) do
    Update p_best and g_best
    Calculate particle velocity
    Update particle position
end
```

FIGURE 7.19 Particle swarm optimization algorithm.

(usually chosen such that $c_1 = c_2 = 2$), and w is a scaling factor. Usually, the velocities of the particles are also clamped to a maximum value v_{max}. The algorithm stops when a suitable stopping criterion is satisfied. A high-level pseudocode of the particle swarm optimization method is presented in Figure 7.19.

The differential evolution method has been shown to outperform the particle swarm optimization method in reconstructing perfectly conducting scatterers limitedly to the specific cases considered (Rekanos 2008). The reader is also referred to studies by Donelli et al. (2006), Donelli and Massa (2005), and Huang and Mohan (2005, 2007) for further details on particle swarm optimization.

7.6 ANT COLONY OPTIMIZATION

The ant colony optimization method is a more recently introduced stochastic optimization algorithm, which was originally inspired by the way ants find the optimal path from nest to food (Dorigo and Gambardella 1997, Dorigo et al. 2006). By exploiting these ideas, the ant colony optimization method was initially developed to solve difficult combinatorial problems, such as the "traveling salesperson" problem but has more recently been extended to continuous domains (Socha and Dorigo 2008).

The ant colony optimization algorithm performs the minimization of a cost function F by iteratively modifying a population $\wp^k$ of P trial solutions, ξ_p^k, $p = 1,\ldots, P$, where k is the iteration number. In the continuous version of ant colony optimization, the elements of the population are real-coded: $\xi_p^k \to \Psi_p^k = [\psi_{p1}^k, \psi_{p2}^k, \ldots, \psi_{pn}^k, \ldots, \psi_{pN}^k]$, with $\psi_{pn}^k \in \Re$. The population is stored in an ordered archive: $F(\xi_1^k) \le F(\xi_2^k) \le \ldots \le F(\xi_P^k)$.

A high-level pseudocode of the ant colony optimization algorithm is shown in Figure 7.20.

The algorithmic steps shown in Figure 7.20 work as follows:

Initialization. An initial set of P randomly chosen trial solutions Ψ_p^0, $i = 1,\ldots, P$, is constructed. In particular, when no a priori information is

```
Initialization
While (convergence check not passed) do
    Ant-based Solution Construction
    Pheromone Update
end
```

FIGURE 7.20 Ant colony optimization algorithm.

available, the components of Ψ_p^0 are obtained by sampling a uniform probability density function (PDF) as in equation (7.4.1).

Ant-Based Solution Construction. At each iteration, a set of Q new trial solutions is created. The N components of the new solution are obtained by sampling a set of N Gaussian kernel PDFs (Socha and Dorigo 2008, Brignone et al. 2008a, 2008b), defined as

$$g_n^k(\psi) = \sum_{p=1}^{P} w_p \frac{1}{\sqrt{2\pi} s_{p_n}^k} e^{-\left(\psi - m_{p_n}^k\right)^2 / 2\left(s_{p_n}^k\right)^2}, \quad n = 1,\ldots,N, \quad (7.6.1)$$

where $m_{p_n}^k$, $s_{p_n}^k$, and w_p, $p = 1,\ldots,P$, $n = 1,\ldots,N$ are parameters of the PDFs, which, at the kth iteration, are given by

$$m_{p_n}^k = \psi_{p_n}^k, \quad p=1,\ldots,P, n=1,\ldots,N, \quad (7.6.2)$$

$$s_{p_n}^k = \xi \sum_{e=1}^{P} \frac{\left|\psi_{e_n}^k - \psi_{p_n}^k\right|}{k-1}, \quad p=1,\ldots,P, n=1,\ldots,,N, \quad (7.6.3)$$

$$w_p = \frac{1}{qP\sqrt{2\pi}} e^{-(p-1)^2/2q^2 P^2}, \quad p=1,\ldots,P, \quad (7.6.4)$$

where ξ (called the *pheromone evaporation rate*) and q are parameters of the ant colony optimization algorithm.

Pheromone Update. The new population $\wp^{k+1}$ is obtained by adding the newly created trial solutions to the ones contained in the current population $\wp^k$ and by discarding the worst Q elements. The algorithm stops when a maximum number of iterations is reached, or when a given stopping criterion is satisfied.

An example application of the ant colony optimization algorithm is described as follows. In particular, the reference configuration introduced in Figure 5.1 is considered. Similarly to the case reported in Section 7.3, the scattering equations are discretized by using the method of moments with spline basis functions and Dirac delta distributions as testing functions. The population elements are the coefficients of the spline expansion of the object function. The cost function considered is the same as that used in Section 7.3 (data equation only). The parameters

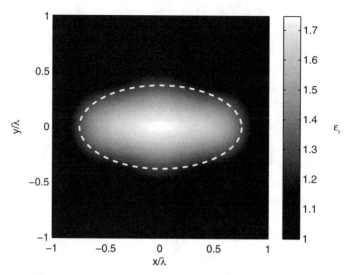

FIGURE 7.21 Reconstructed distribution of the relative dielectric permittivity of an elliptic cylinder, calculated using the ant colony optimization algorithm.

of the ant colony optimization algorithm are $P = 100$, $Q = 10$, $q = 0.1$, $\xi = 0.65$, and $k_{max} = 500$. The algorithm stops when $F(\xi_p^k) \leq F_{th}$, with $F_{th} = 0.1$ or when $k > k_{max}$. Figure 7.21 shows the reconstructed distribution of the relative dielectric permittivity.

Figures 7.22 and 7.23 plot the behavior of the cost function and the reconstruction errors versus the iteration number for the best element of each population. Figures 7.24 and 7.25 compare the ant colony optimization and differential evolution algorithms in terms of cost function values (Fig. 7.24) and relative reconstruction errors (Fig. 7.25) versus the number of evaluations of the functional. As can be seen, the ant colony optimization algorithm requires about half of the cost functional evaluations needed for the differential evolution method, resulting in a significant reduction in the computation time needed to perform the reconstruction.

The second reported example concerns the reconstruction of the two-cylinder configuration used in Section 7.3 to obtain Figure 7.5. The parameters of the ant colony optimization algorithm are $P = 100$, $Q = 10$, $q = 0.1$, $\xi = 0.65$, and $k_{max} = 1000$. The cost function is still the same as that used in the previous examples (data term only). Figure 7.26 shows the reconstructed distribution of the relative dielectric permittivity, whereas Figures 7.27 and 7.28 present the corresponding behaviors of the cost function and reconstruction errors calculated for the best individual of the population. Finally, Figures 7.29 and 7.30 compare the ant colony optimization and the genetic algorithms in terms of the number of evaluations. As can be seen, although it cannot be considered

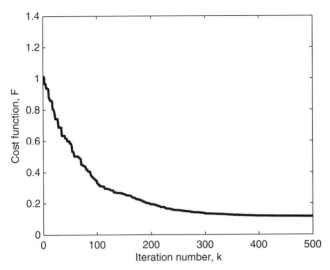

FIGURE 7.22 Cost function of the best individual of the population versus the iteration number for an elliptic cylinder, calculated using the ant colony optimization algorithm.

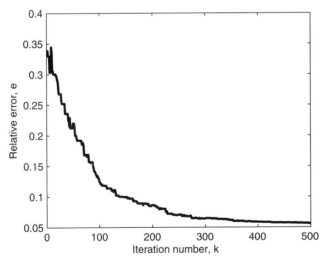

FIGURE 7.23 Relative reconstruction error of the best individual of the population versus the iteration number for an elliptic cylinder, calculated by the ant colony optimization algorithm.

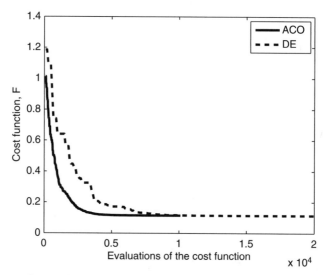

FIGURE 7.24 Comparison between ant colony optimization and differential evolution algorithms, showing cost function values (best individual) versus the number of evaluations of the functional for an elliptic cylinder.

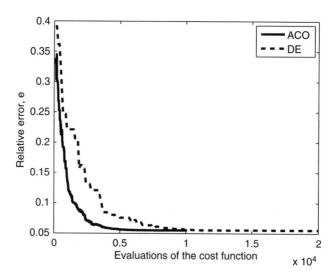

FIGURE 7.25 Comparison between the ant colony optimization and differential evolution algorithms, showing relative reconstruction errors (best individual) versus the number of evaluations of the functional for an elliptic cylinder.

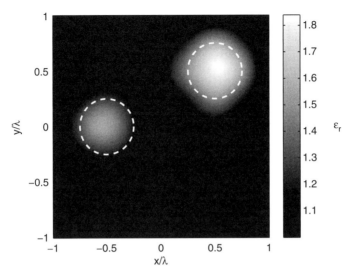

FIGURE 7.26 Reconstructed distribution of the relative dielectric permittivity of two circular cylinders, using the ant colony optimization algorithm.

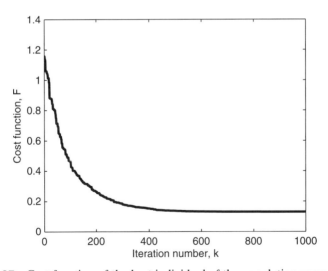

FIGURE 7.27 Cost function of the best individual of the population versus the iteration number for two circular cylinders, calculated using the ant colony optimization algorithm.

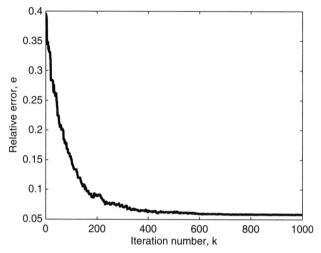

FIGURE 7.28 Relative reconstruction errors of the best individual of the population versus the iteration number for two circular cylinders, calculated using the ant colony optimization algorithm.

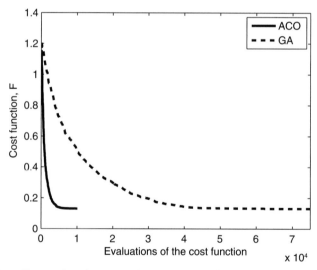

FIGURE 7.29 Comparison between the genetic algorithm and the ant colony optimization algorithm for calculation of cost function values of the best individual versus the number of the evaluations of the functional for two circular cylinders.

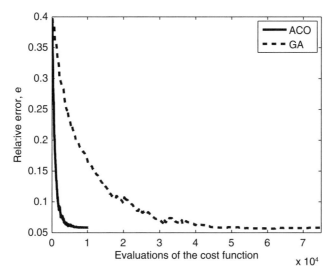

FIGURE 7.30 Comparison between the genetic algorithm and the ant colony optimization algorithm for estimation of relative reconstruction errors of the best individual versus the number of the evaluations of the functional for two circular cylinders.

a general result, for this specific case, the ant colony optimization algorithm outperforms the standard genetic algorithm since it is able to reach the final solution with fewer cost function evaluations.

7.7 CODE PARALLELIZATION

The application of stochastic optimization methods to solve the inverse scattering problem following the formulation described in the previous sections usually results in very expensive computational codes.

To give the reader an idea of the computational load, in a previous study (Pastorino 2007) the author provided data on some stochastic optimization approaches, even though they refer to very different configurations and parameterizations that do not allow for a direct comparison. Such data are presented in Table 7.1 and also pertain to some hybrid techniques, which are described in Chapter 8.

However, when possible, the computation time of the iterative reconstruction can be drastically reduced by using parallel processing. In fact, it is evident that the procedures discussed in Section 7.3–7.6 are intrinsically parallel algorithms since the various operators act on sets of solutions. Consequently, these operators can be applied simultaneously on different chromosomes of the population if more than one processor is available. Moreover, at the kth iteration step, computation of the values of the cost function $F_p^k = F(\xi_p^k)$,

TABLE 7.1 Some Computational Data on Microwave Imaging Methods Based on Stochastic Optimization Algorithms

Stochastic Method	Scatterers	Number of Unknowns	Population Dimension	Number of Iterations	Number of Function Evaluations	Computation Time, min	Computer	Reference
GA[a]	Dielectric	3	126	2	—	NA	NA	Kent and Gunel 1997
GA[b]	Dielectric	4–7	100	57–125	—	≈ 30[c]–438	AMD XP1700+	Franceschini et al. 2005
ACO	Dielectric	6	30	800	—	≈ 61	Pentium 4	Pastorino 2007
MA	Dielectric	9	9	3–15	—	≈ 55	Pentium III	Caorsi et al. 2003
DE	Dielectric/PEC	5	25	30–40	625–750	NA	NA	Michalski 2001
GA[a]	Conductor	10	—	100–200	—	NA	NA	Chiu and Chen 2000
GA[a]	PEC	9	—	4–8	—	≈ 30	Sun Sparc 20	Chiu and Liu 1996
GA[b]	PEC	8	256	35	—	≈ 260	IBM P-133	Qing et al. 2001
μGA	PEC[d]	5–10	—	—	1000–3000	≈ 226–290	Sun Sparc 20	Xiao and Yabe 1998
μGA–DM[e]	PEC[d]	5–10	—	—	402–682	≈ 39–153	Sun Sparc 20	Xiao and Yabe 1998
GA[b]	Dielectric	500	—	6000	—	≈ 20	NA	Donelli et al. 2006
PSO	Dielectric	500	—	6000	—	≈ 12.7	NA	Donelli et al. 2006
GA-CG	PEC	5	200	75[f]–220	—	—	NA	Zhou et al. 2003
GA[a]	Dielectric[g]	900	—	2000	—	≈ 4.5	Pentium II	Caorsi and Pastorino 2000
GA[a]	Dielectric[h]	400	—	4000–8000	—	NA	NA	Caorsi et al. 2001
SA	Dielectric	1600	—	—	—	≈ 60	IBM RISC/6000	Gragnani 2002

[a]Binary-coded genetic algorithm.
[b]Real-coded genetic algorithm.
[c]Parallel processing (16 PCs).
[d]Real data.
[e]Micro-genetic algorithm combined with a deterministic method (hybrid technique).
[f]With the "tabu" list.
[g]First-order Born approximation.
[h]Second-order Born approximation.

Source: Reproduced from M. Pastorino, "Stochastic optimization methods applied to microwave imaging: A review," *IEEE Trans. Anten. Propag.* **55**, 538–548, (March 2007) © 2007 IEEE.

which must be obtained for every chromosome Ψ_p^k, must be parallelized as well. This results in another significant time saving. Discussion of the parallelization of computer codes is beyond the scope of present book; however, it is well known that the use of multiprocessor computers is now quite common and the availability of parallel systems is becoming increasingly widespread.

REFERENCES

Aarts, E. and J. Korst, *Simulated Annealing and Boltzmann Machines*, Wiley, New York, 1990.

Baussard, A., E. L. Miller, and D. Premel, "Adaptive B-spline scheme for solving an inverse scattering problem," *Inverse. Problems.* **20**, 347–365 (2004).

Bort, E., G. Franceschini, A. Massa, and P. Rocca, "Improving the effectiveness of GA-based approaches to microwave imaging through an innovative parabolic crossover," *IEEE Anten. Wireless Propag. Lett.* **4**, 138–142 (2005).

Brignone, M., G. Bozza, A. Randazzo, R. Aramini, M. Piana, and M. Pastorino, "Hybrid approach to the inverse scattering problem by using ant colony optimization and no-sampling linear sampling," *proc. 2008 IEEE Antennas Propagation. Society Int. Symp.*, San Diego, CA, July 5–11, 2008, pp. 1–4 (2008a).

Brignone, M., G. Bozza, A. Randazzo, M. Piana, and M. Pastorino, "A hybrid approach to 3D microwave imaging by using linear sampling and ACO," *IEEE Trans. Anten. Propag.* **56**, 3224–3232 (2008b).

Caorsi, S., A. Costa, and M. Pastorino, "Microwave imaging within the second-order Born approximation: Stochastic optimization by a genetic algorithm," *IEEE Trans. Anten. Propag.* **49**, 22–31 (2001).

Caorsi, S., G. L. Gragnani, S. Medicina, M. Pastorino, and G. Zunino, "Microwave imaging method using a simulated annealing approach," *IEEE Microwave Guided Wave Lett.* **1**, 331–333 (1991).

Caorsi, S., G. L. Gragnani, S. Medicina, M. Pastorino, and G. Zunino, "Microwave imaging based on a Markov random field model," *IEEE Trans. Anten. Propag.* **42**, 293–303 (1994).

Caorsi, S. and M. Pastorino, "Two-dimensional microwave imaging approach based on a genetic algorithm," *IEEE Trans. Anten. Propag.* **48**, 370–373, (2000).

Caorsi, S., M. Pastorino, M. Raffetto, and A. Randazzo, "Electromagnetic detection of buried inhomogeneous elliptic cylinders by a memetic algorithm," *IEEE Trans. Anten. Propagat.* **51**, 2878–2884 (2003).

Chakraborty, U. K., *Advances in Differential Evolution*, Springer, 2008.

Chew, W. C. *Waves and Fields in Inhomogeneous Media*, Van Nostrand Reinhold, New York, 1990.

Chien, W. and C.-C. Chiu, "Using NU-SSGA to reduce the searching time in inverse problem of a buried metallic object," *IEEE Trans. Anten. Propag.* **53**, 3128–3134 (2003).

Chiu, C.-C. and W.-T Chen, "Electromagnetic imaging for an imperfectly conducting cylinder by the genetic algorithm," *IEEE Trans. Microwave Theory Tech.* **48**, 1901–1905 (2000).

Chiu, C.-C. and P. T. Liu, "Image reconstruction of a perfectly conducting cylinder by the genetic algorithm," *IEE Proc. Microwaves Anten. Propag.* **143**, 249–253 (1996).

Clerk, M. and J. Kennedy, "The particle swarm-explosion, stability, and convergence in a multidimensional complex space," *IEEE Trans. Evolut. Comput.* **6**, 58–73 (2002).

Donelli, M., G. Franceschini, A. Martini, and A. Massa, "An integrated multiscaling strategy based on a particle swarm algorithm for inverse scattering problems," *IEEE Trans. Geosci. Remote Sens.* **44**, 298–312 (2006).

Donelli, M. and A. Massa, "Computational approach based on a particle swarm optimizer for microwave imaging of two-dimensional dielectric scatterers," *IEEE Trans. Microwave Theory Tech.* **53**, 1761–1776 (2005).

Dorigo, M., M. Birattari, and T. Stutzle, "Ant colony optimization," *IEEE Comput. Intell. Mag.* **1**, 28–39 (2006).

Dorigo, M. and L. M. Gambardella, "Ant colony system: A cooperative learning approach to the traveling salesman problem," *IEEE Trans. Evolu. Compu.* **1**, 53–66 (1997).

Franceschini, G., D. Franceschini, and A. Massa, "Full-vectorial three-dimensional microwave imaging through the iterative multiscaling strategy—A preliminary assessment," *IEEE Geosci. Remote Sens. Lett.*, **2**, 428–432 (2005).

Garnero, L., A. Franchois, J. P. Hugonin, C. Pichot, and N. Joachimowicz, "Microwave imaging—complex permittivity reconstruction—by simulated annealing," *IEEE Trans. Microwave Theory Tech.* **39**, 1801–1807 (1991).

Geman, S. and D. Geman, "Stochastic relaxation, Gibbs distributions, and the Bayesian restoration of images," *IEEE Trans. Pattern Anal. Machine Intell.* **PAMI-6**, 721–741 (1984).

Goldberg, D. E. *Genetic Algorithms in Search, Optimization, and MachineLearning*, Addison-Wesley, Reading, MA, 1989.

Gragnani, G. L., "Two-dimensional imaging of dielectric scatterers based on Markov random field models: A short review," in *Microwave Nondestructive Evaluation and Imaging*, Trivandrum, Research Signpost, 2002, pp. 121–145.

Hajek, B. "Cooling schedules for optimal annealing," *Math. Oper. Res.* **13**, 311–329 (1988).

Haupt, R. L. "An introduction to genetic algorithms for electromagnetics," *IEEE Anten. Propag. Mag.* **37**, 7–15 (1995).

Haupt, R. L. and S. E. Haupt, *Practical Genetic Algorithms*, Wiley, Hoboken, NJ, 2004.

Holland, J. H. "Genetic algorithms," *Sci. Am.* **267**, 66–72 (1992).

Huang, T. and A. S. Mohan, "Microwave imaging of perfect electrically conducting cylinder by microgenetic algorithm," *Proc. 2004 IEEE Antennas Propagation Society Int. Symp.*, Monterey, CA, 2004, vol. 1, pp. 221–224.

Huang, T. and A. S. Mohan, "Application of particle swarm optimization for microwave imaging of lossy dielectric objects," *Proc. 2005 IEEE Antennas Propagation Society Int. Symp.*, Washington, DC, 2005, vol. 1B, pp. 852–855.

Huang, T. and A. S. Mohan, "A microparticle swarm optimizer for the reconstruction of microwave images," *IEEE Trans. Anten. Propag.* **55**, 568–576 (2007).

Johnson, D. S., C. R. Aragon, L. A. McGeoch, and C. Schevon, "Optimization by simulated annealing: An experimental evaluation. Part I. Graph partitioning," *Oper. Res.* **37**, 365–892 (1989).

Johnson, J. M. and Y. Ramat-Samii, "Genetic algorithms in engineering electromagnetics," *IEEE Anten. Propag. Mag.* **39**, 7–21 (1997).

Kennedy, J. and R. C. Eberhart, "Particle swarm optimization," *Proc. 1995 IEEE Int. Conf. Neural Networks*, 1995, pp. 1942–1948.

Kennedy, J. and R. C. Eberhart, "The particle swarm: Social adaptation in information processing systems," in D. Corne, M. Dorigo, and F. Glover, eds., *New Ideas in Optimization*, McGraw-Hill, London, 1999.

Kent, S. and T. Gunel, "Dielectric permittivity estimation of cylindrical objects using genetic algorithm," *J. Microwave Power Electromagn. Energy.* **32**, 109–113 (1997).

Kirkpatrick, S., C. D. Gelatt, and M. P. Vecchi, "Optimization by simulated annealing," *Science.* **1**, 671–680 (1983).

Krishnakumar, K. "Micro-genetic algorithms for stationary and non-stationary function optimization," *Proc. SPIE Intell. Control Adapt. Syst.* **1196**, 289–296 (1989).

Lin, Y. S., C.-C. Chiu, and Y. C. Chen, "Image reconstruction for a perfectly conducting cylinder buried in a three-layer structure by TE wave illumination," *Proc. 3rd Int. Conf. Computational Electromagnetics and Its Applications (ICCEA 2004)*, Turin, Italy, 2004, pp. 411–414.

Massa, A., M. Pastorino, and A. Randazzo, "Reconstruction of two-dimensional buried objects by a differential evolution method," *Inverse. Problems.* **20**, S135–S150 (2004).

Michalski, K. A. "Electromagnetic imaging of circular-cylindrical conductors and tunnels using a differential evolution algorithm," *Microwave Opt. Technol. Lett.* **27**, 330–334 (2000).

Michalski, K. A. "Electromagnetic imaging of elliptical-cylindrical conductors and tunnels using a differential evolution algorithm," *Microwave Opt. Technol. Lett.* **28**, 164–169 (2001).

Pastorino, M. "Stochastic optimization methods applied to microwave imaging: A review," *IEEE Trans. Anten. Propag.* **55**, 538–548 (2007).

Price, K., R. Storn, and J. Lampinen. *Differential Evolution—a Practical Approach to Global Optimization*, Springer, 2005.

Qing, A. "Electromagnetic inverse scattering of multiple two-dimensional perfectly conducting objects by the differential evolution strategy," *IEEE Trans. Anten. Propag.* **51**, 1251–1262 (2003).

Qing, A. "Dynamic differential evolution strategy and applications in electromagnetic inverse scattering problems," *IEEE Trans. Geosci. Remote Sens.* **44**, 116–125 (2006).

Qing, A. and C. K. Lee, "Microwave imaging of a perfectly conducting cylinder using a real-coded genetic algorithm," *IEE Proc. Microwaves Anten. Propag.* **146**, 421–425 (1999).

Qing, A., C. K. Lee, and L. Jen, "Electromagnetic inverse scattering of two-dimensional perfectly conducting objects by real-coded genetic algorithm," *IEEE Trans. Geosci. Remote Sens.* **39**, 665–676 (2001).

Qing, A. and S. Zhong, "Microwave imaging of two-dimensional perfectly conducting objects using real-coded genetic algorithm," *Proc. 1998 IEEE Antennas Propagat. Society Int. Symp.*, San Diego, CA, 1998, vol. 2, 726–729.

Rahmat-Samii, Y. and E. Michielssen, *Electromagnetic Optimization by Genetic Algorithms*, Wiley, New York, 1999.

Rekanos, I. T. "Shape reconstruction of a perfectly conducting scatterer using differential evolution and the particle swarm optimization," *IEEE Trans. Geosci. Remote Sens.* **46**, 1967–1974 (2008).

Robinson, J. R. and Y. Rahmat-Samii, "Particle swarm optimization in electromagnetics," *IEEE Trans. Anten. Propag.* **52**, 771–778 (2004).

Sekihara, K., H. Haneishi, and N. Ohyama, "Details of simulated annealing algorithm to estimate parameters of multiple current dipoles using biomagnetic data," *IEEE Trans. Med. Imag.* **11**, 293–299 (1992).

Socha, K. and M. Dorigo, "Ant colony optimization for continuous domains," *Eur. J. Oper. Res.* **185**, 1155–1173 (2008).

Storn, R. and K. Price, "Differential evolution—a simple and efficient heuristic for global optimization over continuous spaces," *J. Global Opt.* **11**, 341–359 (1997).

Takenaka, T., Z. Q. Meng, T. Tanaka, and W. C. Chew, "Local shape function combined with genetic algorithm applied to inverse scattering for strips," *Microwave Opt. Technol. Lett.* **16**, 337–341 (1998).

Weile, D. S. and E. Michielssen, "Genetic algorithm optimization applied to electromagnetics: A review," *IEEE Trans. Anten. Propag.* **45**, 343–353 (1997).

Xiao, F. and H. Yabe, "Microwave imaging of perfectly conducting cylinders from real data by micro genetic algorithm couple with deterministic method," *IEICE Trans. Electron.* **E81-C**, 1784–1792 (1998).

Yang, C.-M. and C.-C. Chiu, "Imaging reconstruction of a partially immersed perfect conductor by the genetic algorithm," *Proc. 2002 SBMO/IEEE MTT-S Int. Microwave and Millimeter Wave Technology Conf. (ICMMT 2002)*, 2002, pp. 895–898.

Zhou, Y., J. Li, and H. Ling, "Shape inversion of metallic cavities using hybrid genetic algorithm combined with tabu list," *Electron. Lett.* **39**, 280–281 (2003).

CHAPTER EIGHT

Hybrid Approaches

8.1 INTRODUCTION

As mentioned previously, the fields of industrial, civil, and military engineering continuously need new and efficient diagnostic methods. The same holds in the areas of geophysical applications, buried-object detection, cultural heritage, demining, and medical diagnostics. However, although several excellent results have been obtained by applying the imaging procedures developed in the previous chapters, the need for new application-oriented methods based on hybrid strategies is widely recognized. The idea is to combine different imaging modalities in order to derive profit from the specific features of the different methods and improve the effectiveness of the inspection.

Hybrid methods could include, for example, two-step procedures in which fast qualitative algorithms (Chapter 5) are combined with quantitative methods (Chapters 6 and 7) in order to first derive the supports of the unknown scatterers and successively retrieve the distributions of their dielectric parameters. Other approaches can combine different quantitative procedures (e.g., deterministic and stochastic algorithms can be combined in order to speed up the convergence of slow and computationally expensive stochastic methods near the global solution). Since efficient "general purpose" imaging techniques cannot be developed, these new combined strategies must be closely related to specific applications and should essentially lead to the implementation of model-driven approaches. This chapter discusses some examples of possible hybrid approaches. Although the definition of a hybrid method is again quite an arbitrary task, some examples are reported in this chapter, with reference to proposals appearing in the scientific literature. The first example concerns a case in which iterative methods (Chapter 6) for quantitative imaging of

---

Microwave Imaging, By Matteo Pastorino
Copyright © 2010 John Wiley & Sons, Inc.

strong scatterers require an initial estimate of the distribution of dielectric parameters. As mentioned previously, this initial distribution can be obtained by using a fast qualitative method, such as diffraction tomography (Section 5.11). Quantitative deterministic methods for strong scatterers adopt these strategies almost always if a priori information about the configuration is not available. For example, in the study by Abubakar et al. (2006) the initial solution is obtained by using a linear backpropagation. Another useful approach concerns the use of a qualitative method to determine the shape of the object under test and then, successively, the distribution of its relevant dielectric parameters. An efficient method for deriving the shape of an object is the linear sampling method (Section 5.8), which can be combined with both deterministic and stochastic methods.

The direct application of stochastic optimization algorithms to solve inverse problems in electromagnetics, although highly desirable (as discussed in Chapter 7, these methods are potentially capable of finding the global minimum of a given functional, relatively independently of the properties of the functional itself, e.g., convexity, differentiability), is problematic because of the high computational load required by these methods. In fact, a direct application is possible only for limited parameterizations (see the examples given in Chapter 7). However, the favorable properties of these methods can be kept if they are combined with other faster methods. Hybrid strategies can be devised by using local search methods during the optimization process.

Owing to the versatility of stochastic optimization methods, discussed above, there are evidently several various possibilities of integrating stochastic and deterministic approaches. The simplest way is to start the local search when the stochastic method has reached a reasonable *basin of attraction*, that is, a region of the space solution where the global minimum is likely included. In the first iterations of the stochastic method, the solution space can be explored (starting from arbitrary trial solutions, not necessarily located near the *right* solution) and, possibly, during the minimization process, the trial solutions can escape from local minima. Then, the deterministic procedure starts and the solution is quickly reached. The main difficulty is in devising a criterion for switching from the stochastic method to the determinist one. This approach has been followed in the study by Park and Jeong (1999), where a standard genetic algorithm is combined with a Levenberg–Marquardt algorithm. A similar method has been used in the study by Caorsi et al. (2000), where the genetic algorithm is cascaded with a standard conjugate-gradient method. Another critical point is represented by the modality to be adopted in order to restart the stochastic process if the final solution is not acceptable, that is, if the global minimum is not reached.

Another example of hybridization of a stochastic algorithm with a deterministic one can be found in the paper by Xiao and Yabe (1998). However, the integration of stochastic and deterministic methods can be made stronger. The so-called memetic algorithm represents a significant example.

8.2 THE MEMETIC ALGORITHM

The memetic algorithm is an iterative population-based method (Moscato 1989, Moscato and Norman 1992, Merz and Freisleben 2000). During the iterative evolution, only local minima of the cost function are considered. They are obtained by applying a local minimization method to any element of the population.

Let us consider the classical structure of the genetic algorithm (Section 7.3). After an initialization phase ($k=0$), in which a population $\wp^0$ is constructed (the same considerations made concerning the genetic algorithm still hold true here), before the application of the genetic operators, the chromosomes Ψ_p^k, $p = 1,\ldots, P^0$ are subjected to a local optimization, which provides a set of local optima $\Psi_p^k \to \tilde{\Psi}_p^k, p = 1,\ldots, P^0$. Since the same minimum can be reached more than once, the set of different local minima includes $Q^0 \leq P^0$ elements. This set is used for application of the genetic operators (with the same modalities discussed in Section 7.3 for the genetic algorithm). If a reduced parameterization of the scattering problem can be used, since the algorithm's population is composed of local optima only, the number of individuals involved in the evolution can be chosen very small, even equal to the number of unknowns.

In this way, the memetic algorithm combines, at each iteration, a local search and a stochastic minimization. The use of a local search is aimed at increasing the convergence velocity, whereas the genetic operators are applied to escape from local minima. Consequently, the method combines the advantages of both stochastic and deterministic procedures.

It should be mentioned that when it was introduced, the memetic algorithm was aimed at "emulating the process of exchange of ideas among people" by considering the concept of *meme*, which is a *unit of information* that can be transmitted when people exchange ideas (Merz and Freisleben 2000). Moreover, people process any idea to obtain a personal optimum before propagating it. Accordingly, in the memetic algorithm, each "idea" is an individual and the combined procedure that performs the local/global optimization is inspired by the abovementioned considerations.

Different local search methods can, of course, be considered. Caorsi et al. (2003a) used a standard conjugate gradient together with a real-coded implementation of the genetic operators. Some illustrative examples are presented below. In the first example, the same problem geometry considered by Michalski (2001) is applied. In particular, a borehole configuration is assumed and the target is the reconstruction of an infinite dielectric cylinder under transverse magnetic (TM) illumination conditions. However, Michalski's (2001) approach requires a numerical computation of forward scattering solution needed to calculate the cost function at any iteration. On the contrary, in the study by Caorsi et al. (2003a) the scatterer cross section has been approximated by a multilayer elliptic cylinders, so that the scattered values are computed by using the recursive procedure based on series expansions in Mathieu functions discussed in Section 3.5, according to the assumption of neglecting the effect of the interface (Michalski 2001).

An example is given as follows. The imaging configuration is the one reported in Figure 4.9. The object to be inspected is a void ($\varepsilon_r = 1$) in the soil of cylindrical shape with an elliptic cross section. The center of the cross section is placed at point $\mathbf{r}_t = -0.173\lambda\hat{\mathbf{x}} - 0.865\lambda\hat{\mathbf{y}}$, whereas the semimajor axis of the ellipse is $a = 0.26\lambda$ and the semifocal distance is $d = 0.245\lambda$. The background medium is lossless and characterized by $\varepsilon_b = 12\varepsilon_0$ and $\mu_b = \mu_0$. The source ($S = 1$) is a line current located at point $x_1 = -0.865\lambda$ (position of the transmitting hole) and $y^1 = -1.041\lambda$ (depth of the source), whereas the scattered field is collected at $M = 13$ measurement points such that $x_2 = 0.865\lambda$ (position of the receiving hole) and $y^m = -(m-1)\cdot 0.173\lambda$, $m = 1,\ldots, M$.

The cost function is constituted by the square norm of the residual of the discrete data equation, although amplitude-only data are considered as input data (see Section 11.3 for a brief discussion of this point). Moreover, the memetic algorithm has been applied by using the following parameters (Caorsi et al. 2003b): probability of crossover, $p_c = 0.9$; probability of mutation, $p_m = 0.3$; maximum number of iterations for the conjugate gradient (local search), 20; maximum number of iterations for the genetic operators (global search), 30; threshold for the stopping the local search, F (cost function) $\leq 10^{-5}$; threshold for stopping the application of the genetic operators, $F \leq 10^{-3}$. Moreover, some a priori information is introduced by setting some bounds on the unknowns to be retrieved (some of them are directly related to the geometry of the problem). In particular, the center of the elliptic cross section is searched in the range $-0.865\lambda \leq x \leq 0.865\lambda$ and $-2.1\lambda \leq y \leq 0$, whereas the relative dielectric permittivity of the homogeneous cylinder is assumed in the range $1 \leq \varepsilon_r \leq 120$.

Figure 8.1a shows the behavior of the cost function for three elements of the population. In the figure, CG_{kp} denotes the value of the functional for the pth element of the population at the kth iteration, where $k = 0$ is the initialization phase (the starting solutions are randomly chosen in the abovementioned ranges), whereas MA and the rhombus indicate the values for the best element of the population (i.e., the best value of the kth iteration of the memetic algorithm). Between two iterations of the memetic algorithm (or, as in the present case, between the initialization and the first iteration), the plots in Figure 8.1a concern the local search. As can be seen, just after the first iteration of the memetic algorithm, the convergence threshold is reached. It is worth noting that the local search is trapped in local minima, since the trial solutions in the initialization phase are very far from the actual solution. However, the algorithm, due to the application of the genetic operators, is able to escape from the local minima and explore a different region of the solution space, allowing the convergence of the process. It should be noted that, for the best individual, the starting value of the estimated dielectric permittivity was $\varepsilon_r \approx 40$, whereas in the final solution, the actual value of $\varepsilon_r = 1.0$ has been approximated with an error less than 5%.

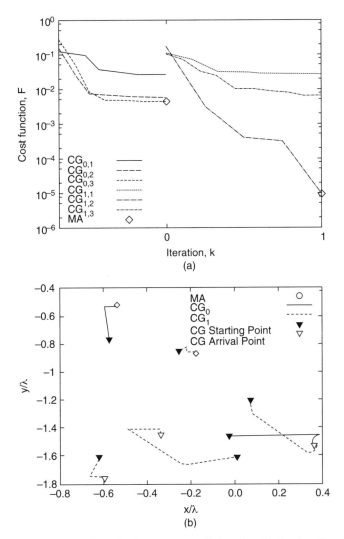

FIGURE 8.1 Reconstruction of a homogenous dielectric cylinder in a borehole configuration using the memetic algorithm: (a) the functional to be minimized for three elements of the population; (b) trajectories of the individuals of the population of the memetic algorithm in the [x,y] plane (coordinates of the center of the reconstructed cross section). [Reproduced from S. Caorsi, A. Massa, M. Pastorino, and A. Randazzo, "Electromagnetic detection of dielectric scatterers using phaseless synthetic and real data and the memetic algorithm," *IEEE Trans. Geosci. Remote Sens.* **41**(12), 2745–2753 (Dec. 2003), © 2003 IEEE.]

Very interesting is Figure 8.1b, which shows the trajectories of the coordinates of center of the elliptic cross section during application of the memetic algorithm. In the figure, the continuous and dashed lines refer to *movements* during initialization and the first iteration, respectively. Furthermore, black and white triangles denote the starting and arrival points during application of the local search, respectively. As in Figure 8.1a, the rhomb denotes the optimum element at each phase. As can be seen, at the beginning of the initialization phase, the trial solutions are quite distant from the true solution (point $r_t = -0.173\lambda\hat{x} - 0.865\lambda\hat{y}$). At the end of this phase, the best solution is represented by the rhomb in the upper left part of the figure. However, all the solutions are not near the true solution. After application of the genetic operators, new solutions are generated (white triangles). One of them is located near the true solution and, starting from this solution, at iteration $k = 1$, the local search is able to rapidly obtain the final optimum solution. On the contrary, the other elements of the population move toward other points in the search space. In particular, the *movement* of the solution in the lowest right part of the figure terminates in a final point coincident with a final point for the initialization phase; that is, the same local minimum is reached.

Another example concerns the reconstruction of a two-layer elliptic dielectric cylinder, which is bounded by two ellipses whose semimajor axes are $a_1 = 0.26\lambda$ and $a_2 = 0.295\lambda$. The relative dielectric permittivities of the two layers are $\varepsilon_{r_1} = 1.0$ (inner layer) and $\varepsilon_{r_2} = 5.0$ (outer layer). The scatterer corresponds essentially to a pipe or a hollowed buried cylinder. The center of the cross section, the semifocal distance, and all the other parameters of the imaging configuration and of the memetic algorithm are the same as in the previous example. Figure 8.2a shows the contours of the reconstructed cross sections at the various iterations corresponding to the best elements of the population. In this case, the final solution is reached after five iterations of the memetic algorithm. The behavior of the cost function during the minimization process is shown in Figure 8.2b, again referred to as the "best element of the population."

In this case, too, at iterations $k = 1$ and $k = 2$, the same contours are retrieved, which correspond to the same local minimum of the cost function. However, at the end of the local search of iteration $k = 5$ a very good solution is reached. The abilities of the memetic algorithm to escape from local minima are clearly shown in Figure 8.2b.

The memetic algorithm has also been used to inspect PEC cylinders. Figure 8.3 provides the results of the reconstruction of a circular cylinder performed by inverting real scattering data that have been measured by a prototype of the microwave imaging system based on the modulated scattering technique and working at 3 GHz (Caorsi et al. 2002). The apparatus is described in Section 9.4.

Finally, it should be mentioned that the memetic algorithm has been also considered a good candidate in other inverse problems in the field of applied electromagnetics. For example, Pisa et al. (2005) used it to compute

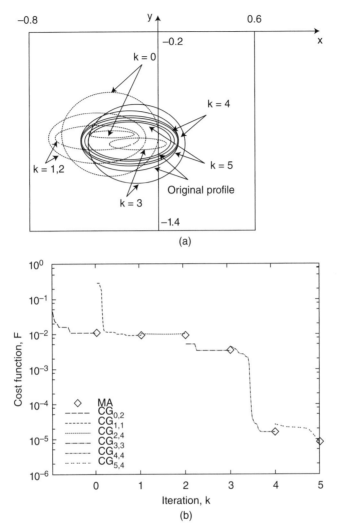

FIGURE 8.2 Reconstruction of a two-layer elliptic cylinder using the memetic algorithm: (a) cross section of the cylinder at various iterations (best elements of the population); (b) behavior of the functional to be minimized (best element of the population). [Reproduced from S. Caorsi, A. Massa, M. Pastorino, and A. Randazzo, "Electromagnetic detection of dielectric scatterers using phaseless synthetic and real data and the memetic algorithm," *IEEE Trans. Geosci. Remote Sens.* **41**(12), 2745–2753 (Dec. 2003), © 2003 IEEE.]

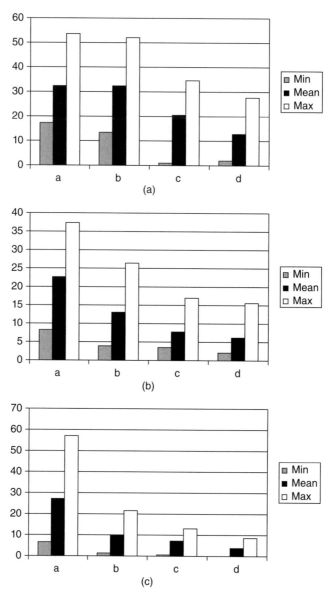

FIGURE 8.3 Errors on the reconstruction on the parameters of a PEC circular cylinder by using real data, memetic algorithm: (a) semimajor axis; (b) x coordinate of the center; (c) radius. In each figure, a, b, c, and d refer to the populations at the beginning of the initial phase, at the end of the initial phase, after the application of the genetic operators, and at the end of the first iteration, respectively. [Reproduced from S. Caorsi, A. Massa, M. Pastorino, and A. Randazzo, "Detection of PEC elliptic cylinders by a memetic algorithm using real data," *Microwave Opt. Technol. Lett.* **43**(4), 271–273 (Nov. 20, 2004), © 2004 Wiley.]

the specific absorption rate inside a human head when subjected to radiation from a cellular phone.

8.3 LINEAR SAMPLING METHOD AND ANT COLONY OPTIMIZATION

As mentioned in Section 8.1, several different hybrid techniques can be devised by combining different approaches. In this section, we consider, for illustration purposes, the combination of a qualitative method, namely, the linear sampling method (Section 5.8), and a quantitative stochastic method, ant colony optimization (Section 7.6). Integration of the two approaches in a single code is straightforward. In fact, the imaging process consists of two stages: (1) starting from the measured values of the scattered electric field in the observation domain, the shapes and positions of the unknown scatterers are determined very rapidly by application of the linear sampling method; and (2) when the support of the scatterer is available, the ant colony optimization method can be applied to retrieve the dielectric properties of the target. In this case, taking into account the a priori information concerning the shape of the target, a relatively small parameterization can be used for the stochastic approach.

This approach has been followed (Brignone et al. 2008a) for two-dimensional configurations and extended to three-dimensional configurations (Brignone et al. 2008b). The hybrid approach combines the robustness and generality of the linear sampling method, which is able to retrieve the contours of a wide class of scatterers (e.g., dielectric and conducting objects, constituting either weak or strong scatterers) with the capability of escaping from local minima of the ant colony optimization algorithm, which has been found to converge within a reasonably small number of iterations.

A reconstruction example is shown in Figure 8.4 (Brignone et al. 2008a). It concerns the inspection of two separated cylinders. The original configuration is shown in Figure 8.4a. The circular and rectangular cylinders are characterized by $\varepsilon_r = 4$ and $\varepsilon_r = 2$, respectively.

The cross sections of the cylinders are located in a square investigation domain of side 0.26 m and discretized into 31×31 square subdomains. The number of incidence and observation angles, uniformly spaced for sake of simplicity, is $S = M = 12$, and the operating frequency is $f = 1\,\text{GHz}$. Each measurement sample is corrupted by a zero mean additive Gaussian noise with variance equal to 7% of its value (Brignone et al. 2008a).

The following parameters have been used the ant colony optimization application (see Section 7.6): $N = 114$, $P = 120$, $q = 0.1$, $Q = 5$, and $\xi = 0.65$. Figure 8.4b gives the profiles detected by the linear sampling methods, whereas Figure 8.4c provides the dielectric permittivity distribution retrieved using the ant colony optimization method (at iteration $k = 300$). As can be seen, the optimization procedure successfully exploits the a priori information on the shape provided by the qualitative approach. It is worth noting that despite

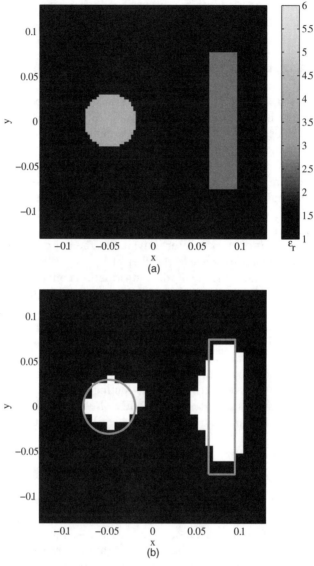

FIGURE 8.4 Imaging of two separated cylinders with circular and rectangular cross sections using the hybrid method of linear sampling and ant colony optimization: (a) actual scatterers; (b) supports provided by the linear sampling method; (c) reconstructed distribution of relative dielectric permittivity using the ant colony optimization algorithm. The gray lines in (b) and (c) represent the true profiles. (Reproduced from M. Brignone, G. Bozza, A. Randazzo, R. Aramini, M. Piana, and M. Pastorino, "Hybrid approach to the inverse scattering problem by using ant colony optimization and no-sampling linear sampling," *Proc. 2008 IEEE Antennas and Propagation Society Int. Symp.*, San Diego, CA, July 5–12, 2008, © 2008 IEEE.)

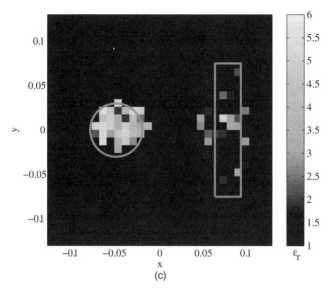

FIGURE 8.4 *Continued*

discretization of the investigation domain into 961 subdomains, the number of unknowns in the stochastic optimization procedure is reduced to only 114 unknowns, due to the useful information provided by the linear sampling method. Finally, some other results obtained using this hybrid technique are discussed in Section 8.3.

Clearly, following this philosophy, the stochastic method can be replaced in a straightforward way with a quantitative deterministic method. This approach has been followed, for example, in the study by Catapano et al. (2007), where the linear sampling method has been used in a two-step procedure together with a solving technique based on the extended Born approximation (Section 4.7) and a gradient procedure (Section 6.8).

REFERENCES

Abubakar, A., P. M. van den Berg, and T. M. Habashy, "An integral equation approach for 2.5-dimensional forward and inverse electromagnetic scattering," *Geophys. J. Int.* **165**, 744–762 (2006).

Brignone, M., G. Bozza, A. Randazzo, R. Aramini, M. Piana, and M. Pastorino, "Hybrid approach to the inverse scattering problem by using ant colony optimization and no-sampling linear sampling," *Proc. 2008 IEEE Antennas Propagation Society Int. Symp.*, San Diego, CA, July 5–12, (2008a).

Brignone, M., G. Bozza, A. Randazzo, R. Aramini, M. Piana, and M. Pastorino, "A hybrid approach to 3D microwave imaging by using linear sampling and ant colony optimization," *IEEE Trans. Anten. Propag.* **56**, 3224–3232 (2008b).

Caorsi, S., M. Donelli, and M. Pastorino, "A passive antenna system for data acquisition in scattering applications," *IEEE Anten. Wireless Propag. Lett.* **1**, 203–206 (2002).

Caorsi, S., A. Massa, and M. Pastorino, "A computational technique based on a real-coded genetic algorithm for microwave imaging purposes," *IEEE Trans. Geosci. Remote Sens.* **38**, 1697–1708 (2000).

Caorsi, S., M. Pastorino, M. Raffetto, and A. Randazzo, "Detection of buried inhomogeneous elliptic cylinders by a memetic algorithm," *IEEE Trans. Anten. Propag.* **51**, 2878–2884 (2003a).

Caorsi, S., A. Massa, M. Pastorino, and A. Randazzo, "Electromagnetic detection of dielectric scatterers using phaseless synthetic and real data and the memetic algorithm," *IEEE Trans. Geosci. Remote Sens.* **41**, 2745–2753 (2003b).

Catapano, I., L. Crocco, M. D'Urso, and T. Isernia, "On the effect of support estimation and of a new model in 2-D inverse scattering problems," *IEEE Trans. Anten. Propag.* **55**, 1895–1899 (2007).

Merz, P. and B. Freisleben, "Fitness landscape analysis and memetic algorithms for the quadratic assignment problem," *IEEE Trans. Evolut. Comput.* **4**, 337–352 (2000).

Michalski, K. A., "Electromagnetic imaging of elliptical-cylindrical conductors and tunnels using a differential evolution algorithm," *Microwave Opt. Technol. Lett.* **28**, 164–169 (2001).

Moscato, P., *On Evolution, Search, Optimization, Genetic Algorithms and Martial Arts toward Memetic Algorithms*, Calif. Institute of Technology, Pasadena, Caltech Concurrent Computation Program Technical Report 826, 1989.

Moscato, P. and M. G. Norman, "A 'memetic' approach for the traveling salesman problem. Implementation of a computational ecology for combinatorial optimization on message-passing systems," in *Parallel Computing and Transputer Applications*, M. Valero et al., eds., Institute of Physics, Amsterdam, 1992.

Park, C.-S. and B.-S. Jeong, "Reconstruction of a high contrast and large object by using the hybrid algorithm combining a Levenberg-Marquart algorithm and a genetic algorithm," *IEEE Trans. Magn.* **35**, 1582–1585 (1999).

Pisa, S., M. Cavagnaro, V. Lopresto, and E. Piuzzi, "A procedure to develop realistic numerical models of cellular phones for an accurate evaluation of SAR distribution in the human head," *IEEE Trans. Microwave Theory Tech.* **53**, 1256–1265 (2005).

Xiao, F. and H. Yabe, "Microwave imaging of perfectly conducting cylinders from real data by micro genetic algorithm couple with deterministic method," *IEICE Trans. Electron.* **E81-C**, 1784–1792 (1998).

CHAPTER NINE

Microwave Imaging Apparatuses and Systems

9.1 INTRODUCTION

As far as instrumentation for microwave imaging is concerned, it should be mentioned that illuminating–receiving systems are essentially based on two different approaches. The first one concerns the use of one or more probes (linear or circular arrays) operating in a real or synthetic mode. Usually, probes are constituted by dipoles or small horns. The second approach involves the use of passive probes that are sequentially modulated. Clearly, the imaging apparatuses must be suitable for use in the measurement configurations described in Chapter 4.

9.2 SCANNING SYSTEMS FOR MICROWAVE TOMOGRAPHY

Scanning systems for tomography are often composed of circular arrays of antennas. Typically, one antenna transmits and the other antennas (or a subset of thereof) are used to receive the signal scattered by the body to be inspected. As mentioned previously, the geometric positions of the transmitting and receiving antennas with respect to the body to be inspected in the various configurations have been described in Chapter 4. To illustrate this approach, we consider the configuration of one of the first proposed imaging systems working at microwaves, namely, the scanning system developed for medical purposes at the Polytechnic University of Catalunya, Barcelona, Spain (Jofre et al. 1990). It consists of a ring of antennas positioned around a tank in which

---

Microwave Imaging, By Matteo Pastorino
Copyright © 2010 John Wiley & Sons, Inc.

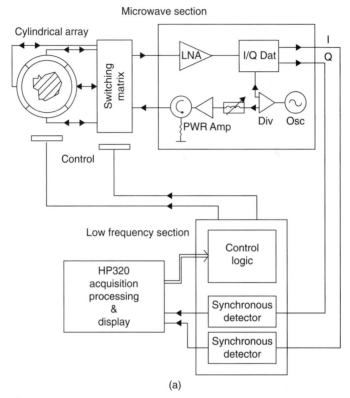

FIGURE 9.1 Prototype of a cylindrical microwave tomographic scanner: (a) scheme and (b) picture of the apparatus. [Reproduced from L. Jofre, M. S. Hawley, A. Broquetas, E. Reyes, M. Ferrando, and A. R. Elias-Fusté, "Medical imaging with a microwave tomographic scanner," *IEEE Trans. Biomed. Eng.* **37**(3), 303–311 (March 1990), © 1990 IEEE.]

the object to be inspected is placed (see Fig. 9.1). In tomographic applications, the transmitting antennas should have a directive beamwidth in the vertical plane and a broad beamwidth in the horizontal plane. The former requirement is due to the need for approximating a cylindrical configuration (see Section 3.3), in which the bodies to be inspected exhibit uniform dielectric properties along the cylinder axis, whereas the latter requirement is related to the need for a uniform illumination of the whole scatterer cross section. In microwave tomography, a uniform illumination is important in order to obtain external measurements containing as much *mutual scattering* information as possible. To accomplish these requirements, in the prototype considered (Jofre et al. 1990), there are 64 water-loaded waveguide antennas, flared only in the vertical plane to form a sectorial horn with a square aperture of side 2.5 cm.

(b)

FIGURE 9.1 *Continued*

The receiving channel is selected by a modulated multiplexing technique, and a double-modulation scheme has been adopted to reduce interferences and crosstalk among the different channels of the illuminating–measurement system.

Similar imaging prototypes and apparatuses have been developed by various research groups for different applications. Despite the similarities, they can exhibit different technical solutions related to any aspect of the design. A few examples are mentioned and related literature sources are cited in the following paragraphs.

The first example involves the imaging system developed for medical imaging by Semenov et al. (2000), which is constituted by antenna elements located inside a chamber of diameter 60 cm and height 40 cm, which is filled with deionized water ($\varepsilon_r = 78 - j0.2$ at 2.36 GHz) or salt solution ($\varepsilon_r = 79 - j21.5$ at 2.36 GHz). The distance between the transmitting antenna and the center of the chamber is 17.9 cm. A receiving antenna is moved around the target at a distance of 11 cm.

Simpler microwave systems can be constructed by using two moving antennas that rotate around the body to be inspected in order to retrieve bistatic scattering data. In this case, the efficiency of the PC-controlled mechanical systems is quite important for fast data acquisition. An example is the prototype of imaging tomograph developed at the Department of Technology and Innovation, University of Applied Sciences of Southern Switzerland (SUPSI), Manno, Switzerland, which is aimed in particular at imaging wooden objects, although other materials with similar dielectric properties could be inspected as well. The choice of operating frequency and power level was optimized for the specific application and considering a standard power levels available from network vector analyzers ($\leq 10\,dBm$). For the first laboratory tests, off-the-shelf horn antennas have been used with typical gains of 8–10 dB, resulting in scattering signals with sufficiently high power levels to be easily detected by a standard network vector analyzers ($-30\,dBm$ or better). Some information about the external surface of the object under test is obtained by a time-domain analysis of the reflected signal S_{11} (single-antenna operation). In another version of the prototype, the laboratory equipment (the network vector analyzer) has been replaced by a low-power short-range microwave radar that generates pulsed or frequency-swept microwave signals (with a center frequency of 5.5 GHz and a signal level of $-30\,dBm$), which are radiated and received by single horn antennas. The result is a graphical representation (obtained by graphical software mapping the time of flight of the reflected radar signal) of the contours of the samples (e.g., wood trunks). The experimental setup is shown in Figure 9.2. Other examples of similar approaches are the prototype shown in Figure 9.3 (Bindu and Mathew 2007) and the scheme for the indirect holography of Figure 9.4 (Elsdon et al. 2006).

Another comparable setup is the one used at the Second University of Naples, Italy, which works in a reflection mode in the X band (Soldovieri et al., 2005). The transmitting and receiving antennas are two pyramidal horns located on two separate linear rails (140 cm long), which can be moved perpendicularly to the cylinder axis when a cylindrical scatterer is to be inspected. The measurement data are collected by using a vector network analyzer.

An important task in imaging systems is the calibration, especially for array-based systems. In fact, the measurements performed by such systems usually needs suitable calibration procedures to compensate for amplitude or phase dispersions in the responses of the various antennas in the array. Different modalities are used to perform the calibration, and the reader is referred to the wide literature concerning microwave measurements and antennas.

For microwave imaging applications, however, the calibration procedure can involve the use of a known *object* (typically, a canonical scatterer) that produces a prescribed scattered field pattern. Since the field that this object should produce at any measurement position is known, it is possible to calculate the weighting coefficients to be applied to each receiving element by comparing the expected field values with the actual ones produced by the calibrating scatterer. As an example of this approach, a metallic circular

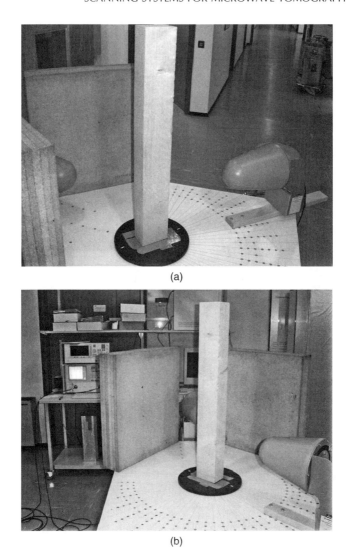

FIGURE 9.2 Experimental setup for inspecting wood objects (including a hollowed wood slab). [Courtesy of A. Salvadé, R. Monleone, and co-workers, Department of Technology and Innovation, University of Applied Sciences of Southern Switzerland (SUPSI), Manno, Switzerland.]

cylinder (diameter 4 cm) positioned at the center of the investigation area was used to calibrate the previously described 64-element array (Joachimowicz et al. 1998).

It should be mentioned that phase retrieval techniques, which have been widely used for diagnostic purposes of antennas, can be also used for imaging applications (Hislop et al. 2007), mainly at the highest frequencies [e.g., in the

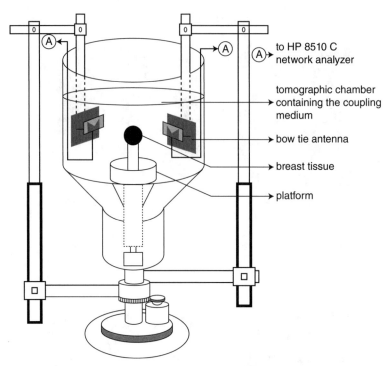

FIGURE 9.3 Experimental setup for two-dimensional microwave tomographic imaging. [Reproduced from G. Bindu and K. T. Mathew, "Characterization of benign and malignant breast tissues using 2-D microwave tomographic imaging," *Microwave Opt. Technol. Lett.* **49**, 2341–2345 (2007), © 2007 Wiley.]

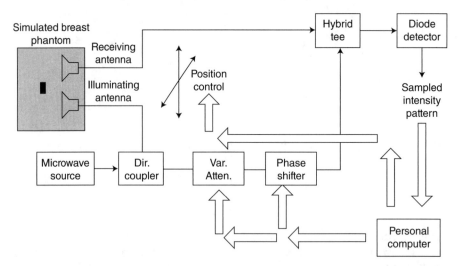

FIGURE 9.4 Experimental arrangement for indirect microwave holographic imaging. [Reproduced from M. Elsdon, D. Smith, M. Leach, and S. J. Foti,"Experimental investigation of breast tumor imaging using indirect microwave holography," *Microwave Opt. Technol. Lett.* **48**(3), 480–482 (2006), © 2006 Wiley.]

recently explored terahertz (THz) band]. By using this technique, one can derive the phase of the field in the observation domain from measurement of the field amplitude. In this way, direct measurement of the phase of the field at the measurement points is not required. The problem is that the phase retrieval process involves the solution of another ill-posed inverse problem. The description of such an inverse problem is outside the scope of the present book, and the reader is referred, for example, to the papers by Isernia and Pierri (1995) and Las-Heras and Sarkar (2002) and references cited therein.

Finally, it must be observed again that many prototypes and measurement systems are currently used for various microwave imaging applications. Detailed descriptions of most of them can be found in the books and papers mentioned in the sections devoted to applications (Chapter 10).

9.3 ANTENNAS FOR MICROWAVE IMAGING

Concerning the transmitting and receiving antennas for imaging systems, the choice depends again on the configuration. Several tomographic systems use dipoles, which ensure a wide beamwith in the azimuthal plane and an acceptably narrow beamwidth in the elevation.

Of particular interest is the case in which wideband systems are used. For example, in the medical field (Section 10.2), the antennas must be small and must operate in an ultrawide range of frequencies in order to achieve good spatial resolutions and suitable penetrations of the field into the biological tissues. Other applications concern the field of nondestructive testing and evaluation. In such cases, one of the most frequently used antennas is the Vivaldi antenna (Fig. 9.5) (Chiappe and Gragnani 2006). A modified version of this antenna is the balanced antipodal Vivaldi antenna (Fig. 9.6) (Bourqui et al. 2007), which has been used even for detecting subsurface bodies. In this area, Vivaldi antennas working in the band 1.3–2.0 GHz have been designed, constructed, and tested (Pichot et al. 2004).

Another widely used antenna for imaging system is the bowtie antenna (Balanis 2005). Resistively loaded bowtie antennas have been proposed (Fear et al. 2003, Kubota et al. 2008), whereas a 2×2 array of bowtie antennas has been designed (Hernández-López et al. 2003) in order to minimize the mutual coupling between the antennas up to a negligible level. For example, Figure 9.7a shows the 2×2 array, in which 2-cm bowtie antennas are printed on a 4-mm-thick lossy substrate ($\varepsilon_r \approx 1$ and $\sigma = 10$ S/m), which minimizes the coupling between the array elements and enhances the broadband characteristics of the antenna. The elements of the array are excited (through coaxial transmission lines) by a signal in the 1.6–11.4 GHz band (the signal is expressed by a time derivative of a Gaussian pulse). Figure 9.7b provides the parameters of the scattering matrix obtained in the two cases in which the antennas are located in free space and when they are printed on the dissipative substrate (Hernández-López et al. 2003). As expected, the presence of the substrate

FIGURE 9.5 Vivaldi antenna. (Courtesy of G. L. Gragnani, University of Genoa, Italy.)

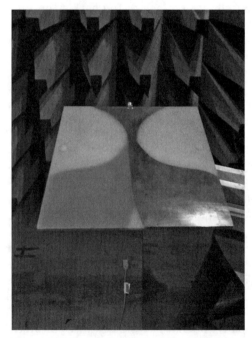

FIGURE 9.6 Antipodal Vivaldi antenna. (Courtesy of G. L. Gragnani, University of Genoa, Italy.)

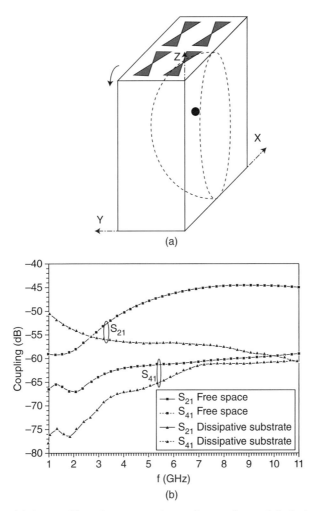

FIGURE 9.7 (a) Array of bowtie antennas located around a model of a human breast: (b) coupling coefficients for the planar array configuration; (c) reflected voltage at the center of the bowtie antenna. [Reproduced from M. A. Hernández-López, M. Quintillán-González, S. González García, A. Rubio Bretones, and R. Gómez Martín, "A rotating array of antennas for confocal microwave breast imaging," *Microwave Opt. Technol. Lett.* **39**, 307–311 (Nov. 20, 2003), © 2003 Wiley.]

significantly reduces the mutual coupling among the antenna elements. Furthermore, the lossy substrate also increases the bandwidth. This can be observed in Figure 9.7c, in which the reflected voltage at the feeding point of one of the array elements is shown for the two cases (free space and dissipative substrate).

Other antennas considered are straight thin-wire dipole antennas (Fear and Stuchly 2000), planar monopoles (Jafari et al. 2007), transverse

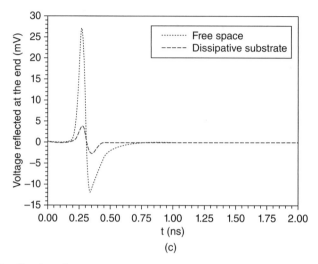

FIGURE 9.7 *Continued*

electromagnetic (TEM) horn antennas, which are traveling-wave structures consisting of two conducting plates (Amineh et al. 2009), "dark eyes" antennas (Kanj and Popovic 2005), and V-antennas (Fernandez Pantoja et al. 2002).

Concerning the important application of microwave imaging in the detection of damage in concrete structures, a significant example is represented by the use of a planar array of microstrip slot antennas shown in Figure 9.8 (Kim et al. 2003). These microstrip slot antennas were selected because they could be directly attached to the concrete surface or the matching cushion.

Patch antennas are proposed even in the most recent imaging systems. One of them is an imaging prototype constituted by 36 patch antennas fabricated on low-loss dielectric material, which has been designed into a "bra-shaped semi-conformal chamber that can be directly attached to the patient's breast" (Stang et al. 2009).

Finally, further information concerning antennas for microwave imaging applications can be obtained in the contributed chapter by C. Furse (Furse 2007).

9.4 THE MODULATED SCATTERING TECHNIQUE AND MICROWAVE CAMERAS

There has been significant interest in applying the *modulated scattering technique* (Richmond 1955), which is as an efficient method for measuring electromagnetic fields not only for imaging applications but also in several other areas, including antenna testing and electromagnetic compatibility. The reader is referred to the text by Bolomey and Gardiol (2001) for a comprehensive

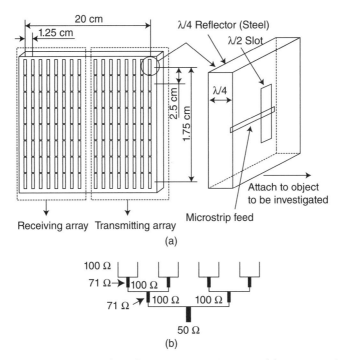

FIGURE 9.8 Planar rectangular microwave antenna array: (a) conceptual design of planar rectangular slot antenna array; (b) eight-element linear array with corporate feed configuration. [Reproduced from Y. J. Kim, L. Jofre, F. De Flaviis, and M. Q. Feng, "Microwave reflection tomographic array for damage detection of civil structures," *IEEE Tran. Anten. Propag.* **51**(11), 3022–3032 (Nov. 2003), © 2003 IEEE.]

description of this technique with in-deepth discussions of any technical issues and applications. The basic principle of the modulated scattering technique is briefly summarized as follows. The basic idea is that a passive probe (or an array of passive probes), when irradiated by an incident electromagnetic field, scatters (under specific loading conditions) a field that is directly related to the unperturbed field at the point where the probe is located.

Collecting this scattered field may result in an efficient measurement technique with a reduced perturbation of the original field distribution to be measured. Moreover, by using a scanning modulation, when an array of probes is used, one can collect a large number of field values with only a microwave receiver with no metallic connection with the probes or any moving mechanical apparatus.

There are two different configurations for implementing the modulated scattering technique: (1) the bistatic configuration, in which the receiving antenna is separated from the transmitting antenna, and (2) the monostatic configuration, in which the two antennas coincide. The passive probe is usually a dipole, but slot antennas have also been used more recently.

Let us first consider the bistatic arrangement for which the key relationship can be justified in terms of the scattering properties of passive antennas (Azaro et al. 1998a,b; Pastorino 2004). To this end, let us consider an incident field $\mathbf{E}_{\text{inc}}$ produced by a given transmitting antenna. A receiving antenna (located at point $\mathbf{r}_{\text{RX}}$) collects the field obtained by the interaction of $\mathbf{E}_{\text{inc}}$ with the scatterer and the probe antenna (located at point $\mathbf{r}_P$), which is assumed here to be a small dipole. The field collected by the receiving antenna can be expressed as

$$\mathbf{E}_1(\mathbf{r}_{\text{RX}}) = \mathbf{E}(\mathbf{r}_{\text{RX}}) + \mathbf{E}^P_{\text{scat}}(\mathbf{r}_{\text{RX}}), \tag{9.4.1}$$

where $\mathbf{E}$ is the total field resulting from the interaction of the incident field with the target and is as given by equation (2.7.2). Essentially, $\mathbf{E}$ represents the *incident* field on the small probe, whereas $\mathbf{E}^P_{\text{scat}}$ is the field scattered by the probe. Since $\mathbf{E}$ is solenoidal, it can be expanded as follows (Stratton 1941)

$$\mathbf{E}(\mathbf{r}) = \sum_p \sum_q \alpha^\gamma_{pq} \mathbf{M}^\gamma_{pq}(\mathbf{r}) + \sum_p \sum_q \beta^\gamma_{pq} \mathbf{N}^\gamma_{pq}(\mathbf{r}), \tag{9.4.2}$$

where $\mathbf{M}^\gamma_{pq}$ and $\mathbf{N}^\gamma_{pq}$ denote spherical vector wavefunctions. The superscript γ indicates that the summation is performed over all the even (when $\gamma = e$) and odd (when $\gamma = o$) modes. Since we assume here that the probe is a small electric dipole polarized along the x axis, it strongly interacts only with the $\mathbf{N}^e_{11}(\mathbf{r})$ mode, which is given by

$$\mathbf{N}^e_{11}(\mathbf{r}) = \frac{2}{k_b r} z_1(k_b r) P^1_1(\cos\theta) \cos\varphi \hat{\mathbf{r}} + \frac{1}{k_b r} \frac{\partial}{\partial r}[r z_1(k_b r)] \frac{\partial P^1_1(\cos\theta)}{\partial r} \cos\varphi \hat{\boldsymbol{\theta}}$$
$$- \frac{1}{k_b r \sin\theta} \frac{\partial}{\partial r}[r z_1(k_b r)] P^1_1(\cos\theta) \sin\varphi \hat{\boldsymbol{\varphi}}, \tag{9.4.3}$$

where $z_1(k_b r)$ is the spherical Bessel function of the first kind and $P^1_1 = \sin\theta$ is the associated Legendre function. If the load Z_L is placed at a point where the impedance looking toward the probe is purely resistive and of value R_P, the power scattered by the probes can be expressed as (Collin 1969, Hansen 1989) follows:

$$P_P = \frac{4\lambda_b^2}{3\pi\eta} |\beta^e_{11}|^2 \left| \frac{Z_L}{R_P + Z_L} \right|^2. \tag{9.4.4}$$

The key idea of the modulated scattering technique is to change, using a modulating signal, the value of the load impedance Z_L. According to equation (9.4.4), the scattered power also changes. In particular, when matched, the dipole scatters as much power as absorbed. In this case, the scattered electric field at the receiving antenna can be expressed as

$$\mathbf{E}_{scat}^{P}(\mathbf{r}_{RX}) = -\frac{2Z_L}{R_P + Z_L}\sqrt{\frac{2\lambda_b^2}{3\pi\eta}}\beta_{11}^{e}\mathbf{N}_{11}^{e}(\mathbf{r}_{RX}). \tag{9.4.5}$$

Finally, the open-circuit output voltage at the receiving-antenna terminals can be expressed as (Balanis 2005)

$$v_{oc} = \mathbf{h}_{RX} \cdot \mathbf{E}_1(\mathbf{r}_{RX}), \tag{9.4.6}$$

where $\mathbf{h}_{RX}$ the complex effective length of the receiving antenna. According to equation (9.4.1), equation (9.4.6) can be immediately rewritten as

$$v_{oc} = \mathbf{h}_{RX} \cdot \mathbf{E}(\mathbf{r}_{RX}) + \mathbf{h}_{RX} \cdot \mathbf{E}_{scat}^{P}(\mathbf{r}_{RX}), \tag{9.4.7}$$

where the term $\mathbf{h}_{RX} \cdot \mathbf{E}_{tot} \cdot (\mathbf{r}_{RX})$ is an unmodulated component of the output voltage, which can be simply removed, whereas the other term, $\mathbf{h}_{RX} \cdot \mathbf{E}_{scat}^{P}(\mathbf{r}_{RX})$, is just proportional, according to equations (9.4.5) and (9.4.2), to the x component of the $\mathbf{E}$ at the probe location $\mathbf{r}_P$.

A different approach is followed for the monostatic configuration, which is related essentially to the imaging modalities defined in Chapter 4. The key relationship of the monostatic configuration is discussed below (Azaro et al. 2003), considering the fact that there are now only one active antenna and the passive probe.

Let us consider Figure 9.9a, in which "P" denotes the probe, loaded by a complex impedance, and "TX" indicates the transmitting antenna (transmission mode). In applying the monostatic approach, the following relation is used (Azaro et al. 2003):

$$\int_S (\mathbf{E}_a \times \mathbf{H}_b - \mathbf{E}_b \times \mathbf{H}_a) \cdot \hat{\mathbf{n}} \, d\mathbf{r} = 0, \tag{9.4.8}$$

This is an expression of the Lorentz reciprocity theorem (Monteath 1973). In equation (9.4.8), S is a surface consisting of S_1 and S_2 (reference surfaces in the transversal planes of the waveguides), S_0 (PEC surface lying outside the antennas and extended on the inner surfaces of the antennas and waveguide up to S_1 and S_2), and S_∞ (the surface at infinity). Moreover, the two fields, denoted by subscripts a and b, represents two different field distributions obtained by imposing two different values to the load Z_L. In particular, a is the field in the case in which $Z_L = \infty$ (open circuit), and b refers to the matching load (Fig. 9.9b).

For a minimum-scattering antenna, the functional state a corresponds to an almost invisible antenna, whereas in the matched-load case, the probe scatters an amount of power equal to that transferred to the load (Collin 1969). According to the notation in Figure 9.9, the following relations hold for incident and reflected currents and voltages in the two states a and b:

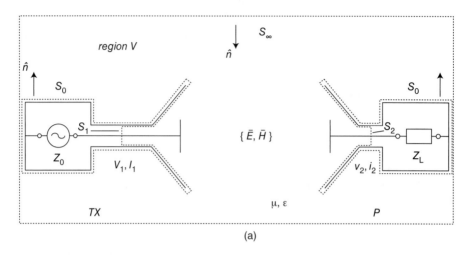

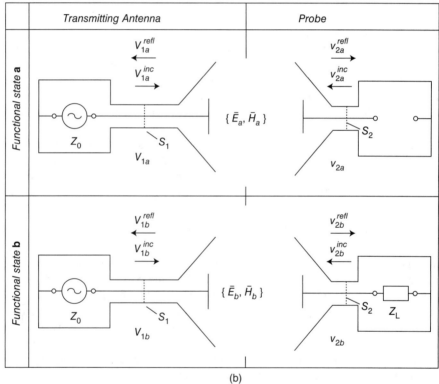

FIGURE 9.9 (a) Schematic representation of the monostatic modulated scattering technique; (b) currents and voltages of the two different functional states. [Reproduced from R. Azaro, S. Caorsi, and M. Pastorino, "On the relationship for the monostatic modulated scattering technique," *Microwave Opt. Technol. Lett.* **38**, 187–190 (2003), © 2003 Wiley.]

THE MODULATED SCATTERING TECHNIQUE AND MICROWAVE CAMERAS   219

$$V_{1ab} = V_{1ab}^{\text{inc}} + V_{1ab}^{\text{ref}}, \qquad (9.4.9)$$

$$I_{1ab} = I_{1ab}^{\text{inc}} + I_{1ab}^{\text{ref}}, \qquad (9.4.10)$$

$$v_{2ab} = v_{2ab}^{\text{inc}} + v_{2ab}^{\text{ref}}, \qquad (9.4.11)$$

$$i_{2ab} = i_{2ab}^{\text{inc}} + i_{2ab}^{\text{ref}}. \qquad (9.4.12)$$

Moreover, assuming the electromagnetic field inside the waveguides to be represented by a single propagation mode, the transversal components of the fields on S_1 and S_2 can be expressed as (Montgomery et al. 1948)

$$\mathbf{E}_t(\mathbf{r}) = V(z)\mathbf{e}_t(\mathbf{r}_t), \qquad (9.4.13)$$

$$\mathbf{H}_t(\mathbf{r}) = I(z)\mathbf{h}_t(\mathbf{r}_t), \qquad (9.4.14)$$

where $\mathbf{e}_t(\mathbf{r}_t)$ and $\mathbf{h}_t(\mathbf{r}_t)$ are real-valued transversal functions such that

$$\int_{S_1, S_1} (\mathbf{e}_t \times \mathbf{h}_t) \cdot \hat{\mathbf{n}} \, d\mathbf{r} = 1. \qquad (9.4.15)$$

Since S_0 is a PEC surface and the fields are vanishing on S_∞, it follows that

$$\int_{S_1} (\mathbf{E}_{ta} \times \mathbf{H}_{tb} - \mathbf{E}_{tb} \times \mathbf{H}_{ta}) \cdot \hat{\mathbf{n}} \, d\mathbf{r} + \int_{S_2} (\mathbf{E}_{ta} \times \mathbf{H}_{tb} - \mathbf{E}_{tb} \times \mathbf{H}_{ta}) \cdot \hat{\mathbf{n}} \, d\mathbf{r} = 0, \qquad (9.4.16)$$

and, by using (9.4.13) and (9.4.14), we obtain

$$\int_{S_1} (V_{1a}\mathbf{e}_{t1} \times I_{1b}\mathbf{h}_{t1} - V_{1b}\mathbf{e}_{t1} \times I_{1a}\mathbf{h}_{t1}) \cdot \hat{\mathbf{n}} \, d\mathbf{r} + \\ \int_{S_2} (v_{2a}\mathbf{e}_{t2} \times i_{2b}\mathbf{h}_{t2} - v_{2b}\mathbf{e}_{t2} \times i_{2a}\mathbf{h}_{t2}) \cdot \hat{\mathbf{n}} \, d\mathbf{r} = 0, \qquad (9.4.17)$$

which can be rewritten as

$$(V_{1a}I_{1b} - V_{1b}I_{1a})\int_{S_1} (\mathbf{e}_{t1} \times \mathbf{h}_{t1}) \cdot \hat{\mathbf{n}} \, d\mathbf{r} + (v_{\hat{n}a}i_{\hat{n}b} - v_{\hat{n}b}i_{\hat{n}a})\int_{S_2} (\mathbf{e}_{t2} \times \mathbf{h}_{t2}) \cdot \hat{\mathbf{n}} \, d\mathbf{r} = 0. \quad (9.4.18)$$

Taking into account (9.4.15), we have

$$V_{1a}I_{1b} - V_{1b}I_{1a} + v_{\hat{n}a}i_{\hat{n}b} - v_{\hat{n}b}i_{\hat{n}a} = 0 \qquad (9.4.19)$$

Let us consider now the functional state a, which is characterized by

$$V_{1a}^{\text{ref}} = 0, \qquad (9.4.20)$$

$$I_{1a}^{\text{ref}} = 0. \qquad (9.4.21)$$

$$v_{2a}^{\text{ref}} = v_{2a}^{\text{inc}}, \qquad (9.4.22)$$

$$i_{2a}^{\text{ref}} = -i_{2a}^{\text{inc}}, \qquad (9.4.23)$$

$$v_{2a} = 2v_{2a}^{\text{inc}}, \tag{9.4.24}$$
$$i_{2a} = 0. \tag{9.4.25}$$

It also follows that

$$V_{1a} = V_{1a}^{\text{inc}}, \tag{9.4.26}$$
$$I_{1a} = \frac{V_{1a}^{\text{inc}}}{Z_0}, \tag{9.4.27}$$

where Z_0 is the characteristic impedance of the feeding waveguide. Analogously, in the functional state b, we have

$$V_{1b}^{\text{inc}} = V_{1a}^{\text{inc}} = V_{1a}, \tag{9.4.28}$$
$$v_{2b} = v_{2b}^{\text{ref}} = v_{2a}^{\text{ref}}, \tag{9.4.29}$$
$$i_{2b} = \frac{v_{2b}^{\text{ref}}}{Z_L} = \frac{v_{2a}^{\text{ref}}}{Z_L}, \tag{9.4.30}$$

and we obtain

$$V_{1b} = V_{1b}^{\text{inc}} + V_{1b}^{\text{ref}} = V_{1a}^{\text{inc}} + V_{1b}^{\text{ref}}, \tag{9.4.31}$$
$$I_{1b} = \frac{V_{1b}^{\text{inc}}}{Z_0} - \frac{V_{1b}^{\text{ref}}}{Z_0} = \frac{V_{1a}^{\text{inc}}}{Z_0} - \frac{V_{1b}^{\text{ref}}}{Z_0}. \tag{9.4.32}$$

From (9.4.19)–(9.4.32) it follows that

$$V_{1b}^{\text{inc}} \left(\frac{V_{1b}^{\text{inc}}}{Z_0} - \frac{V_{1b}^{\text{ref}}}{Z_0} \right) - \left(V_{1b}^{\text{inc}} + V_{1b}^{\text{ref}} \right) \frac{V_{1b}^{\text{inc}}}{Z_0} + 2v_{2a}^{\text{ref}} \left(\frac{v_{2b}^{\text{ref}}}{Z_L} \right) - v_{2a}^{\text{ref}} \cdot 0 = 0, \tag{9.4.33}$$

and

$$V_{1b}^{\text{inc}} V_{1b}^{\text{ref}} = \frac{Z_0}{Z_L} \left(v_{2a}^{\text{ref}} \right)^2 = 0. \tag{9.4.34}$$

If we consider the antenna factor of the small probe, we can express the field at input port of the receiving antenna in functional state b as

$$v_{2b}^{\text{ref}} = \mathbf{h}_{\text{RX}} \cdot \mathbf{E}_1(\mathbf{r}_P). \tag{9.4.35}$$

Finally, we obtain

$$V_{1b}^{\text{ref}} = \frac{Z_0}{Z_L} \frac{1}{V_{1b}^{\text{inc}}} (\mathbf{h}_{RX} \cdot \mathbf{E}_1(\mathbf{r}_P))^2 = 0, \qquad (9.4.36)$$

which is the key relation for the modulated scattering technique in the monostatic mode. By using the modulated scattering technique, microwave cameras can be developed. One of the first proposed and most effective apparatuses of this kind was been developed by Franchois et al. (1983, 1998).

However, a simple scheme of a measurement system based on the modulated scattering technique can be found in Figure 9.10a. The measurement system is composed of three different sections: the first one operates at microwave frequencies and is used to generate the incident electromagnetic field and demodulate the signal received at the receiving antenna (RX); the second and the third ones operate at low frequency. In particular, the second one is used to generate the modulation signal and send it to the field probes. The third one is used to control the system and process the obtained data. The electromagnetic field is generated by means of a sinewave generator connected to an amplifier. The signal is divided by means of a directional coupler.

One part is sent to the transmitting antenna (TX), and the other is used for the reference channel. The transmitter is a $\lambda/4$-dipole antenna (where $\lambda/4$ is the wavelength) with a plane rectangular reflector placed at a distance of $\lambda/4$ from the radiator. Since the distance between the transmitting antenna and the investigation area can be changed, near-field and far-field radiation conditions can be achieved. In the last case, the signal in the investigation area can be assumed as an acceptable approximation of a transverse magnetic planar wave impinging on the scatterer. The passive array is constituted by $\lambda/4$-dipole probes (Fig. 9.10b). Each probe is loaded with a phototransistor, which acts as a nonlinear load. The electronic driving is obtained by sending to the probes an optical signal by means of an optical-fiber connection. The modulation signals are square waves with periods equal to 1 ms. A crystal oscillator generates the modulation signals, which are converted into optical signals by using emitting photodiodes. Since the experimental system is based on the modulated scattering technique in the bistatic configuration, the signals used for the measurements (which are proportional to the field values at probe positions) are finally collected by the receiving antenna, which is an aperture antenna (with a beamwidth of ~120° in the azimuth plane). Because of the low level of the scattering signal, a low-noise microwave amplifier is connected to the receiving-antenna output. The probe array, the receiving antenna, and the optical connections are assembled together by means of a dielectric frame in order to limit the field perturbations (Fig. 9.10b). Figure 9.11 gives an example of measured data in the case of the PEC circular cylinder reconstructed using the memetic algorithm and described in Section 8.2.

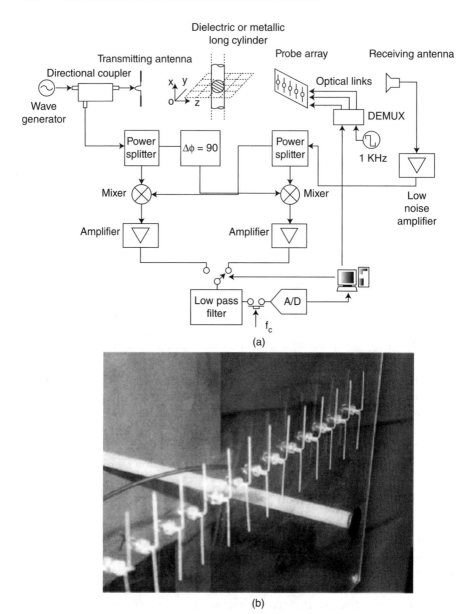

FIGURE 9.10 (a) Experimental setup of an imaging system based on the modulated scattering technique; (b) array of passive probes (dipoles). [Reproduced from S. Caorsi, M. Donelli, and M. Pastorino, "A passive antenna system for data acquisition in scattering applications," *IEEE Anten. Wireless Propag. Lett.* **1**, 203–206 (2002), © 2002 IEEE.]

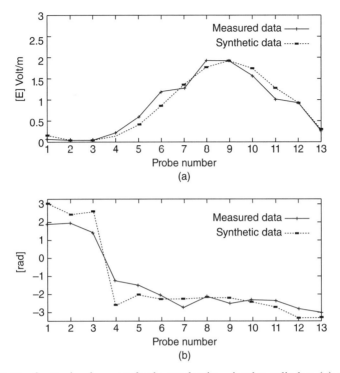

FIGURE 9.11 Scattering by a perfectly conducting circular cylinder: (a) amplitude and (b) phase values of the scattered electric field measured by the 13 probes of the imaging system in Figure 9.10, positioned at x = 0.0. Comparison with analytical data. [Reproduced from S. Caorsi, M. Donelli, and M. Pastorino, "A passive antenna system for data acquisition in scattering applications," *IEEE Anten. Wireless Propag. Lett.* **1**, 203–206 (2002), © 2002 IEEE.]

The modulated scattering technique has also been used to develop a very effective two-dimensional imaging system capable of rapid measurements of amplitude and phase of the electromagnetic field at 24 GHz in the K band (Ghasr et al. 2008). It consists of an array of 30 high-Q resonant slots loaded with PIN diodes. The probes can be modulated sequentially or simultaneously. The element spacing is $\lambda/2$, where λ is the wavelength of the operating frequency.

The key parameters of the systems are the switching speed and the modulation depth (which is the difference between the measured amplitudes of the signal in the two states of the PIN diode). Figure 9.12 shows the measurement setup, whereas Figure 9.13 presents an example of the measured field distribution (amplitude and phase) in the case in which a small PEC sphere is inspected.

224 MICROWAVE IMAGING APPARATUSES AND SYSTEMS

FIGURE 9.12 Measurement setup showing the waveguide transmitter, a small metallic sphere in front of the array, and the slot array. [Reproduced from M. T. Ghasr, M. A. Abou-Khousa, S. Kharkovsky, R. Zoughi, and D. Pommerenke, "A novel 24 GHz one-shot rapid and portable microwave imaging system," *Proc. IEEE Int. Instrumentation and Measurement Technology Conf.* (*I2MTC 2008*), Victoria, Vancouver Island, Canada, May 12–15, 2008, pp. 1798–1802, © 2008 IEEE.]

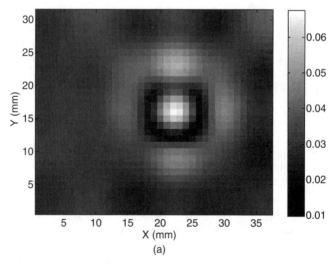

FIGURE 9.13 (a) Magnitude and (b) phase (degree) of measured scattered electric field from a metallic sphere 10 mm in diameter at a distance of 10 mm from the array (Fig. 9.12) in reflection mode. In (c) the data are SAF-processed to produce a precise image of the sphere. [Reproduced from M. T. Ghasr, M. A. Abou-Khousa, S. Kharkovsky, R. Zoughi, and D. Pommerenke, "A novel 24 GHz one-shot rapid and portable microwave imaging system," *Proc. IEEE Int. Instrumentation and Measurement Technology Conf.* (*I2MTC 2008*), Victoria, Vancouver Island, Canada, May 12–15, 2008, pp. 1798–1802, © 2008 IEEE.]

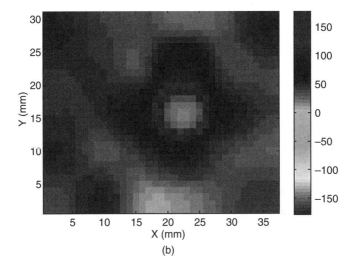

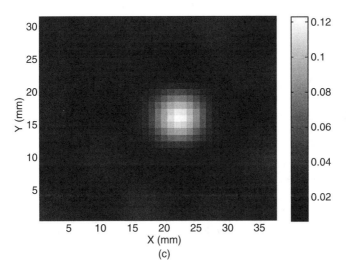

FIGURE 9.13 *Continued*

REFERENCES

Amineh, R. K., A. Trehan, and N. K. Nikolova, "Ultra-wide band TEM horn antenna designed for microwave imaging of the breast," *Proc. 2009 Int. Symp. Antenna Technology and Appl. Electromagn.* (ANTEM/URSI), Banff, Canada, 2009.

Azaro, R., S. Caorsi, and M. Pastorino, "A 3-GHz microwave imaging system based on a modulated scattering technique and on a modified Born approximation," *Int. J. Imag. Syst. Technol.* (Wiley) **9**(5), 395–403 (1998a).

Azaro, R., S. Caorsi, and M. Pastorino, "On the relationship for the bistatic modulated-scattering technique in scattering applications using scattering properties of antennas," *IEEE Trans. Anten. Propag.* **46**(9), 1399–1400 (1998b).

Azaro, R., S. Caorsi, and M. Pastorino, "On the relationship for the monostatic modulated scattering technique," *Microwave Opt. Technol. Lett.* **38**, 187–190 (2003).

Balanis, C. A., *Antenna Theory: Analysis and Design*, Wiley, New York, 2005.

Bindu, G. and K. T. Mathew, "Characterization of benign and malignant breast tissues using 2-D microwave tomographic imaging," *Microwave Opt. Technol. Lett.* **49**, 2341–2345 (2007).

Bolomey, J.-C. and F. E. Gardiol, *Engineering Applications of the Modulated Scatterer Technique*, Artech House, 2001.

Bourqui, J., M. Okoniewski, and E. C. Fear, "Balanced antipodal Vivaldi antenna for breast cancer detection," *Proc. 2nd Eur. Conf. Antennas and Propagation*, Edinburgh, UK, 2007.

Chiappe, M. and G. L. Gragnani, "Vivaldi antennas for microwave imaging: Theoretical analysis and design considerations," *IEEE Trans. Instrum. Meas.* **55**, 1885–1891 (2006).

Collin, R. E., "The receiving antenna," in *Antenna Theory*, Part 1, R. E. Collin and F. J. Zucker, eds., McGraw-Hill, New York, 1969.

Elsdon, M., D. Smith, M. Leach, and S. J. Foti, "Experimental investigation of breast tumor imaging using indirect microwave holography," *Microwave Opt. Technol. Lett.* **48**, 480–482 (2006).

Fear, E. C., P. M. Meaney, and M. A. Stuchly, "Microwaves for breast cancer detection?" *IEEE Potentials* **22**, 12–18 (2003).

Fear, E. C. and M. A. Stuchly, "Microwave detection of breast cancer," *IEEE Trans. Microwave Theory Tech.* **48**, 1854–1863 (2000).

Fernandez Pantoja, M., S. Gonzalez Garcia, M. A. Hernandez-Lopez, A. Rubio Bretones, and R. Gomez Martin, "Design of an ultra-broadband V-antenna for microwave detection of breast tumors," *Microwave Opt. Technol. Lett.* **34**, 164–166 (2002).

Franchois, A., A. Joisel, N. Joachimowicz, J.-C. Bolomey, and Ch. Pichot, "Nonlinear iterative reconstruction algorithm for a 2.45-GHz planar microwave camera," *Proc. 2nd Int. Scientific Meeting IEEE-URSI Microwaves in Medicine*, Rome, Italy, 1983, pp. 263–266.

Franchois, A., A. Joisel, Ch. Pichot, and J.-C. Bolomey, "Quantitative microwave imaging with a 2.45-GHz planar microwave camera," *IEEE Trans. Med. Imag.* **17**, 550–561 (1998).

Furse, C., "Antennas for medical applications," in *Antenna Engineering Handbook*, J. Volakis, ed., McGraw-Hill, New York, 2007.

Ghasr, M. T., M. A. Abou-Khousa, S. Kharkovsky, R. Zoughi, and D. Pommerenke, "A novel 24 GHz one-shot, rapid and portable microwave imaging system," *Proc. IEEE Instrumentation and Measurement Technology Conf.*, Victoria, Vancouver Island, Canada, 2008, pp. 1798–1802.

Hansen, R. C. "Relationships between antennas as scatterers and as radiators," *Proc. IEEE* **77**, 659–662 (1989).

Hernández-López, M. A., M. Quintillán-González, S. González García, A. Rubio Bretones, and R. Gómez Martín," A rotating array of antennas for confocal microwave breast imaging," *Microwave Opt. Technol. Lett.* **39**, 307–311 (2003).

Hislop, G., G. C. James, and A. Hellicar, "Phase retrieval of scattered fields," *IEEE Trans. Anten. Propag.* **55**, 2332–2341 (2007).

Isernia, T. and R. Pierri, "Phase retrieval of radiated fields," *Inverse Problems* **11**, 183–203 (1995).

Jafari, H. M., M. J. Deen, S. Hranilovic, and N. K. Nikolova, "A study of ultrawideband antennas for near-field imaging," *IEEE Trans. Anten. Propag.* **55**, 1184–1188 (2007).

Joachimowicz, N., J. J. Mallorqui. J. C. Bolomey, and A. Broquetas, "Convergence and stability assessment of Newton-Kantorovich reconstruction algorithms for microwave tomography," *IEEE Trans. Med. Imag.* **17**, 562–570 (1998).

Jofre, L., M. S. Hawley, A. Broquetas, E. de los Reyes, M. Ferrando, and A. R. Elias-Fusté, "Medical imaging with a microwave tomographic scanner," *IEEE Trans. Biomed. Eng.* **37**, 303–311 (1990).

Kanj H. and M. Popovic, "Miniaturized microstrip-fed "dark-eyes" antenna for near field microwave sensing," *IEEE Anten. Wireless Propag. Lett.* **4**, 397–401 (2005).

Kim, Y., J. L. Jofre, F. De Flaviis, and M. Q. Feng, "Microwave reflection tomographic array for damage detection of civil structures," *IEEE Trans. Anten. Propag.* **51**, 3022–3032 (2003).

Kubota, S., X. Xiao, N. Sasaki, K. Kimoto, W. Moriyama, and T. Kikkawa, "Experimental confocal imaging for breast cancer detection using silicon on-chip UWB microantenna array," *Proc. 2008 Antennas Propagation Society Int. Symp.*, San Diego, CA, 2008.

Las Heras F., and T. Sarkar, "A direct optimization approach for source reconstruction and NF-FF transformation using amplitude-only data," *IEEE Trans. Anten. Propag.* **50**, 500–510 (2002).

Monteath, G. D., *Applications of the Electromagnetic Reciprocity Principle*, Pergamon Press, New York, 1973.

Montgomery, G., R. H. Dicke, and E. M. Purcell, *Principles of Microwave Circuits*, McGraw-Hill, New York, 1948.

Pastorino, M., "Recent inversion procedures for microwave imaging in biomedical, subsurface detection and nondestructive evaluation," *Measurement* **36**, 257–269 (2004).

Pichot, C., J. Y. Dauvignac, C. Dourthe, I. Aliferis, and E. Guillanton, "Inversion algorithms and measurement systems for microwave tomography of buried objects," *Proc. IEEE Int. Workshop on Imaging Systems and Techniques*, Stresa, Italy, 2004.

Richmond, J. H., "A modulated scattering technique for measurement of field distributions," *IRE Trans. Microwave Theory Tech.* **3**, 13–15 (1955).

Semenov, S. Y. et al., "Spatial resolution of microwave tomography for detection of the myocardial ischemia and infarction. Experimental study on two-dimensional models", *IEEE Trans. Microwave Theory Tech.* **48**, 538–544 (2000).

Soldovieri, F., A. Brancaccio., G. Leone, and R. Pierri, "Shape reconstruction of perfectly conducting objects by multiview experimental data," *IEEE Trans. Geosci. Remote Sens.* **32**, 65–71 (2005).

Stang, J. P. et al., "A tapered microstrip patch antenna array for use in breast cancer screening via 3D active microwave imaging," *Proc. 2009 Antennas Propagation Society Int. Symp.*, Charleston, SC, 2009.

Stratton, J. A., *Electromagnetic Theory*, McGraw-Hill, New York, 1941.

CHAPTER TEN

Applications of Microwave Imaging

10.1 CIVIL AND INDUSTRIAL APPLICATIONS

Microwave diagnostic techniques are widely used in civil and industrial applications, especially because of their ability of penetrate inside dielectric materials. The simplest way to understand this phenomenon is to consider the propagation of a plane wave through a homogeneous medium. To this end, let us assume that a plane wave [equation (2.4.24)] is propagating along the z axis inside a material characterized by $\varepsilon = \varepsilon' - j\varepsilon''$ [equation (2.3.14)], and μ (real-valued). Thus

$$\mathbf{k} = k\hat{\mathbf{z}} = (\beta - j\alpha)\hat{\mathbf{z}}, \tag{10.1.1}$$

where

$$\beta = \omega \left\{ \frac{\mu\varepsilon'}{2} \left[\sqrt{1 + \left(\frac{\varepsilon''}{\varepsilon'}\right)^2} + 1 \right] \right\}^{1/2} \quad \text{(phase constant)}, \tag{10.1.2}$$

$$\alpha = \omega \left\{ \frac{\mu\varepsilon'}{2} \left[\sqrt{1 + \left(\frac{\varepsilon''}{\varepsilon'}\right)^2} - 1 \right] \right\}^{1/2} \quad \text{(attenuation constant)}. \tag{10.1.3}$$

Accordingly, equation (2.4.24) can be rewritten as

$$\mathbf{E}(\mathbf{r}) = -\eta \hat{\mathbf{z}} \times \mathbf{H}(\mathbf{r}) = E_p(z)\hat{\mathbf{p}} = E_0 e^{-jkz}\hat{\mathbf{p}} = E_0 e^{-\alpha z} e^{-j\beta z}\hat{\mathbf{p}}. \tag{10.1.4}$$

Microwave Imaging, By Matteo Pastorino
Copyright © 2010 John Wiley & Sons, Inc.

For ideal dielectrics, $\alpha = 0$; consequently, from equation (10.1.4) it results that $|E_p(z)| = |E_0|$ for any z; that is, the electromagnetic field is not attenuated. For low-loss scatterers ($\varepsilon'' \ll \varepsilon'$), we have

$$\beta \approx \omega\sqrt{\mu\varepsilon'} \quad \text{(phase constant)}, \tag{10.1.5}$$

$$\alpha \approx \frac{\omega\varepsilon''}{2}\sqrt{\frac{\mu}{\varepsilon'}} \quad \text{(attenuation constant)}, \tag{10.1.6}$$

$$\eta \approx \sqrt{\frac{\mu}{\varepsilon'}}\left(1 + j\frac{\varepsilon''}{2\varepsilon'}\right), \tag{10.1.7}$$

whereas for good conducting materials, for which the conduction effect is dominant and $\varepsilon'' \approx \sigma/\omega \gg \varepsilon$ [see again equation (2.3.14)], we obtain

$$\beta = \alpha \approx \sqrt{\frac{\mu\sigma\omega}{2}}, \tag{10.1.8}$$

$$\eta \approx (1 + j)\alpha. \tag{10.1.9}$$

In both cases, since $\alpha > 0$, it follows from equation (10.1.4) that $|E_p(z)| = |E_0|e^{-\alpha z}$. Then the field attenuates as it propagates through the material. Since the material is homogeneous, after any length Δz, the attenuation is such that

$$\frac{|E_p(z+\Delta z)|}{|E_p(z)|} = e^{-\alpha \Delta z} \tag{10.1.10}$$

If we choose $\Delta z = \delta$ such that the ratio shown above is equal to $1/e$, it follows immediately that

$$\delta = \frac{1}{\alpha}, \tag{10.1.11}$$

which is the *penetration depth* (m) and is used to quantify the degree to which an electromagnetic wave can propagate inside a material, although in real cases the targets are not homogeneous, and the incident wave cannot always be approximated by a plane wave. It should be noted that for PEC objects, we have $\sigma \to \infty$, and, from (10.1.8) and (10.1.11), $\delta \to 0$, with the well-known result that a time-varying field is unable to penetrate inside a PEC object. For an idea about the penetration depth, it should be noted that, due to the exponential decay of the field (except for ideal dielectrics and PEC objects), after a length equal to the penetration depth $\Delta z = \delta$, the field amplitude is reduced to about 37% of its initial value; after a length $\Delta z = 2\delta$, we have a reduction to about 13.5%, whereas after $\Delta z = 3\delta$, the field is only about 5% of the initial value. For the sake of illustration, some results concerning materials that can be inspected by microwave techniques will be provided below.

As mentioned earlier, microwave techniques are widely used in civil and industrial engineering, such as for material characterization and crack detection. However, these techniques usually are based on transmission and reflection concepts. There is usually one transmitting antenna (e.g., an open waveguide or a small horn) which may or may not be in contact with the bodies to be inspected and that generates the incident wave. The scattered wave is then collected by the same antenna (reflection mode) or by another antenna (the receiving antenna) located on the other side of the object (transmission mode). Care must be exercised in determining the dielectric properties of the bodies by measuring the reflection or transmission coefficients (e.g., the [S] parameters of the system), and the effects of the antenna elements have to be taken into account. Despite these additional difficulties, the approach is essentially one-dimensional. Procedures for solving the one-dimensional inverse problem have been reported (e.g., Mittra 1973, Bolomey et al. 1981), and reviewed (Chew, 1990). Moreover, several applications of microwave techniques in nondestructive testing and evaluation (NDT&E), based mainly on reflection/transmission concepts, have been reviewed in detail (Kharkovsky and Zoughi 2007). It is quite straightforward that many of the applications for which these techniques are effective can probably also benefit from image-based inspection methods, such as those described in the previous chapters.

The first area of interest for microwave techniques is that of material characterization, in which microwaves are used to draw information about the main properties of the inspected materials, such as the determination of constituents, evaluation of porosity, and assessment of the curing state. It is evident that this goal can be achieved by microwave techniques if there is a precise correlation between the properties of the materials under test and the values of the macroscopic dielectric parameters that can be detected by microwave inspection techniques, working under both classical transmission/reflection conditions and by possibly using imaging techniques in order to derive maps of the inspected objects.

A large number of studies correlate physical and chemical properties of materials to their dielectric parameters [see, e.g., the text by Zoughi (2000) and references cited therein]. Measurement of the dielectric parameters at microwave frequencies is a classical topic in applied electromagnetics, and the scientific literature on the subject is wide. The reader can refer to the book by Chang (2005) and other specialized publications. In most cases the dielectric parameters are calculated by inserting samples of the material to be characterized inside a waveguide often closed on a short circuit. Incident progressive waves impinging on the samples create reflected waves that can be measured by means of a vector network analyzer. The values of the reflection coefficients can be related to the dielectric properties of the samples constituting the *load* of the waveguide. Cavity-based measurement systems are also used, as well as free-space measurement approaches.

However, retrieving the values of the dielectric parameters from measured values of the reflection/transmission coefficients is essentially another *inverse problem*. Several techniques can be adopted to *invert* this relation, and often various samples of the same material with different thicknesses are used. Such experiments and the related *inversions* allow one to deduce the relation among the values of the dielectric parameters and the specific properties of the material, for a characterization of the material that can be used to extrapolate the specific properties of interest when an *unknown* material is inspected by microwave techniques able to retrieve the distributions of the dielectric parameters.

For illustration purposes, we consider here an example of the extensive measurements of dielectric properties of rubber that has been performed by Zoughi (2000) using a waveguide measurement system (working as outlined previously) with an accuracy of less than 1% on the measurements of the real part of the dielectric permittivity and less than 3.5% for the imaginary part (lossy factor). It was observed that rubber dielectric properties significantly depend on the compound constituents, the volume fraction of these constituents, and chemical reactions among them.

For example, it has been found that the real part of the dielectric permittivity changes almost monotonically from $\varepsilon'_r \approx 5$ to $\varepsilon'_r \approx 27$ for a percentage of carbon block volume ranging from about 8% to about 32% at 5 GHz. It is quite evident that these ranges of permittivity and conductivity values are compatible with the *resolution* expected from the microwave imaging techniques discussed in the previous chapters. For an idea of the capability of penetration of microwaves inside this material, we can evaluate, using equation (10.1.11), the penetration depths of microwave signals at the operating frequency and for the extremes of the specified range of variation for the percentage of the carbon block volume. The results are $\delta \approx 3$ cm for 8% and $\delta \approx 0.35$ cm for 32%.

Another key aspect is represented by the capabilities of microwave techniques in detecting cracks and defects inside dielectric structures (Pastorino et al. 2002). One of the most interesting applications is in the civil engineering field, in particular the evaluation of concrete structures and other materials (Park and Nguyen 2005; Xu et al. 2006). For example, the linear sampling method–Green's function approach (Sections 5.8 and 4.10) can be applied for the nondestructive assessment of a concrete structure. To this end, let us consider the numerical simulation reported in Figures 10.1 and 10.2, which refer to the detection of a void crack inside a cement pillar, modeled with dielectric parameters similar to those encountered in the area of civil engineering. The investigation domain is illuminated by $M = 36$ line-current sources uniformly distributed on a circumference of radius a, and the scattered electric field is collected at $M = 36$ measurement points uniformly distributed on a circumference of radius b. The values of a and b are chosen such that the antennas are in the far-field region of the investigation domain. The *measured* data to be inverted have been simulated by using a direct solver based on the method of

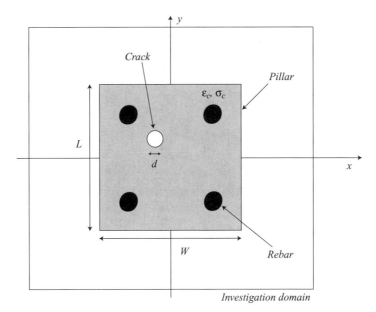

FIGURE 10.1 Model of a concrete pillar with four rebars (cross section).

moments. A Gaussian noise with zero mean value and variance corresponding to a signal-to-noise ratio of 25 dB has been added to the simulated data. In the inversion stage, the far-field matrix has been discretized by using $N = 3969$ pulse basis functions and $M = 36$ testing points, corresponding to the M measurement locations.

The concrete pillar, whose cross section is shown in Figure 10.1, is modeled as a cylinder with square cross section made from homogeneous cement paste with dielectric permittivity $\varepsilon_c = 2.37\varepsilon_0$ and electric conductivity $\sigma_c = 5.7 \times 10^4$ (Hayt and Buck 2001). The pillar is located in the center of the investigation domain and has sides $W = 3\lambda_0$ and $L = 3\lambda_0$, respectively. Inside the pillar, four rebars, centered at points (λ_0, λ_0), $(\lambda_0, -\lambda_0)$, $(-\lambda_0, \lambda_0)$, and $(-\lambda_0, -\lambda_0)$, are present. This configuration represents the *unperturbed* structure used to compute the inhomogeneous Green function, according to the formulation described in Section 4.10. The crack is modeled as a circular void cylinder with diameter $d = 0.6\lambda_0$ and centered at point $\mathbf{r}_d = (-0.6\lambda_0, 0.6\lambda_0)$. The reconstruction results are reported in Figure 10.2. In particular, Figure 10.2a shows the normalized indicator function Ψ, introduced in Section 5.8. As can be seen, a *peak* is present exactly where the crack is located. The binary map obtained by thresholding the normalized indicator function at $\Psi_{th} = 0.4$ is provided in Figure 10.2b, allowing for a correct identification of the support of the defect.

It is quite evident that microwave techniques are able to detect concealed metallic object, too. See for example the result in Figure 10.3, which has been

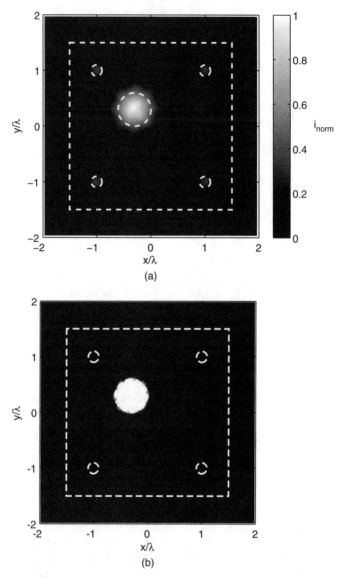

FIGURE 10.2 (a) Normalized indicator function obtained by applying the linear sampling method, with detection of a crack inside the concrete pillar of Figure 10.1; (b) reconstructed support of a crack inside the concrete pillar with four rebars of Figure 10.1. (Simulations performed by A. Randazzo, University of Genoa, Italy.)

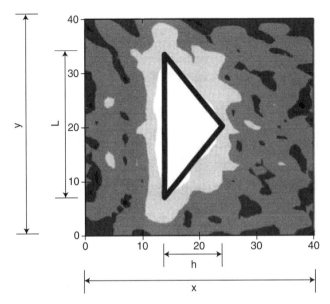

FIGURE 10.3 Reconstructed image of a concealed metal object (10 dB/contour) with $x = y = 40$ cm, $L = 36$ cm, and $h = 10$ cm. [reproduced from M. Elsdon, D. Smith, M. Leach, and S. Foti, "Microwave imaging of concealed metal objects using a novel indirect holographic method," *Microwave Opt. Technol. Lett.* **47**(6), pp. 536–537 (Dec. 20, 2005), © 2005 Wiley.]

obtained by the microwave holography system in Figure 9.4. Other accurate images of defects and voids in concrete structures based on experimental data have been obtained by using the planar microstrip slot array shown in Figure 9.8 and are shown in Figure 10.4 (Kim et al. 2003).

It must be noted that civil structures have been addressed by using inverse scattering techniques in other studies, including that by Catapano et al. (2006), who used a tomographic technique to invert experimental laboratory-controlled data on a multilayer configuration, consisting of air, bricks, and concrete. Further discussion of civil applications can be found elsewhere in the literature (e.g., Mikhnev and Vainikainen 2002, Chang 2005, Zeng et al. 2006).

In the industrial field, another interesting application is related to the inspection of new materials such as ceramics (Bahr, 1978), plastics, and other composites, which can be suitably evaluated at microwave frequencies. In particular, it has been recognized that "the ever expanding materials technology by which lighter, stiffer, stronger and more durable electrically insulating composites are replacing metals in many applications demands alternative inspection techniques to existing NDT approaches. This is due to the fact that the existing standard and well established NDT techniques (developed primarily for inspecting metallic structures) may not always be capable of inspecting these composites" (Zoughi 2000, p. 2). Moreover, the inspection of

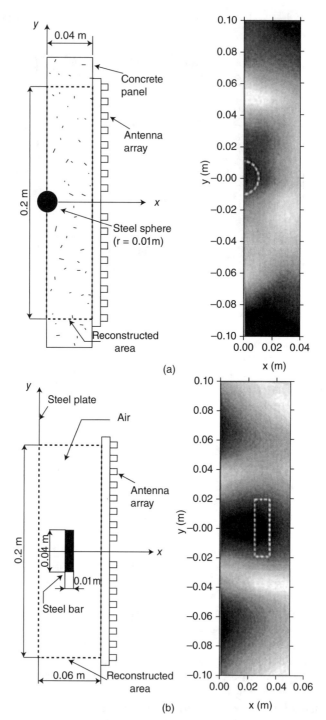

FIGURE 10.4 Reconstructions obtained using the planar array shown in Figure 9.8 in different cases. [Reproduced from Y. J. Kim, L. Jofre, F. De Flaviis, and M. Q. Feng, "Microwave reflection tomographic array for damage detection of civil structures," *IEEE Trans. Anten. Propag.* **51**(11), 3022–3032 (Nov. 2003), © 2003 IEEE].

CIVIL AND INDUSTRIAL APPLICATIONS   237

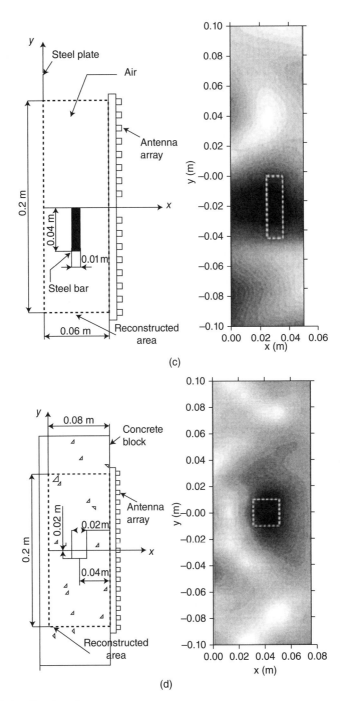

(c)

(d)

FIGURE 10.4 *Continued*

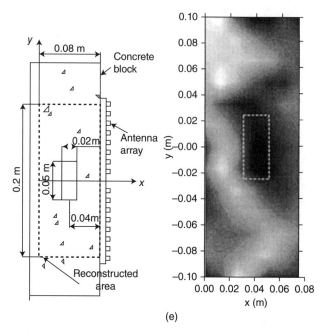

(e)

FIGURE 10.4 *Continued*

low-density fiberglass composites, which are widely used as insulating and coating materials, can be very efficiently performed by microwave imaging methods, which can also indirectly monitoring the curing process. To this end, let us consider just one of the results reported (Zoughi 2000), in which the dielectric properties of a fresh liquid resin binder and the same material after 12 days (at ambient temperature) have been studied. It has been found that, at a frequency of 5.5 GHz, ε'_r varies from 16.0 to 18.7. This change in the real part of the dielectric permittivity, even for a limited curing at room temperature, is significant and fully compatible with the inspection capabilities of the approaches considered throughout this book.

Another interesting field for microwave imaging techniques is represented by the wood-processing industry (Bacur 2003), where inverse-scattering-based diagnostic techniques can be used together with microwave inspection methods based on reflection or transmission in order to characterize wood materials and find defects and knots inside slabs and trunks (Lhiaubet et al. 1992, Shen et al. 1994, Eskelinen and Harju 1998, Kaestner and Baath 2005, Salvadé et al. 2007; Pastorino et al. 2007). An example of the dielectric reconstructions obtained by using the tomograph prototype shown in Figure 9.2 is given in Figure 10.5. It involves the nondestructive evaluation of a hole in a wood slab. The slab has a rectangular cross section (dimensions 11.7×7.7 cm), and a rectangular hole (dimensions 6×3 cm) has been drilled in the sample. The measured scattered data are collected in the L band, and the actual values of

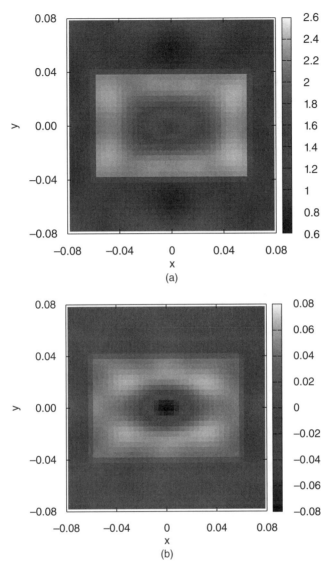

FIGURE 10.5 Distributions of (a) the reconstructed relative dielectric permittivity and (b) electric conductivity. [Reproduced from A. Salvadé, M. Pastorino, R. Monleone, T. Bartesaghi, G. Bozza, and A. Randazzo, "Microwave detection of dielectric structures by using a tomographic approach," Proc. IEEE Int. Instrumentation and Measurement Technology Conf. (I2MTC 2008), Victoria, Vancouver Island, Canada, May 12–15, 2008, pp. 1300–1305, © 2008 IEEE.]

the dielectric parameters are $\varepsilon_w \approx 2.2\varepsilon_0$ and electric conductivity $\sigma_w \approx 0.04\,\text{S/m}$, which correspond to a penetration depth of $\delta \approx 6.5\,\text{cm}$ at $f = 1.5\,\text{GHz}$ and $\delta \approx 4.2\,\text{cm}$ at $f = 3.5\,\text{GHz}$. In order to invert the experimental data, a truncated singular value decomposition has been used. The slab cross section is embedded in a square investigation domain of side 0.16 m and discretized into $N = 1600$ subdomains.

In the wood-processing field, microwave imaging techniques can also be of interest for the detection of foreign bodies inside wood trunks, such as bullets, screws, or other debris, which can severely damage carpentry machinery if they are not detected and removed prior to processing (Salvadé et al. 2008).

Detection of inclusions inside plastic materials has also been performed by using a waveguide apparatus (Brovko et al. 2008). In particular, the position and the size of a spherical inclusion in a dielectric sample have been deduced by using an inverse problem by means of an artificial neural network (ANN). The network has been trained by using measured complex reflection and transmission coefficients (which are related to the scattered data) obtained by solving the direct scattering problem by means of a full-wave three-dimensional finite-difference time-domain (FDTD) code. The operating frequency was 915 MHz and allowed the sizes and positions of glass and air spheres of more than 15 mm diameter to be accurately detected in a Teflon block (with relative errors in the coordinates and radius of the center of the inclusion of the order of 0.9–2.2%).

Further applications in the field of material characterization may involve evaluation of the porosity in polymers, ceramics, and other materials, which can become very weak if the porosity level is too high. It has been observed that "porosity often concentrates at specific locations in composite materials (either between plies or at the fiber/matrix interface) and can dramatically lower flexural and shear performance" (Zoughi 2000). Consequently, porosity can render inhomogeneous a given structure, and imaging techniques able to inspect inhomogeneous scatterers can be applied. Since *porosity* means the presence of air inside the material, the dielectric permittivity and, mainly, the electric conductivity significantly decrease if the porosity level increases. For example, at 10 GHz, a 36% reduction of ε'_r has been measured for a specific polymer when the air content was about 49%.

According to Zoughi (2000), microwave techniques can be good candidate techniques for inspection of composites, in particular for the accurate thickness measurement of dielectric layers, such as coatings or layered composites (Mudanyal et al. 2008); for evaluation of disbond and delamination in half-space and stratified materials (Zoughi and Bakhtiari 1990); for the detection of voids, rust, and corrosion (Qaddoumi et al. 2007) under paint and stratified coatings, for the inspection of thick plastics and glass-reinforced composites; and for several other diagnostic objectives (Bozza et al. 2007a,b) that can be performed both during the material production and during the service or use. In the former case, the inspection is strictly related to the procedures for process control (Zoughi, 2000).

Finally, an additional proposed application is in the food industry, such as detection of the sugar content in fruits. In particular, the use of the chirp radar technique (briefly mentioned in Section 5.9) has been exploited. Watanabe et al. (2006) considered some fruits (e.g., pears and apples), whose sugar content varies between 8% and 22%. As a result, the sugar content directly affects the value of the complex dielectric permittivity, and, consequently, can be detected by using microwaves. The final image is a gray-level image [the correlation coefficients between the sugar content and the corresponding gray level of the final images result in the range 0.36–0.73 for Japanese pears and less than 0.7 for other pears (Watanabe et al. 2006)]. In the authors' opinion (Watanabe et al. 2006) "It seems to be difficult to visualize the sugar content distribution with its absolute value. Generally speaking, a fruit contains sugar as various chemical substances and those substances change their structure according to the growth. However, it will not be difficult to designate the sugar content of the same kind of fruits by comparing the gray levels."

10.2 MEDICAL APPLICATIONS OF MICROWAVE IMAGING

The preliminary ideas concerning the application of microwave techniques to medical imaging have been summarized in a book by Larsen and Jacobi (1986). When that pioneering work appeared, it seemed that techniques based on interrogating microwaves would have provided in a short time new and powerful tools for medical diagnostics. However, more than 20 years later, microwave medical imaging is often still considered an *emerging technique*, since accurate and effective dielectric reconstructions are still very difficult to obtain. One should consider, for comparison, the developments in the area of X-ray computerized tomography. Essentially, the basic formulation of computerized tomography, specifically Radon transform, considered to be useless for some decades, was *rediscovered* in the 1970s and, a few years later, tomographs were available in several advanced diagnostic units of major hospitals; now, computerized tomography has become a routine diagnostic methodology. The *history* of medical microwave imaging is clearly far different. Nevertheless, some quite interesting results using microwaves have been obtained, particularly in the field of breast cancer detection. The important fact is that some of the factors that have so far limited the real applicability of microwave medical imaging tend to further reduce their impact on the development of imaging systems. This delineates an *optimistic perspective*, which is supported by the significant achievements in both the design and realization of efficient illumination/measurement systems and the development of fast, effective, and reliable reconstruction procedures, some of them covered by this monograph. At the same time, the medical community is presently looking for new diagnostic tools with high sensitivity and specificity. It is a well-known fact that the early-stage detection of cancer is the fundamental issue not only for long-term survival of patients but also for their life quality.

This is particularly true for breast cancers, as this disease is one of the main causes of death among women. Unfortunately, mammography, which is the main actual diagnostic modality, still exhibits a relatively high false-negative rate, resulting in a large number of unnecessary biopsies (Huynh et al. 1998, Li et al. 2005). Incidentally, it should be added that mammography is usually rather uncomfortable for patients, due to breast compression.

In this scenario, the potentialities of microwave imaging for medical diagnosis are widely recognized, since this technique exhibits some unique features from which medical diagnostic units should realize special benefits.

First, microwaves can directly provide the distributions of the dielectric parameters (dielectric permittivity and electric conductivity). This information cannot be obtained using any other different technique. In addition, the retrieved values of the dielectric parameters can be directly related not only to the various biological tissues but also to specific pathologies (Lazebnik et al. 2007). For example, some studies indicate that the contrast in dielectric properties of malignant and normal breast tissues is greater than 2:1 in the radiofrequency–microwave frequency range, due to the increased water content in neoplastic tissues resulting from the protein hydration and vascularization of the cancerous tissues. On the contrary, mammography, as well as magnetic resonance imaging (MRI), are judged not sensitive or specific enough and too operator-dependent (Foster and Schwan 1989; Li et al. 2005).

Moreover, it has been argued that microwaves "can exploit strong indicators of malignancy associated with physical or physiological factors of clinical interest, such as water content, vascularization/angiogenesis, blood-flow rate, and temperature" (Li et al. 2005, p. 20).

Significant changes in dielectric properties have been noted in other cases, for example, for normal bones versus leukemic marrow (Colton and Monk 1995). In addition, considering again the contrast in biological tissues (in terms of the values of the dielectric permittivity and the electric conductivity), the use of gas-filled microbubbles, metallic nanoparticles, and single-walled carbon nanotubes as contrast agents has recently been proposed for microwave imaging, since it seems possible that their accumulation in a tumor enhances the contrast between normal and malignant tissues (Mashal et al. 2009, Shea et al. 2009).

Another fundamental aspect is that microwaves are nonionizing radiations and interrogating signals with very low levels of power can safely be applied for diagnostic purposes. This is safe for both patients and operators and could also allow for quasi-continuous monitoring. In addition, the costs of the apparatuses can be very limited as compared with those needed for the most consolidated actual imaging systems.

Furthermore, although microwave imaging exhibits a limited spatial resolution when compared with X-ray-based diagnostic techniques or other advanced imaging techniques currently adopted, the modern development of focusing procedures, as well as of tomographic approaches based on nonlinear formulations, has definitely proved the capabilities of subwavelength resolution of microwave imaging techniques. Incidentally, microwaves can in principle be applied in conjunction with other techniques, combining good resolution with high specificity.

It has also been observed that contrast between normal and malignant dielectric parameters is higher at frequencies different from those of microwaves (Lazebnik et al. 2008) and that other electromagnetic techniques can be applied in the medical field (e.g. electrical impedance tomography operating at lower frequency or near-infrared tomography, which can exploit the higher contrast due to higher absorption in DNA, protein, and hemoglobin of malignant tissues). Nevertheless, the microwave range represents a good compromise between resolution and penetration depth (Li et al. 2005, Lin, 1985).

Among the various diagnostic applications, microwave imaging has been proposed, for example, for the detection of myocardial ischemia and infarction (Semenov et al. 2000). It has been shown that changes in myocardial dielectric properties of ~10% arise in cases of ischemia as well as in chronic infarction. These results are mainly related to canine and pig studies at frequencies in the 0.9–2.36 GHz band, irradiated by waves corresponding to 1 W of transmitted power. A circular configuration has been used to image a phantom of the heart ($\varepsilon_r = 70 - j9$ at 2.36 GHz), in which some inhomogeneities have been inserted in order to simulate specific pathological situations (in particular, an acute ischemic zone and a chronic infarcted zone). Moreover, a situation in which infarcted and ischemic tissues are present has also been considered, since a chronic infarcted zone surrounded by an ischemic area may cause arrhythmic problems (Semenov et al. 2000). Consequently, the phantom structure under test and related studies are important in the light of potential arrhythmia detection as well.

Let us now provide some examples of the assumptions made in evaluating the various proposed methodologies. These examples are provided only for illustration purposes, and the reader can refer to the cited literature for details. In applying the confocal method to breast cancer detection, the patient can be in a supine position in order to better detect tumors near the chest wall or in a prone position (Xie et al. 2006). In the latter case, holes are present in the treatment table in order to access to the full breast volume. Various models of the breast, both two-dimensional and three-dimensional ones, have been considered in the literature for simulation and testing purposes. In the case in which the patient is lying in a supine position, a commonly considered two-dimensional model assumes the breast profile to be approximated by a semicircle, whereas the corresponding three-dimensional model considers a hemispherical breast. Concerning the imaging approach, some improved procedures have been proposed with respect to the basic formulation. In particular, the use of a matched filter for estimation of propagation times from the elements of the array and the skin–breast interface has been proposed, which is used to *shape* the interface and compensate for the response of this interface "which is at least one order of magnitude larger that any tumor response" (Li et al. 2005, p. 20).

In the study by Xie et al. (2006), for example, the hemispherical model has a radius of 10 cm and includes randomly distributed glandular and fatty breast tissues. In addition, 2-mm-thick skin and 4- and 6-mm spherical models of the tumors to be detected are assumed.

It should be mentioned that more realistic models of the human body have also been developed, not only for imaging purposes but also for other applications, such as numerical computation of the specific absorption rate (SAR) due to the exposure to electromagnetic field (this computations requires the solution of a direct scattering problem). Zastrow et al. (2008) anatomically constructed realistic breast phantoms of different shape and sizes, starting with MRI images of prone patients and taking into account the dispersivity of the biological tissues.

Concerning the dielectric properties of the biological tissues, the models are based on extensive studies based on dielectric measurements (Gabriel 1996, Lazebnik et al. 2007). For example, Xie et al. (2006) chose the following values for various biological tissues: external medium ($\varepsilon_r = 9$, $\sigma = 0$ S/m), chest wall ($\varepsilon_r = 50$, $\sigma = 7$ S/m), skin ($\varepsilon_r = 36$, $\sigma = 4$ S/m), fatty breast tissue ($\varepsilon_r = 9$, $\sigma = 0.4$ S/m), nipple ($\varepsilon_r = 45$, $\sigma = 5$ S/m), glandular tissue ($\varepsilon_r = 11-15$, $\sigma = 0.4-0.5$ S/m), and tumor ($\varepsilon_r = 50$. $\sigma = 4$ S/m). As mentioned, working in a wide band of frequencies, dispersion of the biological tissues must be taken into account. Usually, a first-order Debye-type dispersion model is assumed (see Chapter 2), but higher-order models can also be considered.

In most cases the transmitting and receiving antennas are simulated by line-current sources (two-dimensional problems) or point sources (three-dimensional problems), although more accurate numerical descriptions are possible. For example, the imaging configuration adopted by Xie et al. (2006) consists of 72 elements (point sources), located on six circles with different radii, such that the distance from the skin is always 1 cm. The centers of the circles are on a line (the symmetry axis of the hemispherical breast model), and 12 antennas are located on each circle. Along the symmetry axis, the circles are spaced at 0.5 cm. The 72 elements emit sequentially and, for each emission, the scattered field is collected by all the 72 elements. A Gaussian pulse of about 120 ps is used, with a maximum signal at 5 GHz. At this frequency, the penetration depth [equation (10.1.11)] in the breast tissue is $\delta \approx 1.1$ cm.

Preliminary experiments concerning the confocal method described in Section 5.9 have also been performed by using a multilayer phantom made of materials with dielectric properties similar to those of the normal breast (soybean oil, with dielectric properties similar to those of fat with very low water content, $\varepsilon_r = 2.6$. $\sigma = 0.05$ S/m at 6 GHz) and skin tissues (1.5-mm-thick FR4 glass epoxy), with a small synthetic tumor (diecetin–water solution), suspended in liquid (Li et al. 2004). The antenna used for the ultrawideband data acquisition was a miniaturized pyramidal horn with a single ridge and curved launching plane terminated with Chip resistors (Li et al. 2003). A square planar 6×6 cm array was synthesized by moving the horn antenna. The measurements were performed by using a network analyzed with a frequency sweep in the range 1–11 GHz (201 samples). The "signal to clutter ratio" (the ratio of the maximum received energy to the energy received in the tumor-free case) was 8.4 dB, sufficient for allowing tumor localization.

In order to reduce the coupling loss due to dielectric discontinuity represented by the biological body, the antennas and the target to be inspected are usually assumed to be immersed in a medium with a dielectric permittivity similar to that of the breast (Xie et al. 2006, Chen et al. 2008). Li et al. (2004), for example, used the same liquid for modeling the breast tissue.

Moreover, in experimental reconstructions, the biological body is often immersed in water. In this case, it must be understood that the dielectric parameters of the water depends not only on the frequency of the incident field but also on the temperature. In the experiments performed using the circular array described in Section 9.2, the temperature uniformity is maintained within ±1 °C (Joachimowicz et al. 1998). The sensitivity of the dielectric permittivity versus the temperature have been assumed (Joachimowicz et al. 1998) to be governed by temperature coefficients such as around the ambient temperature the real and imaginary parts of the relative dielectric permittivity changes of about −0.5 % and +2 % for any degree of temperature increment.

An example of a reconstruction obtained by using the bowtie-type antenna array discussed in Chapter 9 (Fig. 9.7) is shown in Figure 10.6 (Hernandez-Lopez et al. 2003). The images have been constructed using a beamforming approach and starting with a numerical model of the breast similar to the one just discussed. It consists of a 4-cm-radius half-sphere attached to an infinite half-space characterized by the same dielectric parameters of the healthy tissue. Moreover, a layer of skin tissue is added. The tumor is modeled by a sphere located at an arbitrary position inside the breast. Both healthy and cancerous breast tissues are characterized by dispersive Debye models. In the simulation shown in Figure 10.6, the distance between the center of the breast and the center of the tumor is about 2.1 cm. The transmitting/receiving array is placed 1 mm above the skin. The system rotates around the breast (see Fig. 9.7), and four perpendicular measurement positions are considered, which correspond to 64 measured signals.

Figure 10.6 provides a three-dimensional distribution of the (normalized) reconstructed intensities. In particular, three sections are shown, in which the light patches indicates the tumor regions, which are located quite accurately. In particular, the distance between the original and reconstructed centers of the tumor is 2.25 mm, which must be compared with the maximum error due to the discretized mesh. In this case, the numerical forward simulations have been performed by means of a FDTD code with a mesh consisting of $96 \times 96 \times 58$ (cubic) Yee cells of side $\Delta = 1$ mm, and a temporal step assumed equal to 1.83 ps. Accordingly, the maximum discretization error is equal to $2\sqrt{3}\Delta \approx 3.5$ mm.

Sensitivity issues of microwave imaging systems are also very important. Essentially, the question is as follow: Considering the current generation of illumination/measurement systems, how small can a discontinuity be inside a biological body while still allowing the detection of the field that it scatters?

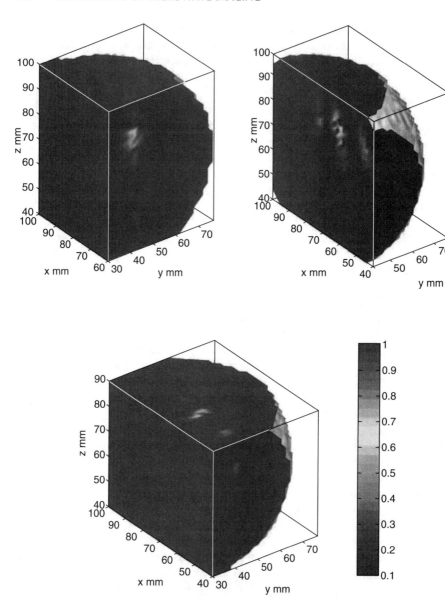

FIGURE 10.6 An example of a reconstructed model of a spherical tumor, obtained by considering the array of bowtie antennas shown in Figure 9.7 [reproduced from M. A. Hernández-López, M. Quintillán-González, S. González García, A. Rubio Bretones, and R. Gómez Martín, "A rotating array of antennas for confocal microwave breast imaging," *Microwave Opt. Technol. Lett.* **39**, 307–311 (Nov. 20, 2003), © 2003 Wiley.]

This topic, which deserves further investigation, has been addressed, for example, by Zhang et al. (2003) by means of numerical simulations of the forward scattering problem and considering a three-dimensional model of the breast. The direct scattering problem has been solved by using the method of moments with a solver based on a biconjugate-gradient procedure. In their opinion, considering the changes in the scattered field outside the biological body and assuming a dynamic range of a network analyzer of about -100 dB or better, tumors with a diameter ≥6 mm at 800 MHz and ≥2 mm at 6 GHz are detectable in ideal situation, although Zhang et al. (2003) state that "in reality, it may be too optimistic […] due to the spurious scattering in the environment." Finally, it should be noted that if the contrast between malignant and normal tissues decreases, the minimum detectable volume increases, since the scattered signal, according to equation (2.7.2), for a small scattering target of volume ΔV is essentially proportional to $\tau \Delta V$, where τ is proportional to the contrast value (Zhang et al. 2003).

The idea of combining different imaging modalities (discussed in Chapter 8) can also be a good solution in the medical field. For example, the beamforming technique briefly described in Chapter 5, which does not provide the distribution of the dielectric parameters of the biological tissues, can be used to detect the positions of the most relevant scatterers inside a biological body and provide an estimation of their spatial dimensions. Successively, a quantitative imaging technique (Chapter 6 and 7) can be applied. In the study by Sabouni et al. (2006), after the scatterers (e.g., tumors) were located, the detected areas were inspected by using an iterative procedure based on a genetic algorithm (see Section 7.3), whereas the forward scattering problem was solved at each iteration by using a FDTD code (including a first-order dispersion relation of Debye type). By using this method, a 0.9-cm circular tumor ($\varepsilon_r = 54$, $\sigma = 0.7$ S/m) was satisfactory reconstructed in a quantitative way in a two-dimensional model (Sabouni et al. 2006).

The detection of tumors has been also experimentally pursued using the prototype shown in Figure 9.3. Excised specimens of biological tissues are reconstructed starting with the inversion of measured data. The distorted Born iterative method (Section 6.6) has been used for this task. Figures 10.7 and 10.8 provide two examples of the reconstructions obtained (Bindu and Mathew 2007). An experimental model, consisting of a bath of vegetable oil ($\varepsilon_r \approx 54$), has been used to represent both the immersion liquid and the healthy tissue in the simplified breast phantom developed by (Elsdon et al. 2006) and inspected by using the indirect holography schematized in Figure 9.4 (frequency: $f = 8.59$ GHz, scanning aperture: 40×40 cm. The recorded data and the reconstructed image of the dielectric discontinuity are presented in Figure 10.9.

Other methods have been applied to detect scatterers representing tumor models. One of them is the previously mentioned level-set method (references for this method are cited at the end of Section 5.8), which has been combined with the MUSIC algorithm (Irishina et al. 2006). An image of the retrieved dielectric permittivity distribution is shown in Figure 10.10. The

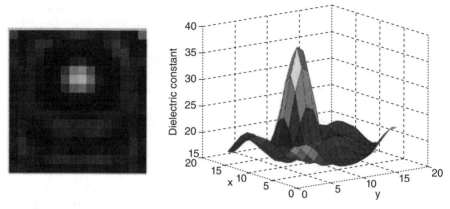

FIGURE 10.7 2D tomographic image and permittivity profile of benign breast tissue of radius ≈0.5 cm inserted in normal breast tissue sample of radius ≈1 cm, of patient 2. [Reproduced from G. Bindu and K. T. Mathew, "Characterization of benign and malignant breast tissues using 2-D microwave tomographic imaging," *Microwave Opt. Technol. Lett.* **49**, 2341–2345 (2007), © 2007 Wiley.]

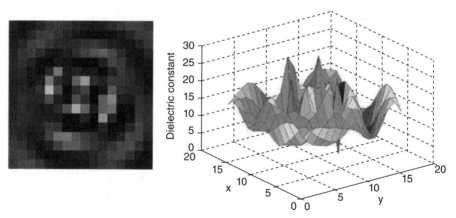

FIGURE 10.8 2D tomographic image and permittivity profile of malignant breast tissue of radius ≈0.5 cm inserted in normal breast tissue sample of radius ≈1 cm, of patient 3. [Reproduced from G. Bindu and K. T. Mathew, "Characterization of benign and malignant breast tissues using 2-D microwave tomographic imaging," *Microwave Opt. Technol. Lett.* **49**, 2341–2345 (2007), © 2007 Wiley].

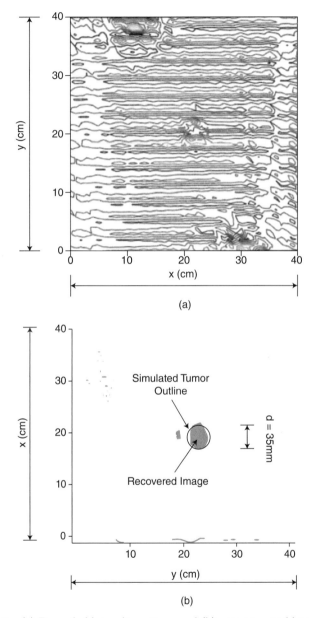

FIGURE 10.9 (a) Recorded intensity pattern and (b) reconstructed image of a simulated breast tumor ($x = y = 40$ cm, $d = 35$ mm). [Reproduced from M. Elsdon, D. Smith, M. Leach, and S. J. Foti, "Experimental investigation of breast tumor imaging using indirect microwave holography," *Microwave Opt. Technol. Lett.* **48**, 480–482 (2006) © 2006 Wiley.]

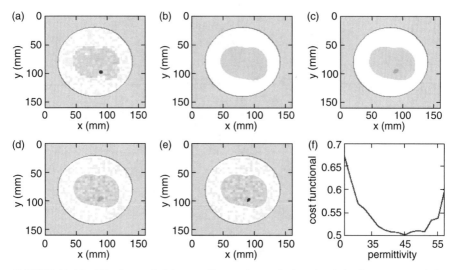

FIGURE 10.10 Final permittivity profiles at the end of each of the four stages of the algorithm (level-set method). (a) true permittivity profile of the breast model with tumor of $\varepsilon_s = 50$; (b) final profile of stage 1; (c) final profile of stage 2; (d) final profile of stage 3; (e) final profile of our reconstruction algorithm, which is the profile corresponding to the minimum of the curve shown in image (f) (here at permittivity value equal to 45); (f) final cost versus tumor permittivity (stage 4 result). [Reproduced from N. Irishina, O. Dorn, and M. Moscoso, "Level-set techniques for microwave medical imaging," *Proc. Appl. Math. Mech.* **7** (1), 1151601–1151602 (Dec. 2007).]

linear sampling method (Section 5.8) has also been applied. Figure 10.11 gives an example of the results obtained by using this qualitative technique in conjunction with the inhomogeneous Green function approach described in Section 4.10, and considering the detailed three-dimensional breast model developed by Zastrow et al. (2008). In particular, Figures 10.11a,b provide the distributions of the relative dielectric permittivity and of the electric conductivity in a two-dimensional section of the model (used for the computation of Green's function at a frequency $f = 3\,\text{GHz}$). Figure 10.11c gives the amplitude distribution of the indicator function Ψ (see Section 5.8), which has been obtained after severely perturbing the input data (the computed values of the scattered electric field in the observation domain) by an additive random noise of 5%. The indication function clearly allows for the localization of a discontinuity (a circular model of a tumor with a radius equal to 0.2 cm). In particular, the error in the estimation of the tumor center is about 0.1 cm.

Finally, it should be noted that other excellent reconstruction results have been obtained by using beamforming and confocal imaging techniques. As mentioned at the end of Section 5.9, the scientific literature devoted to these approaches is now really wide. Consequently, the reader can refer to the

MEDICAL APPLICATIONS OF MICROWAVE IMAGING 251

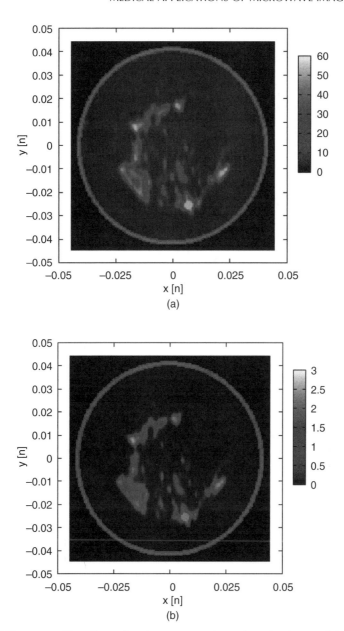

FIGURES 10.11 Two-dimensional reconstruction of a circular model of a tumor (radius 0.2 cm) inside a slide of an accurate numerical breast phantom (Zastrow et al. 2008); linear sampling method/inhomogeneous Green's function. (a), (b) original distributions of the relative dielectric permittivity and of the electric conductivity (S/m); (c) indicator function. Additive random noise (5%). (Simulation performed by G. Bozza, University of Genoa, Italy.)

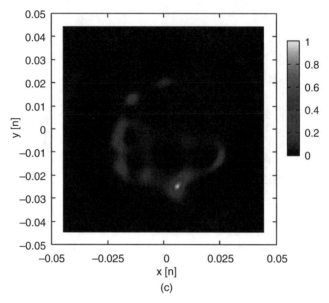

FIGURE 10.11 *Continued*

numerous papers cited throughout this section and in Section 5.9, whereas other important considerations concerning medical applications of microwave imaging can be found elsewhere in the literature (Semenov et al. 2002, Fear et al. 2003, Paulsen et al. 2005, Bertero and M. Piana 2006, Williams et al. 2006, Arunachalam et al. 2008, Lim et al. 2008, Meaney et al. 2007, Irishina et al. 2009, Salvador and Vecchi 2009).

10.3　SHALLOW SUBSURFACE IMAGING

The detection and identification of buried objects is very important in geophysical prospections, as well as in humanitarian applications and in the inspection of archeological sites. In particular, electromagnetic techniques are able to detect, for example, voids, tunnels, pipes, cables, and other civil or military structures. The demining activity is also another very important application. It is well known that ground-penetrating radar (GPR) represents a very efficient tool for this kind of applications. More recently, however, in order to improve the detection capabilities of the original GPR approach based on a direct evaluation of the so-called radiograms, which requires specific experience in the interpretation of final images, several inversion procedures have been proposed. In implementing these procedures, the localization and identification processes are treated as inverse scattering problems, essentially as described in the previous sections (D'Errico et al. 1993, Cassidy 2007,

Bozza et al. 2007c). In particular, at higher frequencies, the imaging is limited to shallow subsurface detection, which is of paramount importance in several cases (e.g., demining or pollution evaluations).

Concerning the adopted antennas, in order to radiate and receive wideband signals, according to the discussion in Section 9.3, they are essentially the same as those used in radar applications. In particular, for subsurface detection purposes, the most frequently used are resistively loaded dipoles, biconical, Vivaldi, and bowtie antennas. Bowtie antennas have been used, for example, in a prototype of an imaging system for buried-object detection, which has been developed by Pichot et al. (2004). In that apparatus, coaxial fed bow-tie microstrip antennas are used as transmitting antennas in the frequency band 0.3–1.3 GHz, with a linear polarization and a low reflection coefficient with a standing-wave ratio (SWR) < 2.

At lower frequencies (which actually should be outside the frequency *range* considered in the present monograph), similar approaches can be considered. One of them is the *very-early-time electromagnetic* (VETEM) system (Cui et al. 2000, 2001), which operates from several tens of kilohertz to several tens of megahertz and has potential environmental applications that include "delineation of the boundaries of waste-burial pits and trenches, characterization of the contents of those pits and trenches, measurement of the thickness and assessment of the integrity of clay caps, mapping liquid-spill plumes, ground-injection monitoring, and so on" (Cui et al. 2000, p. 17).

Two examples of buried-object reconstruction are now provided. They concern far different techniques. The first example is related to the inversion of real data collected in a cross-borehole configuration (see Section 4.5 and Fig. 4.9). The reconstruction results are presented in Figures 10.12 and 10.13 (Tarnus et al. 2004).

The second example involves the monitoring of fluid movements in a reservoir. The configuration is still of borehole type, but both single-hole and cross-borehole imaging modalities are considered. The results provided refer to numerical simulations in which a 2.5-dimensional problem is assumed (Abubakar et al. 2006). In the cross-well inversion experiment, there are 29 sources uniformly spaced inside the left well bore ($x = -25$ m), from $y = -70$ m up to $y = 70$ m. For each source, 30 receivers are used to measure the scattered field. The receivers are located inside the right well bore ($x = 25$ m), from $y = -72.5$ m up to $y = 72.5$ m. The source and receiver wells are denoted by dashed lines in Figure 10.14. The operating frequency is 500 Hz, and the rectangular investigation domain (subdivided into 32×48 square pixels) has dimensions of 80 m in the x direction and 120 m in the y direction. It should be noted that the discretization step considered (2.5 m) represents about 5% of the well spacing. For this configuration, the final reconstructed images are reported in Figures 10.14–10.16.

Another important application is represented by estimation of the water lost by pipes of aqueducts, which has a strong social impact in regions where the availability of water is limited (Catapano et al. 2006). In this case, since

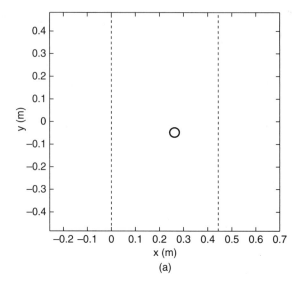

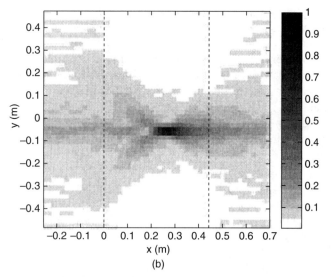

FIGURE 10.12 (a) Geometry of a transverse magnetic tomography with a 4-cm cylinder; (b) numerical and (c) experimental results for nine frequencies (400–480 MHz). [Reproduced from R. Tarnus, X. Dérobert, and Ch. Pichot," Multifrequency microwave tomography between boreholes," *Microwave Opt. Technol. Lette.* **42**(1), 4–8 (July 2004), © 2004 Wiley.]

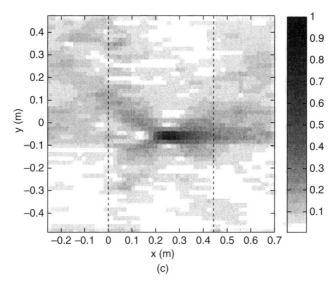

(c)

FIGURE 10.12 *Continued*

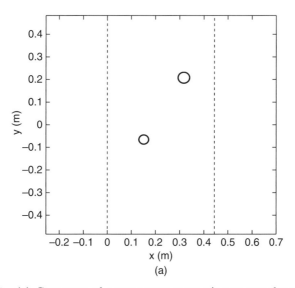

FIGURE 10.13 (a) Geometry of a transverse magnetic tomography with two 4-cm cylinders; (b) numerical and (c) experimental results for nine frequencies (400–480 MHz). [Reproduced from R. Tarnus, X. Dérobert, and Ch. Pichot," Multifrequency microwave tomography between boreholes," *Microwave Opt. Technol. Lette.* **42**(1), 4–8 (July 2004), © 2004 Wiley.]

256 APPLICATIONS OF MICROWAVE IMAGING

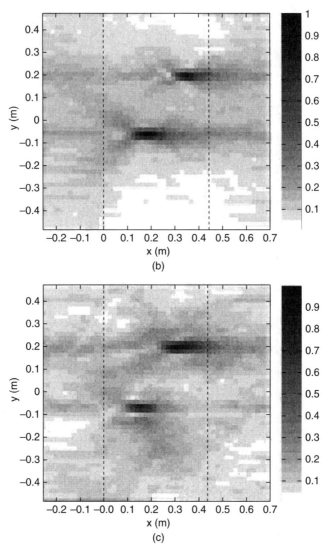

FIGURE 10.13 *Continued*

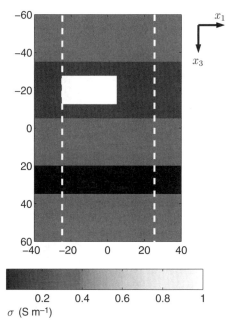

FIGURE 10.14 The conductivity distribution of the test configuration. The well bores are denoted by the dashed lines and are located at x = −25 m and x = 25 m. [Reproduced from A. Abubakar, P. M. van den Berg, and T. M. Habashy, "An integral equation approach for 2.5-dimensional forward and inverse electromagnetic scattering," Geophys. J. Int. **165**(3), 744–762 (2006), © 2006 Wiley.]

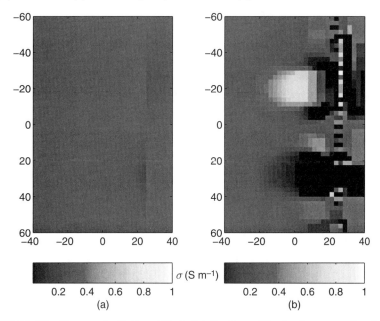

FIGURE 10.15 The conductivity σ (S/m) distribution of the backpropagation (a) and the full inversion (b) from cross-well data without invoking reciprocity. [Reproduced from A. Abubakar, P. M. van den Berg, and T. M. Habashy, "An integral equation approach for 2.5-dimensional forward and inverse electromagnetic scattering," Geophys. J. Int., **165**(3), 744–762 (2006), © 2006 Wiley.]

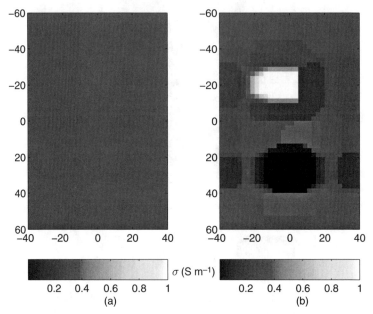

FIGURE 10.16 The conductivity σ (S/m) distribution of the backpropagation (a) and the full inversion (b) from cross-well data invoking reciprocity. [Reproduced from A. Abubakar, P. M. van den Berg, and T. M. Habashy, "An integral equation approach for 2.5-dimensional forward and inverse electromagnetic scattering," *Geophys. J. Int.* **165**(3), 744–762 (2006), © 2006 Wiley.]

the water is usually leaking under the pipe, it is quite difficult to detect it from common radiograms. Tomographic imaging techniques can represent a good solution. Moreover, since the position and dimensions of the pipe can be known to some extent, it can be fruitful to consider an approach like the one described in Section 4.10. In particular, the Green function for the unperturbed structure (i.e., soil and pipe) can be computed. Consequently, the medium "soil and pipe" is now the propagation medium for electromagnetic waves. This approach has been followed in the study by Catapano et al. (2006), where, after neglecting the effects of the air–soil interface and assuming the pipe to be approximated by an ideal circular cylinder, Green's function has been analytically computed.

REFERENCES

Abubakar, A., P. M. van den Berg, and T. M. Habashy, "An integral equation approach for 2.5-dimensional forward and inverse electromagnetic scattering," *Geophys. J. Int.* **165**, 744–762 (2006).

Arunachalam, K., L. Udpa, and S. S. Udpa, "A computational investigation of microwave breast imaging using deformable reflector," *IEEE Trans. Biomed. Eng.* **55**, 554–562 (2008).

Bacur, V., *Nondestructive Characterization and Imaging of Wood*, Springer, Berlin, 2003.

Bahr, A. J., "Nondestructive microwave evaluation of ceramics," *IEEE Trans. Microwave Theory Tech.* **26**, 676–683 (1978).

Bertero, M. and M. Piana, "Inverse problems in biomedical imaging: Modeling and methods of solution," in *Complex Systems in Biomedicine*, A. Quarteroni, L. Formaggia, and A. Veneziani, eds., Springer, Milan, 2006.

Bindu, G. and K. T. Mathew, "Characterization of benign and malignant breast tissues using 2-D microwave tomographic imaging," *Microwave Opt. Technol. Lett.* **49**, 2341–2345 (2007).

Bolomey, J.-C., D. Lesselier, C. Pichot, and W. Tabbara, "Spectral and time domain approaches to some inverse scattering problems," *IEEE Trans. Anten. Propag.* **29**, 206–212 (1981).

Bozza, G., C. Estatico, M. Pastorino, and A. Randazzo, "Electromagnetic imaging for non-intrusive evaluation in civil engineering," *Proc. 2007 IEEE Int. Workshop on Imaging Systems and Techniques*, Krakow, Poland, 2007 (2007a).

Bozza, G., C. Estatico, M. Pastorino, and A. Randazzo, "Microwave imaging for non-destructive testing of dielectric structures: Numerical simulations by using an Inexact-Newton method," *Mater. Eval.* **65**, 917–922 (2007) (2007b).

Bozza, G., C. Estatico, M. Pastorino, and A. Randazzo, "Application of an inexact-Newton method within the second-order Born approximation to buried objects," *IEEE Geosci. Remote Sens. Lett.* **4**, 51–55, (2007) (2007c).

Brovko, A. V., E. K. Murphy, M. Rother, H. P. Schuchmann, and V. V. Yakovlev, "Waveguide microwave imaging: Spherical inclusion in a dielectric sample," *IEEE Microwave Wireless Compon. Lett.* **18**, 647–649 (2008).

Cassidy, N. J., "A review of practical numerical modeling methods for the advanced interpretation of ground-penetrating radar in near-surface environments," *Near-Surface Geophys.* **5**, 5–22 (2007).

Catapano, I., L. Crocco, R. Persico, M. Pieraccini, and F. Soldovieri, "Linear and nonlinear microwave tomography approaches for subsurface prospecting: Validation on real data," *IEEE Anten. Wireless Propag. Lett.* **5**, 49–53 (2006).

Chang, K., ed., *Encyclopedia of RF and Microwave Engineering*, Wiley, New York, 2005.

Chen, Y., E. Gunawan, K. S. Low, S.-C. Wang, C. B. Soh, and T. C. Putti, "Effect of lesion morphology on microwave signature in 2-D ultra-wideband breast imaging," *IEEE Trans. Biomed. Eng.* **55**, 2011–2021 (2008).

Chew, W. C., *Waves and Fields in Inhomogeneous Media*, Van Nostrand Reinhold, New York, 1990.

Colton, D. and P. Monk, "The detection and monitoring of leukemia using electromagnetic waves: Numerical analysis," *Inverse Problems* **11**, 329–342 (1995).

Cui, T. J. et al., "Numerical modeling of an enhanced very early time electromagnetic (VETEM) prototype system," *IEEE Anten. Propag. Mag.* **42**, 17–27 (2000).

Cui, T. J., W. C. Chew, A. A. Aydiner, D. L. Wright, and D. V. Smith, "Detection of buried targets using a new enhanced very early time electromagnetic (VETEM) prototype system," *IEEE Trans. Geosci. Remote Sens.* **39**, 2702–2712 (2001).

D'Errico, M. S., B. L. Douglas, and H. Lee, "Subsurface microwave imaging for nondestructive evaluation of civil structures," *Proc. IEEE Int. Conf. Acoustics, Speech, and Signal Processing*, Minneapolis, MN, USA, 1993, vol. 5, pp. 453–456.

Elsdon, M., D. Smith, M. Leach, and S. Foti, "Microwave imaging of concealed metal objects using a novel indirect holographic method," *Microwave Opt. Technol. Lett.* **47**, 536–537 (2005).

Elsdon, M., D. Smith, M. Leach, and S. J. Foti, "Experimental investigation of breast tumor imaging using indirect microwave holography," *Microwave Opt. Technol. Lett.* **48**, 480–482 (2006).

Eskelinen, P. and P. Harju, "Characterizing wood by microwaves," *IEEE Aerospace Electron. Syst. Mag.* **13**, 34–35 (1998).

Fear, E. C., J. Sill, and M. A. Stuchly, "Experimental feasibility study of confocal microwave imaging for breast tumor detection," *IEEE Trans. Microwave Theory Tech.* **51**, 887–892 (2003).

Foster, K. R. and H. P. Schwan, "Dielectric properties of tissues and biological materials: A critical review," *Crit. Rev. Biomed. Eng.* **17**, 25–104 (1989).

Gabriel, C., Compilation of the Dielectric Properties of Body Tissues at RF and Microwave Frequencies, Report AL/OE-TR-1996–0037, Radiofrequency Radiation Division, Brooks Air Force Base, Texas, 1996.

Hayt, W. H. and J. A. Buck, *Engineering Electromagnetics*, McGraw-Hill, New York, 2001.

Hernández López, M. A., M. Quintillán González, S. González García, A. Rubio Bretones, and R. Gómez Martín, "A rotating array of antennas for confocal microwave breast imaging," *Microwave Opt. Technol. Lett.* **39**, 307–311 (2003).

Huynh, P. T., A. M. Jarolimek, and S. Daye, "The false-negative mammogram," *Radio Graphics* **18**, 1137–1154 (1998).

Irishina, N., O. Dorn, and M. Moscoso, "Level-set techniques for microwave medical imaging," *Proc. Appl. Math. Mech.* **7**, 1151601–1151602 (2007).

Irishina, N., M. Moscoso, and O. Dorn, "Detection of small tumors in microwave medical imaging using level sets and MUSIC," *Proc. 2006 Progress in Electromagnetics Research Symp.* Cambridge, MA, USA, 2006, pp. 43–47.

Irishina, N., M. Moscoso, and O. Dorn, "Microwave imaging for early breast cancer detection using a shape-based strategy," *IEEE Trans. Biomed. Eng.* **56**, 1143–1153 (2009).

Joachimowicz, N., J. J. Mallorqui, J. C. Bolomey, and A. Broquetas, "Convergence and stability assessment of Newton-Kantorovich reconstruction algorithms for microwave tomography," *IEEE Trans. Med. Imag.* **17**, 562–570 (1998).

Kaestner, A. P. and L. B. Baath, "Microwave polarimetry tomography of wood," *IEEE Sensors J.* **5**, 209–215 (2005).

Kharkovsky, S. and R. Zoughi, "Microwave and millimeter wave nondestructive testing and evaluation—overview and recent advances," *IEEE Instrum. Meas. Mag.* **10**, 26–38 (2007).

REFERENCES

Kim, Y. J., L. Jocre, F. De Flaviis, and M. Q. Feng, "Microwave reflection tomographic array damage detection of civil structures," *IEEE Trans. Anten. Propag.* **51**(11), 3022–3032 (2003).

Larsen, L. E. and J. H. Jacobi, *Medical Applications of Microwave Imaging*, IEEE Press, Piscataway, NJ, 1986.

Lazebnik, M. et al., "A large-scale study of the ultrawideband microwave dielectric properties of normal, benign, and malignant breast tissues obtained from cancer surgeries," *Phys. Med. Biol.* **52**, 6093–6115 (2007).

Lazebnik, M. et al., "Electromagnetic spectroscopy of normal breast tissue specimens obtained from reduction surgeries: Comparison of optical and microwave properties," *IEEE Trans. Biomed. Eng.* **55**, 2444–2451 (2008).

Lhiaubet, C., G. Cottard, J. Ciccotelli, J. F. Portala, and J. C. Bolomey, "On-line control in wood and paper industries by means of rapid microwave linear sensors," *Proc. 1992 Eur. Microwave Conf.*, 1992, vol. 2, pp. 1037–1040.

Li, X., E. J. Bond, B. D. Van Veen, and S. C. Hagness, "An overview of ultra-wideband microwave imaging via space-time beamforming for early-stage breast-cancer detection," *IEEE Anten. Propag. Mag.* **47**, 19–34 (2005).

Li, X., S. K. Davis, S. C. Hagness, D. W. van der Weide, and B. D. Van Veen, "Microwave imaging via space-time beamforming: experimental investigation of tumor detection in multilayer breast phantoms," *IEEE Trans. Microwave Theory Tech.* **52**, 1856–1865 (2004).

Li, X., S. C. Hagness, M. K. Choi, and D. W. van der Weide, "Numerical and experimental investigation of an ultrawideband ridged pyramidal horn antenna with curved launching plane for pulse radiation," *Antennas Wireless Propagat. Lett.* **2**, 259–262 (2003).

Lim, H. B., N. T. Nhung, E. Li, and N. D. Thang, "Confocal microwave imaging for breast cancer detection: Delay-multiply-and-sum image reconstruction algorithm," *IEEE Trans. Biomed. Eng.* **55**, 1697–1704 (2008).

Lin, J. C., "Frequency optimization for microwave imaging of biological tissues," *IEEE Proc.* **73**, 374–375 (1985).

Mashal, A., J. H. Booske, and S. C. Hagness, "Towards contrast-enhanced microwave induced thermoacoustic imaging of breast cancer: An experimental study of the effects of microbubbles on simple thermoacoustic targets," *Phys. Med. Biol.* **54**, 641–650 (2009).

Meaney, P. M. et al., "Initial clinical experience with microwave breast imaging in women with normal mammography," *Acad. Radiol.* **14**, 207–218 (2007).

Mikhnev, V. and P. Vainikainen, "Microwave imaging of layered structures in civil engineering," *Proc. 2002 Eur. Microwave Conf.*, Milan, Italy, 2002, pp. 717–720.

Mittra, R. "Inverse scattering and remote probing" in *Computer Techniques for Electromagnetics*, R. Mittra, ed., Pergamon, New York, 1973, pp. 372–392.

Mudanyal, O., S. Yldz, O. Semerci, A. Yapar, and I. Akduman, "A microwave tomographic approach for nondestructive testing of dielectric coated metallic surfaces," *IEEE Geosci Remote Sens. Lett.* **5**, 180–184 (2008).

Park, J. and C. Nguyen, "An ultrawide-band microwave radar sensor for nondestructive evaluation of pavement subsurface," *IEEE Sensor J.* **5**, 942–949 (2005).

Pastorino, M., A. Massa, and S. Caorsi, "A global optimization technique for microwave nondestructive evaluation," *IEEE Trans. Instrum. Meas.* **51**, 666–673 (2002).

Pastorino, M., A. Salvadé, R. Monleone, T. Bartesaghi, G. Bozza, and A. Randazzo, "Detection of defects in wood slabs by using a microwave imaging technique," *Proc. 2007 IEEE Instrum entation and Measurement Technology Conf.*, Warsaw, Poland, 2007.

Paulsen, K. D., P. M. Meaney, and L. Gilman, eds., *Alternative Breast Imaging. Four Model-Based Approaches*, Springer, Berlin, 2005.

Pichot, C., J. Y. Dauvignac, I. Aliferis, E. Le Brusq, R. Ferrayé, and V. Chatelée, "Recent nonlinear inversion methods and measurement system for microwave imaging," *Proc. 2004 IEEE Int. Workshop on Imaging Systems and Techniques*, Stresa, Italy, 2004, pp. 95–99.

Qaddoumi, N. N., W. M Saleh, and M. Abou-Khousa, "Innovative near-field microwave nondestructive testing of corroded metallic structures utilizing open-ended rectangular waveguide probes," *IEEE Trans. Instrum. Meas.* **56**, 1961–1966 (2007).

Sabouni, A., D. Flores-Tapia, S. Noghanian, G. Thomas, and S. Pistorius, "Hybrid microwave tomography technique for breast cancer imaging," *Proc. 28th IEEE EMBS Annual Int. Conf.*, New York, 2006, pp. 4273–4276.

Salvadè, A. et al., "Microwave imaging of foreign bodies inside wood trunks," *Proc. 2008 IEEE Int. Workshop on Imaging Systems and Techniques*, Chania, Greece, 2008, pp. 88–93.

Salvadè, A., M. Pastorino, R. Monleone, G. Bozza, A. Randazzo, and T. Bartesaghi, "A noninvasive microwave method for the inspection of wood beams," *Proc. 3rd Int. Conf. Electromagnetic Near-Field Characterization & Imaging*, St. Louis, MO, 2007, pp. 395–400.

Salvador, S. M. and G. Vecchi, "Experimental tests of microwave breast cancer detection on phantoms," *IEEE Trans. Anten. Propagat.* **57**, 1705–1712 (2009).

Semenov, S. Y. et al., "Spatial resolution of microwave tomography for detection of the myocardial ischemia and infarction. Experimental study on two-dimensional models," *IEEE Trans. Microwave Theory Tech.* **48**, 538–544 (2000).

Semenov, S. Y. et al., "Three-dimensional microwave tomography: Initial experimental imaging of animals," *IEEE Trans. Biomed. Eng.* **49**, 55–63 (2002).

Shea, J. D., P. Kosmas, S. C. Hagness, and B. D. Van Veen, "Contrast-enhanced microwave breast imaging," *Proc. 2009 Int. Symp. Antenna Technology and Appl. Electromagn. (ANTEM/URSI)*, Banff, Canada, 2009.

Shen, J., G. Schajer, and R. Parker, "Theory and practice in measuring wood grain angle using microwaves," *IEEE Trans. Instrum. Meas.* **43**, 803–809 (1994).

Tarnus, R. X. Dérobert, and Cr. Pichot, "Multifrequency microwave tomography between boreholes," *Microwave Opt. Technol. Lett.* **42**, 4–8 (2004).

Watanabe, M., M. Miyakawa, Y. Miyazaki, N. Ishii, and M. Bertero, "Image restoration of CP-MCT for sugar distribution imaging inside a fruit," *Proc. 36th Eur. Microwave Conf.*, Manchester, UK, 2006, pp. 1248–1251.

Williams, T. C., E. C. Fear, and D. T. Westwick, "Tissue sensing adaptive radar for breast cancer detection—investigation of an improved skin-sensing method," *IEEE Trans. Microwave Theory Tech.* **54**, 1308–1414 (2006).

Xie, Y., B. Guo, L. Xu, J. Li, and P. Stoica, "Multistatic adaptive microwave imaging for early breast cancer detection," *IEEE Trans. Biomed. Eng.* **53**, 1647–1657 (2006).

Xu, H., T. Li, and Y. Sun, "The application research of microwave imaging in nondestructive testing of concrete wall," *Proc. 6th World Congress on Intelligent Control and Automation*, Dalian, China 2006, pp. 5157–5161.

Zastrow, E., S. K. Davis, M. Lazebnik, F. Kelcz, B. D. Van Veen, and S. C. Hagness, "Development of anatomically realistic numerical breast phantoms with accurate dielectric properties for modeling microwave interactions with the human breast," *IEEE Trans Biomed. Eng.* **55**, 2792–2800 (2008).

Zeng, Z., K. Arunachalam, C. Lu, B. Shanker, and L. Udpa, "Element-free Galerkin method in modeling microwave inspection of civil structures," *Proc. 12th Biennial IEEE Conf. Electromagnetic. Field Computation*, Miami, FL, USA, 2006, p. 268.

Zhang, Z. Q., Q. H. Liu, C. Xiao, E. Ward, G. Ybarra, and W. T. Joines, "Microwave breast imaging: 3D forward scattering simulation," *IEEE Trans. Biomed. Eng.* **50**, 1180–1189 (2003).

Zoughi, R., *Microwave Non-Destructive Testing and Evaluation*, Kluwer Academic Publishers, Dordrecht, 2000.

Zoughi, R. and S. Bakhtiari, "Microwave nondestructive detection and evaluation of disbonding and delamination in layered-dielectric slabs," *IEEE Trans. Instrum. Meas.* **39**, 1059–1063 (1990).

CHAPTER ELEVEN

Microwave Imaging Strategies, Emerging Techniques, and Future Trends

11.1 INTRODUCTION

Research activity on microwave imaging has evolved rapidly, and it is almost impossible to mention the plethora of new ideas, methods, and strategies that continuously appear in the scientific literature and are proposed by a number of very active research teams working around the world and discussed at many conferences devoted to applied electromagnetics. Nevertheless, some more recent ideas concerning possible new imaging strategies and emerging solutions can be summarized, in an attempt to delineate the trend in scientific research in this field.

One trend is certainly represented by the development of hybrid techniques, which allow utilization of the specific features of the different methods applied in combination. This topic has been discussed in Chapter 8. It should be mentioned once again that the ability to devise specific techniques for any particular application is of paramount importance from an engineering perspective, since it is absolutely evident that *general-purpose* approaches cannot be developed.

However, besides hybrid techniques, other strategies, approaches, and particular applications have been proposed by the research community. Some of them are discussed in the following sections, without any claim of completeness.

Microwave Imaging, By Matteo Pastorino
Copyright © 2010 John Wiley & Sons, Inc.

11.2 POTENTIALITIES AND LIMITATIONS OF THREE-DIMENSIONAL MICROWAVE IMAGING

Most of the approaches described in the previous sections concern two-dimensional imaging configurations, although it has been mentioned that, theoretically, most of them can be immediately extended to directly accommodate three-dimensional configurations.

Although three-dimensional inversion techniques were proposed in the 1980s or so, the limitations due to the computer powers made it possible to inspect only very simple scatterers with very rough discretizations and reduced meshes (Ghodgaonkar et al. 1983, Guo and Guo 1987). As an example, the pseudoinverse matrix [equation (5.5.11)] was used to *invert* a discretized form of the data equation [equation (3.2.1)] in order to reconstruct dielectric cubes (Ney et al. 1984, Caorsi et al. 1988). After these rather naive attempts, the microwave imaging community focused attention on two-dimensional approaches that allow spending all the available computer resources for obtaining acceptably fine discretizations. Concerning this point, also of interest is the discussion in the paper by Hagmann and Levin (1990), in which some of the previously reported results concerning three-dimensional configurations are judged to be "overly optimistic," because "sizable errors in the forward solution were largely cancelled by errors in the inverse calculations"; that is, the so-called *inverse crime* perhaps could not be avoided.

Although research activity on microwave imaging was previously focused, as mentioned, on two-dimensional methods, some other techniques based on approximations have been extended to handle three-dimensional problems. Among them, we can mention the distorted Born iterative method (Section 6.6), which has been applied together with a solving procedure based on a biconjugate-gradient method and fast Fourier transform (FFT) by Gan and Chew (1995), whereas the extension of diffraction tomography (based on the Born and Rytov approximations) to three-dimensional problems has also been proposed (Vouldis et al. 2006). Moreover, three-dimensional reconstructions of buried objects have also been obtained (at low frequencies) by using the very-early time electromagnetic (VETEM) system mentioned in Section 10.3 (Cui et al. 2003).

Obviously, when the objective of the investigation is simply the shape of the targets, qualitative methods can be applied without difficulties to three-dimensional problems, due to the reduced computational complexity (see Chapter 5). The linear sampling method, for example, has been used for three-dimensional scatterers several times (Giebermann et al. 1999; Brignone et al. 2009) and, recently, the level-set method has been extended to shaping three-dimensional metallic objects (Ferrayé et al. 2009).

Very recently, however, the direct application of nonlinear inspection techniques to reconstruct three-dimensional object has been reconsidered. The main reason is the terrific improvement in the computer power of modern PCs and also the wide availability of multiprocessor computers. In addition, it should be consider that an improved capability of handling the ill-posedness of the inverse scattering problem had a strong impact on this new research direction. In fact, the dramatic increase in the number of unknowns in three-dimensional problems, with respect to two-dimensional problems, is reflected in a significant increase in the ill-posedness of the inverse problem.

As an example of *full-wave* three-dimensional imaging, a quantitative iterative approach based on a Gauss–Newton method (see Chapter 6) has been applied to inspect a cube and a sphere by using both synthetic and real data (De Zaeytijd et al. 2007), in which the total electric field has been computed with a fast-forward solver at each iteration, whereas in the inverse solution, only the values of the dielectric permittivities have been considered as unknowns. Quite good results have been obtained with a three-dimensional mesh of $15 \times 15 \times 15$ cells, for the inverse problem, and with a finer mesh for the (well-posed) direct computation ($30 \times 30 \times 30$ cells) (De Zaeytijd et al. 2007). A quasi-Newton minimization has also been proposed for reconstruction of the three-dimensional distribution of electrical conductivity for applications related to the detection of oil reservoirs (which is the same problem considered in Figs. 10.14–10.16). For these applications, imaging techniques can suitably exploit the high contrast in resistivity between saline-filled rocks and hydrocarbons (Abubakar et al. 2009). An iterative Newton method has also been used (Rubaek et al. 2009), in particular for processing the data collected by a cylindrical multistatic antenna setup constituted by 32 horizontally oriented antennas (see Chapter 9). In this approach, a multiplicative functional has been assumed.

In the medical field, an interesting application of the distorted Born iterative method has been proposed in the study by Winters et al. (2009), where, in order to solve a complex three-dimensional inverse problem, the use of patient-specific basis functions is proposed for discretization of the continuous model (Section 3.4). In this way, a reduced parameterization of the problem is obtained with a significant computational saving. Such an approach confirms once again the importance of developing specific application-oriented systems and reconstruction techniques.

It should be also mentioned that, although computationally expensive, stochastic optimization methods have been proposed for three-dimensional imaging as well. An example is represented by the application of the micro-genetic algorithm (Huang and Mohan 2005), which involves a limited population of individuals (Section 7.3).

The perspective of three-dimensional imaging can, of course, be of interest for research on hybrid methods (described in Chapter 8), since the reduction in numerical complexity is one of the main reasons for developing hybrid imaging approaches. An example of hybrid approaches for two-dimensional problems successively extended to three-dimensional configurations is represented by the combined strategy of linear sampling method and ant colony optimization (Section 8.3). It must be noted that the use of stochastic optimization methods (quite computationally heavy), although combined with fast qualitative procedures, still requires some additional assumptions in order to treat realistic configurations. For this reason, the bodies to be inspected are assumed, for example, to be homogeneous (Brignone et al. 2008). Under this hypothesis, the proposed hybrid method provided very accurate reconstructions, as can be seen in Figure 11.1.

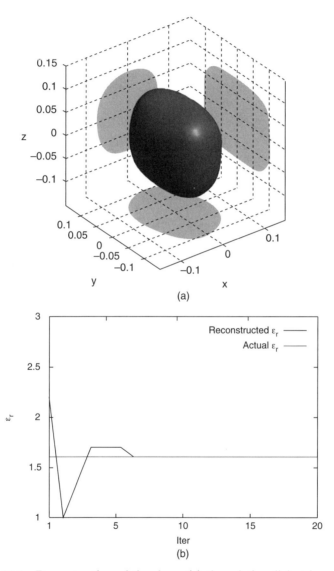

FIGURE 11.1 Reconstruction of the shape (a) the relative dielectric permittivities (b) and electric conductivity (c) versus the iteration number for a homogeneous dielectric parallelepiped; (d) number of cost function evaluations. [Reproduced from M. Brignone, G. Bozza, A. Randazzo, R. Aramini, M. Piana, and M. Pastorino, "A hybrid approach to 3D microwave imaging by using linear sampling and ant colony optimization," *IEEE Trans. Anten. Propag.* **56**(10), 3224–3232 (Oct. 2008), © 2008 IEEE.]

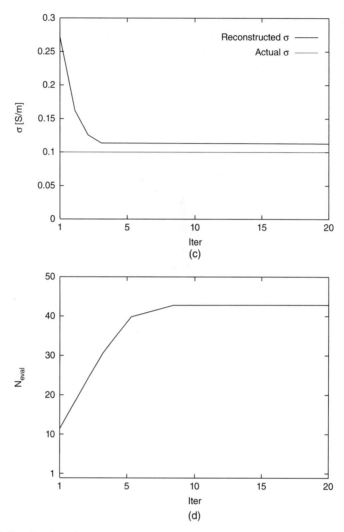

FIGURE 11.1 *Continued*

It should be recalled that qualitative approaches, which are able to provide the support of the scatterer, can be also combined with deterministic quantitative techniques (Section 8.3). An example is the method proposed by Morabito et al. (2009), who combined the linear sampling method with a technique based on the extended Born approximation (Section 4.7) and applied it to a contrast source formulation (Section 3.2).

11.3 AMPLITUDE-ONLY METHODS

Although measurement of the phase of the electromagnetic field at microwave frequencies no longer represents a technical problem, there is still significant interest in developing imaging approaches based on the amplitude of the scattered field only. This assumption results in less expensive instrumentation, which is particularly suitable at the highest frequencies of the microwave range and at millimeter waves. There are essentially two possible conceptual approaches:

1. Application of phase retrieval techniques that are able to compute the phase of the field in the observation domain starting from measured values of the field amplitude (Hislop et al. 2007). Once amplitudes and phases of the scattered field at the measurement points are available, one of the methods described in Chapters 5–8 can be directly applied. This approach has been previously mentioned at the end of Section 9.2 and some references are provided there.
2. Exploring inversion procedures that reconstruct the scatterers directly, starting with amplitude-only data. In principle, a function similar to the one reported in Section 6.7 can be constructed on the bases of the amplitudes of the fields. In particular, the *residual* [equation (6.7.2)] between the measured data and the computed data (i.e., the data calculated at each iteration on the basis of the current trial solution) can be formed by using the difference of the amplitudes between the two fields, instead of their complex values (amplitudes and phases). An approach of this kind was followed by Caorsi et al. (2003), who optimized a *phaseless* functional by using the memetic algorithm (Section 8.2).

Finally, it should be mentioned that amplitude-only data have been inverted by extending other classical methods. For example, diffraction tomography (Section 5.11) has been used (Devaney 1992), an approach based on the Rytov approximation (Section 4.8) has been proposed (Zhang et al. 2009), and a so-called multiscale approach has been followed (Franceschini et al. 2006).

11.4 SUPPORT VECTOR MACHINES

In the previous sections, it was mentioned several times that the *image* of the target is not always the final objective of the inspection process. In some cases, sufficient information is represented by the knowledge of some useful parameters of the target, such as, position or dimensions. In certain applications, the object under test can be a canonical object or can be approximated by a canonical object. As a priori information, one can also assume, in specific cases, that the target is homogeneous or that its dielectric parameters exhibit a specific dependence on the spatial coordinates. Fast and simplified techniques can be applied in these cases, as described in Chapter 5. Moreover, when few parameters have to be detected, the global optimization methods discussed in Chapter 7 can be suitably applied. Beside these approaches, the determination of few parameters of a scattering configuration can be efficiently addressed by using neural network approaches as well.

Conceptually, the network must first be *trained*, in particular by using, as input data, two sets of data. The first one includes several arrays of parameters, each of them able to describe one of the known scatterers belonging to a given set. The second set of data contains the values of the scattered field (at the measurement points) produced by the previously considered known scatterers. Both sets constitute the *training data* of the network. The previously considered known scatterers might be canonical objects (e.g., the ones mentioned in Section 3.5), for which the produced scattered field can be calculated in a simple and fast way. Once the offline *training phase* is completed, the network should be able to almost instantaneously retrieve the key parameters of an *unknown* scatterer just starting from the values of the field that it scatters. This constitutes the *test phase*.

The use of neural networks has been proposed for inverse scattering purposes in several papers (e.g., Rekanos 2002, Bermani et al. 2002, and references cited therein). However, more recently, support vector machines (SVMs) (Vapnik 1998) have attracted notable attention due to their excellent generalization capabilities (i.e., the ability of dealing, in the test phase, with data quite different from those considered in the training phase) in several applications, such as pattern recognition, time prediction, and regression. Support vector machines have been also successfully applied in other areas of electromagnetics (e.g., for the determination of the arrival angles of incident waves (Hines et al. 2008)).

In order to retrieve an unknown target from scattering data, the functional to be minimized, F [equation (6.7.1)] can be successfully approximated by using a support vector regression (SVR) approach. To this end, let $\hat{\xi}^l$, $l = 1,\ldots, L$ be the array of data characterizing the lth known scatterer of the training set (which includes L known targets). In addition, $\hat{\mathbf{z}}^l, l = 1,\ldots, L$, is the array containing the samples of the scattered electric field at the measurement points. These field samples are produced by solving a direct scattering problem involving the lth known scatterer. Thus, the training set Ω_{train} is given by

$$\Omega_{\text{train}} = \left\{ \left(\hat{\mathbf{z}}^1, \hat{\xi}^1\right), \ldots, \left(\hat{\mathbf{z}}^l, \hat{\xi}^l\right), \ldots, \left(\hat{\mathbf{z}}^L, \hat{\xi}^L\right) \right\}. \quad (11.4.1)$$

The values contained in the training set Ω_{train} are used to evaluate the functional $\tilde{F}$ that approximates the unknown actual functional F. In particular, the approximating relationship is given by

$$\tilde{F}(\mathbf{z}) = \langle \mathbf{w}, \varphi(\mathbf{z}) \rangle + b, \quad (11.4.2)$$

where φ is a nonlinear function used to transform the input data $\mathbf{z}$ from their original space, namely, Σ, to a higher-dimensional space H (usually called *feature space*) and $\langle \cdot , \cdot \rangle$ denotes the inner product of H; $\mathbf{w}$ and b are parameters whose optimal values are obtained by minimizing the so-called regression risk function associated with SVR, which is given by

$$R_{\text{reg}} = C\sum_{l=1}^{L} c\left(\hat{\mathbf{z}}^l, \hat{\xi}^l\right) + \frac{1}{2}\|\mathbf{w}\|^2, \quad (11.4.3)$$

where C is a constant and $c\left(\hat{\mathbf{z}}^l, \hat{\xi}^l\right)$ is the ε-insensitive loss function, that is

$$c\left(\hat{\mathbf{z}}^l, \hat{\xi}^l\right) = \begin{cases} 0 & \text{if } \left|\hat{\xi}^l - \tilde{F}(\hat{\mathbf{z}}^l)\right| \leq \varepsilon, \\ \left|\hat{\xi}^l - \tilde{F}(\hat{\mathbf{z}}^l)\right| - \varepsilon & \text{otherwise,} \end{cases} \quad (11.4.4)$$

Where ε is the allowed error during the training phase. According to Vapnik (1998), the regression risk is minimized by using a dual formulation. In particular, a standard dualization method utilizing Lagrange multipliers is employed. The following quadratic optimization problem, to be solved with respect to the Lagrange multipliers α_i and α'_i, is obtained:

$$\max\left\{-\varepsilon\sum_{l=1}^{L}(\alpha'_l + \alpha_l) + \sum_{l=1}^{L}\hat{\xi}^l(\alpha'_l - \alpha_l) - \frac{1}{2}\sum_{l,i=1}^{L}(\alpha'_l + \alpha_l)(\alpha'_l - \alpha_l)\kappa(\hat{\mathbf{z}}^l, \hat{\mathbf{z}}^i)\right\}, \quad (11.4.5)$$

subject to

$$\sum_{l=1}^{L}(\alpha'_l - \alpha_l) = 0, \quad (11.4.6)$$

In both equations (11.4.5) and (11.4.6), $\alpha'_l, \alpha_l \in [0, C]$. The parameter C [equation (11.4.3)] has been found to control the generalization properties of the support vector machine (Smola and Schölkopf 2004). In particular, high values of C produce small estimation errors for configurations equal to those contained in the training set, but they result in a reduced generalization capability. Finally, in equation (11.4.5), κ is the *kernel function*, working on the original space Σ and defined as

$$\kappa(\hat{\mathbf{z}}^l, \mathbf{z}) = \langle \varphi(\hat{\mathbf{z}}^l), \varphi(\mathbf{z}) \rangle. \quad (11.4.7)$$

A commonly used kernel is the Gaussian kernel (Smola and Schölkopf 2004, Pastorino and Randazzo 2005), which is defined as

$$\kappa(\hat{\mathbf{z}}^l, \mathbf{z}) = e^{-\gamma\|\mathbf{z} - \hat{\mathbf{z}}^l\|^2}, \quad (11.4.8)$$

where γ is a constant to be selected by the user. By using the *dual-optimization problem*, one can express $\mathbf{w}$ in terms of the input data $\mathbf{z}$ as follows:

$$\mathbf{w} = \sum_{l=1}^{L}(\alpha_l - \alpha'_l)\varphi(\hat{\mathbf{z}}^l). \quad (11.4.9)$$

By substituting (11.4.9) into (11.4.2), one can rewrite $\tilde{F}$ as

$$\tilde{F}(\mathbf{z}) = \sum_{l=1}^{L}(\alpha_l - \alpha'_l)\langle\varphi(\hat{\mathbf{z}}^l), \varphi(\mathbf{z})\rangle + b$$
$$= \sum_{l=1}^{L}(\alpha_l - \alpha'_l)\kappa(\hat{\mathbf{z}}^l, \mathbf{z}) + b. \quad (11.4.10)$$

The resulting quadratic optimization problem can be efficiently solved by applying, for example, the sequential minimal optimization (SMO) algorithm (Keerthi et al. 2001). It is worth noting that only a subset of the Lagrange multipliers have nonzero values (i.e., those satisfying the condition $\left|\hat{\xi}^l - \tilde{F}(\hat{\mathbf{z}}^l)\right| \geq \varepsilon$); thus a sparse solution is obtained. The samples associated with the nonzero Lagrange multipliers are called *support vectors* (Vapnik 1998).

A very preliminary example of the use of support vector machines for detecting a discontinuity in a biological body is presented in Figure 11.2. In particular, a quite rough two-dimensional model of a human abdomen (Caorsi et al. 2004) includes a circular discontinuity in one of the tissues. The support vector machine has been trained by using various elliptic scatterers (Section 3.5), and the forward problem has been solved by using the method of moments. The final image is provided in Figure 11.2b. As can be seen, for this simple configuration, the mass in the abdomen is correctly located and shaped.

The application of support vector machines has been also proposed by Zhang et al. (2007), with some differences in implementation for the localization of passive scatterers. In particular, the application considered concerns detection of the dielectric parameters (problem A) and, separately, of the position (problem B) of an infinite circular cylinder in free space starting from measurements of the scattered electric field. The reconstruction errors for the two problems are briefly summarized in Tables 11.1 and 11.2.

11.5 METAMATERIALS FOR IMAGING APPLICATIONS

Another challenge is use of metamaterials for microwave imaging applications, since important steps have been taken toward the development of devices involving such materials. In particular, left-handed metamaterials, which were theoretically studied for the first time by the Russian scientist V. G. Veselago (Veselago 1968), have more recently been considered in some advanced technical areas of electromagnetics.

In certain frequency bands, these materials have a negative dielectric permittivity and a negative magnetic permeability. This property results in a number of significant effects and consequences. However, in terms of imaging applications, the most relevant aspect is the possibility of realizing focusing lens made by flat slabs. Because of the negative values of ε_r and μ_r, a microwave field, generated by a localized source (ideally, a point source), can be focused on a specific point by a flat slab. In this way it is possible to scan the

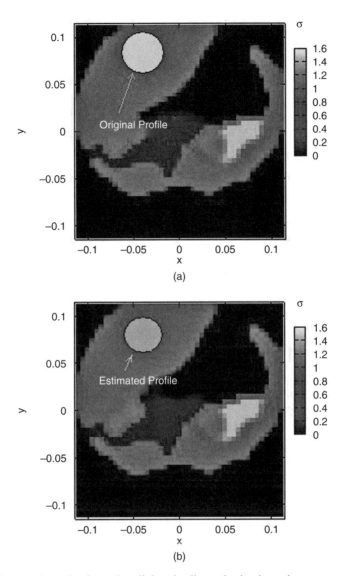

FIGURE 11.2 Localization of a dielectric discontinuity by using a support vector machine: (a) original configuration (rough model of a human abdomen); (b) final image. (Simulation performed by A. Randazzo, University of Genoa, Italy.)

TABLE 11.1 Error Analysis for Problem A

Error	Average, %	Maximum, %	Minimum, %
RelErr(ε_r)	0.630	2.500	0.011
RelErr(σ)	0.840	5.810	0.042

Source: Reproduced from Q.-H. Zhang, B.-X. Xiao, and G.-Q. Zhu, "Inverse scattering by dielectric circular cylinder using support vector machine approach," *Microwave Opt. Technol. Lett.* **49**(2), 372–375 (Feb. 2007), © 2007 Wiley.

TABLE 11.2 Error Analysis for Problem B

Error	Average, %	Maximum, %	Minimum, %
RelErr(x_c)	3.020	17.470	0.031
RelErr(y_c)	3.018	18.320	0.007

Source: Reproduced from Q.-H. Zhang, B.-X. Xiao, and G.-Q. Zhu, "Inverse scattering by dielectric circular cylinder using support vector machine approach," *Microwave Opt. Technol. Lett.* **49**(2), 372–375 (Feb. 2007), © 2007 Wiley.

focal point inside a target to be inspected by simply moving the source (in front of the slab) in the lateral and longitudinal directions (Wang et al. 2007).

11.6 THROUGH-WALL IMAGING

Recently there has also been increasing interest in *through-wall imaging* (Chang et al. 2009), in which a target is to be localized and reconstructed by using an imaging system positioned behind a wall or another blinding structure. The main purpose is for security, but other surveillance applications are possible. Radar techniques have been proposed mainly for this purpose, but the field can benefit from the development of inverse scattering-based imaging methods, as well. From a theoretical perspective, the presence of the hidden obstacle can be detected by using the proper Green function for the inhomogeneous structure, which can be assumed to be approximately known. Other approaches are also possible. For example, a combination of 2D images obtained by using a linear approach based on the Kirchhoff approximation (Section 4.9) and solved by a truncated singular value decomposition (SVD) (Chapter 5) has been proposed (Solimene et al. 2007, Kidera et al. 2008).

REFERENCES

Abubakar, A., J. Liu, T. Habashy, M. Zaslavsky, V. Druskin, and G. Pan, "3D multiplicative regularized Gauss-Newton inversion," *Proc. 2009 Antennas Propagation Society Int. Symp.* Charleston, SC, USA, 2009.

Bermani, E., S. Caorsi, and M. Raffetto, "Microwave detection and dielectric characterization of cylindrical objects from amplitude-only data by means of neural networks," *IEEE Trans. Anten. Propag.* **50**, 1309–1314, (2002).

Brignone, M., G. Bozza, R. Aramini, M. Pastorino, and M. Piana, "A fully no-sampling formulation of the linear sampling method for three-dimensional inverse electromagnetic scattering problems," *Inverse Problems* **25**, 015014 (2009).

Brignone, M., G. Bozza, A. Randazzo, R. Aramini, M. Piana, and M. Pastorino, "A hybrid approach to 3D microwave imaging by using linear sampling and ant colony optimization," *IEEE Trans. Anten. Propag.* **56**, 3224–3232 (2008).

Caorsi, S., G. L. Gragnani, and M. Pastorino, "Electromagnetic vision-oriented numerical solution to three-dimensional inverse scattering," *Radio Sci.* **23**, 1094–1106 (1988).

Caorsi, S., A. Massa, M. Pastorino, and A. Randazzo, "Electromagnetic detection of dielectric scatterers using phaseless synthetic and real data and the memetic algorithm," *IEEE Trans. Geosci. Remote Sens.* **41**, 2745–2753 (2003).

Caorsi, S., A. Massa, M. Pastorino, and A. Rosani, "Microwave medical imaging: Potentialities and limitations of a stochastic optimization technique," *IEEE Trans. Microwave Theory Tech.* **52**, 1909–1916 (2004).

Chang, P. C., R. J. Burkholder, J. L. Volakis, R. J. Marhefka, and Y. Bayram, "High-frequency EM characterization of through-wall building imaging," *IEEE Trans. Geosci. Remote Sens.* **47**, 1375–1387 (2009).

Cui, T. J., A. A. Aydiner, W. C. Chew, D. L. Wright, and D. V. Smith, "Three-dimensional imaging of buried objects in very lossy earth by inversion of VETEM data," *IEEE Trans. Geosci. Remote Sens.* **41**, 2197–2210 (2003).

Devaney, A. J., "Diffraction tomographic reconstruction from intensity data," *IEEE Trans. Image Proc.* **1**, 221–228 (1992).

De Zaeytijd, J., A. Franchois, C. Eyraud, and J.-M. Geffrin, "Full-wave three-dimensional microwave imaging with a regularized Guass-Newton method—theory and experiment," *IEEE Trans. Anten. Propag.* **55**, 3279–3292 (2007).

Ferrayé, R. et al., "Shape-gradient optimization applied to the reconstruction of 2-D and 3-D metallic objects," *Proc. 2009 Antennas Propagation Society Int. Symp.*, Charleston, SC, USA, 2009.

Franceschini, G., M. Donelli, R. Azaro, and A. Massa, "Inversion of phaseless total field data using a two-step strategy based on the iterative multiscaling approach," *IEEE Trans. Geosci. Remote Sens.* **44**, 3527–3539 (2006).

Gan, H. and W. C. Chew, "3D inhomogeneous inversion for microwave imaging using distorted Born iterative method and BCG-FFT," *Proc. 1995 Antennas Propagation Society Int. Symp.*, Newport Beach, CA, USA, 1995, vol. 3, pp. 1614–1617.

Ghodgaonkar, D. K., O. P., Gandhi, and M. J. Hagmann, "Estimation of complex permittivities of three-dimensional inhomogeneous biological bodies," *IEEE Trans. Microwave Theory Tech.* **31**, 442–446 (1983).

Giebermann, K., D. Colton, and P. Monk, "The linear sampling method for three-dimensional inverse scattering problems," *Proc. Computational Techniques and Applications Conf.*, Canberra, Australia, 1999.

Guo, T. C. and W. W. Guo, "Physics of image formation by microwave scattering," *SPIE Proc.* **767**, 30–39 (1987).

Hagmann, M. J. and R. L. Levin, "Procedures for noninvasive electromagnetic property and dosimetry measurements," *IEEE Trans. Anten. Propag.* **38**, 99–106 (1990).

Hines, E. et al., eds., *Intelligent Systems: Techniques and Applications*, Shaker Verlag, Maastricht, The Netherlands, 2008.

Hislop, G., G. C. James, and A. Hellicar, "Phase retrieval of scattered fields," *IEEE Trans. Anten. Propag.* **55**, 2332–2341 (2007).

Huang, T. and A. S. Mohan, "Microwave imaging of perfect electrically conducting cylinder by microgenetic algorithm," *Proc. Antennas Propagation. Society. Int. Symp.*, Washington, DC, USA, 2005, vol. 1, pp. 221–224.

Keerthi, S. S., S. Shevade, C. Bhattacharyya, and K. R. K. Murthy, "Improvements to Platt's SMO algorithm for SVM classifier design," *Neural Comput.* **13**, 637–649 (2001).

Kidera, S., T. Sakamoto, and T. Sato, "High-resolution and real-time three-dimensional imaging algorithm with envelopes of spheres for UWB radars," *IEEE Trans. Geosci. Remote Sens.* **46**, 3503–3513 (2008).

Morabito, A., I. Catapano, M. D'Urso, L. Crocco, and T. Isernia, "A stepwise approach for quantitative 3D imaging: Rationale and experimental results," *Proc. 2009 Antennas Propagation Society Int. Symp.*, Charleston, SC, USA, 2009.

Ney, M. M., A. M. Smith, and S. S. Stuchly, "A solution of electromagnetic imaging using pseudoinverse transformation," *IEEE Trans. Med. Imag.* **3**, 155–162 (1984).

Pastorino, M. and A. Randazzo, "A smart antenna system for direction of arrival estimation based on a support vector regression," *IEEE Trans. Anten. Propag.* **53**, 2161–2168 (2005).

Rekanos, I. T., "A neural-network-based inverse-scattering technique for online microwave medical imaging," *IEEE Trans. Magn.* **38**, 1061–1064 (2002).

Rubaek, T., O. S. Kim, and P. Meincke, "Computational validation of a 3-D microwave imaging system for breast-cancer screening," *IEEE Trans. Anten. Propag.* **57**, 2105–2115 (2009).

Smola, A. J. and B. Schölkopf, "A tutorial on support vector regression," *Stat. Comput.* **14**, 199–222 (2004).

Solimene, R., F. Soldovieri, G. Prisco, and R. Pierri, "Three-dimensional microwave tomography by a 2-D slice-based reconstruction algorithm," *IEEE Geosci. Remote Sens. Lett.* **4**, 556–560 (2007).

Vapnik, V., *Statistical Learning Theory*, Wiley, New York, 1998.

Veselago, V. G., "The electrodynamics of substances with simultaneously negative values of ε and μ," *Sov. Phys. Usp.* **10**, 509–514 (1968).

Vouldis, A. T., C. N. Kechribaris, T. A. Maniatis, K. S. Nikita, and N. K. Uzunoglu, "Three-dimensional diffraction tomography using filtered backpropagation and multiple illumination planes," *IEEE Trans. Instrum. Meas.* **55**, 1975–1984 (2006).

Wang, G., J. Fang, and X. Dong, "Resolution of near-field microwave target detection and imaging by using flat LHM lens," *IEEE Trans. Anten. Propag.* **55**, 3534–3541 (2007).

Winters, D. W., J. D. Shea, P. Kosmas, B. D. Van Veen, and S. C. Hagness, "Three-dimensional microwave breast imaging: Dispersive dielectric properties estimation using patient-specific basis functions," *IEEE Trans. Med. Imag.* **28**, 969–981 (2009).

Zhang, W., L. Li, and F. Li, "Multifrequency imaging from intensity-only data using the phaseless data distorted Rytov iterative method," *IEEE Trans. Anten. Propag.* **57**, 290–295 (2009).

Zhang, Q.-H., B.-X. Xiao, and G.-Q. Zhu, "Inverse scattering by dielectric circular cylinder using support vector machine approach," *Microwave Opt. Technol. Lett.* **49**, 372–375 (2007).

INDEX

a priori information 16, 22, 79, 102, 107, 117, 118, 119, 124, 126, 147, 154, 156, 158, 171, 179, 180, 194, 196, 201, 269
acceleration coefficient 179
adjoint matrix 86
adjoint operator 81
air 11, 65, 102, 235, 240, 258
aircraft 265
algebraic equation 29, 76
algebraic reconstruction technique 108
amplitude-only data 196, 269
amplitude-only method 269
angiogenesis 242
angular coefficient 63
ant colony optimization 154, 180, 181, 182, 187, 201, 266
antenna 12, 61, 101, 102, 207, 208, 211, 214, 215, 216, 217, 220, 221, 231, 244
antipodal Vivaldi antenna 211
apple 241
applied electromagnetics 198, 231, 264
aqueduct 253
archeological sites 252
arrhythmia 243
associate Legendre function 216
asymptotic expressions 14, 47
atmospheric sounding 20
attenuation constant 229, 230
attraction basin 173
azimuthal plane 211

background 48, 58, 92, 96, 109, 117, 118, 123, 142, 196
backscattering 103
base vector 172
basis function 28, 139, 155, 164, 181, 233, 266
beamforming 95, 101, 102, 245, 247, 250
beamwidth 61, 101, 206, 211, 221
benchmark 142
Bessel function 13, 216
biconjugate stabilized gradient method 36, 150
biopsy 242
bistatic configuration 215, 221
blood 242
bone 242
bore-hole configuration 65
Born approximation 66, 67, 68, 72, 73, 75, 79, 107, 108, 109, 117, 118, 119, 136, 142, 167
boundary conditions 6, 10, 43
bow-tie antenna 211, 253
breast 214, 241, 242, 243, 244, 245, 247, 250
brick 235
bullet 240
buried object 65, 173, 193, 253, 265

cable 252
calibration 147, 208

Microwave Imaging, By Matteo Pastorino
Copyright © 2010 John Wiley & Sons, Inc.

cancer 241, 242, 243
carbon 232, 242
Cartesian coordinates 25, 104
ceramic 235, 240
chest wall 243, 244
chirp pulse 102, 103
chromosome 158, 159, 160, 161, 164, 167, 168, 187, 189, 195
circular cylinder 43, 48, 142, 144, 162, 198, 209, 221, 258, 272
civil engineering 232
clutter 244
coating 240
coaxial transmission line 211
compact operator 22
complex dielectric permittivity 9, 74, 155, 241
complex effective length 217
complex phase function 70
complex vectors 5
computational load 30, 154, 167, 173, 187, 194
computerized tomography 103, 108, 241
concrete 214, 232, 233, 235
condition number 89, 90
conducting media 9
confocal method 243, 244
conjugate gradient 83, 125, 147, 148, 149, 150, 194, 195, 196
constitutive equations 7
continuous monitoring 242
continuous operator 22
contrast function 35, 123, 139
contrast source 23, 27, 58, 59, 123, 136, 137, 139, 148, 150, 268
corrosion 240
cost function 149, 153, 154, 155, 156, 157, 158, 160, 162, 164, 166, 168, 173, 174, 177, 179, 180, 181, 182, 187, 195, 196, 198
cross section 25, 26, 27, 41, 48, 53, 59, 62, 67, 72, 95, 96, 102, 107, 156, 167, 168, 173, 195, 196, 198, 201, 206, 233, 238, 240
crossover 153, 158, 159, 161, 162, 167, 168, 170, 172
crosstalk 207
crystal oscillator 221

curing process 238
current density 68, 71, 73, 105, 106, 107, 118
cylinder axis 41, 206, 208
cylindrical scatterer 67, 173, 208
cylindrical waves 41

data acquisition 208, 244
data equation 22, 24, 26, 27, 58, 59, 117, 155, 156, 173, 181, 196, 265
debris 240
Debye model 244, 245
defect 16, 57, 74, 75, 154, 232, 233, 235, 238
deterministic method 126, 154, 162, 194, 203
diagnostic applications 11, 243
diagnostic method 193, 241
diecetin 244
dielectric permittivity 8, 9, 34, 72, 74, 79, 109, 142, 144, 155, 156, 162, 164, 174, 177, 182, 196, 201, 232, 233, 238, 240, 241, 242, 245, 247, 250, 272
dielectric permittivity tensor 8
differential evolution 154, 170, 171, 172, 173, 174, 177, 180, 182
diffraction tomography 26, 103, 104, 106, 107, 108, 118, 126, 194, 265, 269
dipole 205, 213, 215, 216, 221, 253
Dirac delta function 29
direct problem 20, 21
direct scattering problem 15, 18, 66, 67, 90, 124, 164, 240, 244, 247, 270
directions of arrival 57
directivity 61
discretization 8, 30, 48, 89, 90, 108, 203, 245, 253, 266
dispersion 102, 127, 244, 247
dissipative substrate 211
distorted Born iterative method 74, 142, 144, 247, 265, 266
DNA 243

eigenfunction 41, 48, 96
eigenvalue 84
electric charge 4
electric conductivity 9, 14, 79, 233, 240
electric current 4, 6, 11, 12, 17

electric field 4, 7, 13, 15, 17, 18, 19, 23, 25, 27, 43, 66, 67, 96, 105, 106, 109, 124, 155, 162, 164, 168, 216, 232, 266
electric flux density 4, 7
electromagnetic compatibility 214
electromagnetic field 4, 6, 10, 13, 16, 19, 41, 70, 215, 219, 221, 223, 230, 244, 269
electromagnetic wave 230, 258
elementary source 10
elitism 159, 161, 173
elliptic coordinates 41
elliptic cylinder 41, 42, 48, 53, 95, 96, 109, 139, 195
energy 63, 101, 119, 156, 157, 158, 245
environment 83, 247
equivalent current 17, 104
equivalent current density 18, 19, 23, 27, 71, 118, 119
equivalent magnetic current 17, 18
Euler equation 83, 90
evolutionary algorithm 168
existence 21, 80, 81
extended Born approximation 68, 69, 70, 203, 268

false-negative rate 242
far field 12, 13, 14, 47, 91, 92, 232
fast Fourier transform 107, 131, 265
fat 244
feeding point 213
finite element method 28
finite-difference approximation 33
first order approximation 67
Fletcher-Reeves approach 149
Fourier diffraction theorem 106
Fourier transform 105, 106, 107
Fréchet derivative 125, 130, 131, 137, 139
Fredholm equation 18, 23, 92, 109
free space 10, 11, 12, 14, 15, 17, 26, 41, 73, 95, 104, 211, 213, 231, 272
frequency 5, 8, 13, 73, 92, 102, 103, 106, 107, 166, 223, 232, 238, 240, 242, 243, 244, 245, 253, 272
frequency hopping 126
fruit 241

Galerkin's method 43
Gaussian kernel 181, 271
Gaussian noise 109, 164, 167, 201, 233
Gaussian pulse 211, 244
gene 158, 159, 161, 162
generalized discrepancy principle 94
generalized inverse 80, 82, 83, 88
generalized solution 80, 81, 82, 83, 87, 88, 89, 127, 128
genetic algorithm 126, 153, 154, 158, 161, 162, 164, 166, 167, 168, 169, 170, 171, 173, 174, 177, 179, 182, 187, 194, 195, 247, 266
geometrical optics 73
glandular tissue 244
glass 240, 244
global minimum 153, 155, 194
gray level 241
Green's dyadic tensor 10, 37
Green's function 11, 26, 58, 72, 73, 74, 75, 76, 92, 93, 118, 142, 232, 233, 250, 258, 274
ground penetrating radar 252

half space 11, 73
Hankel function 13, 26, 43
heart 243
heat conduction 20
Helmholtz equation 68, 71
hemoglobine 243
Hilbert space 80, 127
hole 65, 196, 238, 253
holography 208, 247
homogeneous medium 9, 229
homogeneous sphere 37, 41
horizontal plane 42, 206
horn antenna 208, 214, 244
humanitarian applications 252
hybrid method 154, 193, 266, 267

ill-posed problem 22, 80, 82, 83, 89
ill-posedness 21, 22, 124, 137, 266
image 28, 57, 63, 86, 102, 108, 144, 167, 231, 235, 241, 244, 245, 247, 252, 253, 269, 272, 274
imaging 2, 3, 12, 21, 22, 24, 28, 29, 41, 57, 58, 59, 67, 73, 74, 75, 79, 92, 96, 101, 102, 103, 106, 118, 125, 126, 136, 142, 149, 153, 154, 155, 167, 170, 173, 193, 196, 198, 201, 205, 207, 208, 209, 211, 214, 217, 223, 231, 240, 241, 242, 243, 244, 247, 253, 258, 264, 265, 266, 269, 272, 274

imaging applications 12, 21, 22, 28, 155, 170, 209, 214, 272
imaging method 75, 238, 274
impedance 216, 217, 220, 243
impressed current 9
impressed source 10, 68, 71
incident direction 42, 62, 95
incident field 15, 17, 19, 21, 23, 25, 27, 37, 41, 58, 60, 61, 62, 66, 70, 71, 73, 74, 75, 92, 93, 95, 108, 109, 118, 130, 216, 245
indicator function 93, 96, 233, 250
induced current density 9
industrial applications 229
inexact Newton method 129, 125, 130, 136, 139
infarction 243
inhomogeneous Green's function 74, 76, 142, 233
initialization 117, 126, 142, 155, 157, 158, 171, 179, 180, 195, 196, 198
injectivity 21
inner product 28, 29, 271
insulating material 238
integral equation 18, 19, 22, 23, 24, 26, 30, 58, 69, 70, 72, 76, 92, 104, 109, 118, 129, 130, 155
integral operator 10, 37
integral representation 26
integrodifferential equation 30
interface conditions 6
interference 207
interrogating fields 13
intrinsic impedance 10
inverse crime 265
inverse operator 21
inverse problem 20, 21, 22, 23, 67, 89, 108, 123, 124, 125, 137, 147, 148, 153, 162, 164, 168, 173, 194, 198, 211, 231, 232, 240, 266
inverse scattering problem 15, 18, 19, 20, 23, 24, 26, 66, 118, 123, 124, 125, 130, 136, 139, 147, 155, 187, 252, 266
inverse source problem 23, 118
investigation area 27, 72, 95, 96, 109, 114, 174, 209, 221
investigation domain 24, 28, 58, 61, 66, 71, 74, 75, 79, 96, 104, 117, 123, 130, 131, 136, 139, 156, 162, 201, 203, 232, 233

ischemia 243
isorefractive material 53
isotropic media 8

Kirchhoff approximation 73
knot 238

Lagrange multiplier 271, 272
Landweber method 84, 125, 127, 128, 129, 142
least-square solution 81, 82
leukemic marrow 242
level set method 91, 96, 247
Levenberg-Marquardt algorithm 194
linear materials 7
linear sampling method 80, 91, 92, 93, 94, 96, 194, 201, 203, 232, 250, 266, 268
liquid 238, 244, 245, 253
local minima 126, 148, 153, 155, 173, 194, 195, 196, 198, 201
local search method 162, 194, 195
localization 57, 91, 245, 250, 252, 272
Lorentz reciprocity theorem 217

magnetic field 4, 6, 7, 12, 17
magnetic flux density 4, 7
magnetic permeability 8, 9, 79, 272
magnetic permeability tensor 8
main lobe 61
malignant tissue 242, 243
mammography 242
matching load 217
Mathieu functions 42, 43, 44, 96, 195
matrix equation 44, 48
Maxwell's equations 4, 5, 6, 7, 9, 16
measured data 22, 101, 162, 221, 232, 247, 269
measurement error 21
measurement point 28, 29, 58, 62, 63, 65, 96, 106, 109, 132, 139, 148, 167, 196, 211, 232, 269, 270
measurement system 11, 24, 80, 106, 207, 211, 221, 231, 232, 241, 245
mechanical system 101
medical imaging 20, 102, 207, 241
meme 195
memetic algorithm 194, 195, 196, 198, 221, 269
metallic strip 168

method of moments 28, 155, 164, 181, 233, 247, 272
Metropolis criterion 154, 155, 156, 161
micro genetic algorithm 169, 266
microbubble 242
microstrip 214, 235, 253
microwave amplifier 221
microwave camera 214, 221
microwave imaging 9, 13, 20, 23, 24, 58, 67, 75, 79, 80, 101, 102, 107, 108, 118, 125, 126, 136, 149, 154, 164, 167, 170, 198, 205, 208, 211, 214, 229, 232, 238, 240, 241, 242, 243, 245, 252, 264, 265, 272
microwave tomography 59, 205, 206
minimization process 153, 155, 156, 179, 194, 198
modulated scattering technique 198, 214, 215, 216, 221, 223
modulation 207, 215, 221, 223
modulation signal 221
monostatic configuration 63, 215, 217
Morozov principle 91
multifrequency 148
multiillumination 58, 119, 136, 142, 148
multilayer circular cylinder 48
multilayer cylinder 53
MUSIC 247
mutant vector 172
mutation 153, 158, 159, 162, 167, 168, 172, 173, 196
mutual coupling 101, 211, 213

nanoparticle 242
nanotube 242
near-infrared tomography 243
network analyzer 208, 231, 240, 244, 247, 270
neural network 240, 270
Newton method 80, 125, 142, 266
Newton-Kantorovich method 142
nipple 244
noise 21, 22, 80, 82, 83, 91, 94, 96, 101, 109, 119, 125, 132, 139, 162, 164, 167, 201, 221, 233
nondestructive evaluation 102, 238
nondestructive testing 16, 57, 154, 211, 231
noninvasive analysis 154

nonlinear load 221
nonlinear operator 22
nonradiating source 118, 119
normed space 21
numerical code 37, 48, 53
numerical method 18, 28, 30, 35, 150

object function 18, 23, 58, 66, 67, 72, 74, 95, 109, 117, 118, 124, 136, 181
observation domain 22, 23, 27, 58, 59, 117, 118, 130, 201, 211, 269
offspring 171
Ohm's law 9
open circuit 217
open-circuit output voltage 217
operator 7, 10, 21, 22, 31, 33, 80, 81, 82, 83, 89, 107, 130, 131, 136, 137, 139, 153, 159, 161, 170, 172, 179, 187, 195, 196, 198, 242
optical signal 221
optical-fiber connection 221
optimization 119, 124, 126, 147, 148, 149, 153, 154, 155, 158, 160, 161, 167, 168, 170, 177, 180, 181, 182, 187, 194, 195, 201, 203, 266, 270, 271, 272
optimum solution 93, 157, 198
orthonormal basis 86, 88

paint 240
parallel computing 154
parallelization 187, 189
particle 179
particle swarm optimization 154, 177, 179, 180
passive probe 205, 215, 217
patch antenna 214
patient 102, 214, 241, 243, 266
pear 241
PEC 6, 7, 19, 27, 53, 73, 166, 167, 168, 174, 198, 217, 219, 221, 223, 230
penetration depth 230, 232, 240, 243, 244
perfect electric conductor 6
perfectly matched layer 48
phantom 243, 244, 247
phase change 67, 73, 102
phase constant 229, 230
phase retrieval 209, 211, 269
phaseless 269

pheromone 181
phototransistor 221
physical optics 73
piecewise constant representation 28
pillar 232, 233
PIN diode 223
pipe 173, 198, 252, 258
pits 253
plane wave 13, 14, 37, 41, 42, 43, 60, 67, 93, 104, 166, 229, 230
plastic 253, 240
Polak-Ribière approach 149
polar coordinates 60, 96, 109
polarization vector 13
polymer 240
population 153, 154, 158, 159, 160, 161, 162, 167, 168, 169, 171, 172, 173, 174, 176, 177, 179, 180, 181, 182, 187, 195, 196, 198, 266
porosity 231, 240
position 616, 57, 63, 64, 65, 93, 101, 102, 132, 168, 179, 196, 201, 205, 208, 240, 243, 247, 258, 269, 272
position vector 4, 8, 104
principal value 11
probability 155, 156, 158, 160, 161, 162, 167, 168, 181, 196
probability density function 181
probe 22, 27, 29, 63, 64, 65, 101, 102, 205, 215, 216, 217, 220, 221, 223
probing line 27, 62, 63, 65, 104, 105, 106, 107, 166
projection 89, 103, 106
propagation medium 10, 13, 14, 16, 65, 74, 102, 258
propagation vector 13, 92
proportional selection 160
protein 242, 243
prototype 198, 206, 207, 208, 211, 214, 238, 247, 253
pseudoinverse matrix 88, 265
pseudosolution 81
pseuodoinverse 129
pulsation 5
pulse duration 102
pyramidal horn 208, 244

qualitative methods 79, 80, 124, 126, 265
quality control 57

quantitative methods 80, 124, 147, 193
quantum theory 20

radar 57, 91, 101, 102, 208, 241, 252, 253, 274
radiation pattern 13, 60
radiogram 252, 258
Radon transform 104, 241
ray propagation 73, 103, 106, 108
ray tracing 108
rebar 233
receiving antenna 102, 103, 108, 205, 207, 211, 216, 231
receiving channel 206
rectangular reflector 221
reflected signal 63, 101, 102, 103, 208
reflection 63, 102, 208, 231, 232, 240, 253
regularization 22, 80, 82, 83, 84, 91, 93, 94, 124, 125, 147, 167
relaxation time 8
reproduction 160, 161
reservoir 253, 266
resin 238
resistor 244
resonant slot 223
Riccati equation 71
roulette wheel 160
rubber 232
rust 240
Rytov approximation 70, 72, 73, 80, 108, 265, 269

scanning configuration 58, 63
scanning system 205
scattered field 15, 17, 21, 23, 28, 41, 47, 57, 63, 71, 73, 92, 104, 107, 130, 196, 208, 215, 244, 247, 253, 269, 270
scattered wave 43, 63, 231
scattering matrix 211
scattering potential 18, 30, 68, 70, 117, 130, 136, 139, 142
scattering signature 101
screw 240
second order approximation 67
sectorial horn 206
seismology 20
selective pressure 160
semiconvergence 126
semi-focal distance 48, 196, 198

semi-major axes 41, 48, 95, 173, 198
semi-minor axes 41
sensitivity 89, 241, 245
shape 16, 24, 25, 57, 92, 96, 102, 154, 166, 168, 176, 194, 201, 243
shape function 166, 168
short-range imaging 20, 73, 91
short-range inspection 13
Silver's formula 61
Silver-Müller radiation conditions 10
simulated annealing 153, 154, 155, 156, 157, 158, 161
singular system 86, 94, 128
singular value 80, 83, 84, 85, 87, 89, 90, 94, 109, 125, 127, 129
singular value decomposition 80, 83, 84, 85, 87, 90, 94, 109, 125
singular vector 86, 87, 90, 94
skin 102, 243, 244, 245
slab 238, 240, 272
smart antenna 57
soil 11, 65, 196, 258
soybean oil 244
spatial distribution 57
spatial resolution 101, 107
spherical Bessel function 216
spline basis function 162, 164, 174, 181
state equation 22, 24, 27, 58, 59, 117
stationary condition 157
stochastic method 194, 201, 203
stochastic optimization 153, 154, 180, 203
stopping criterion 117, 118, 155, 157, 160, 173, 180, 181
stopping rule 131
strong scatterer 123, 126
sugar 241
support of the scatterer 24, 73, 201
surface charge 6
surface current density 6, 19, 28
surjectivity 21
swarm 179
synthetic aperture focusing 63
synthetic focusing 101, 102

tabu list 170
tangential vector 13
target vector 172, 173

teflon 240
temperature 155, 156, 157, 161, 238, 242, 245
test region 24, 27
testing function 29
three-dimensional configuration 24, 70, 266
three-dimensional scattering 20, 30
Tikhonov method 83, 94, 125
tilt angle 173
time dispersiveness 8
time-harmonic fields 4, 5, 6
tomograph 208, 238, 241
tomographic applications 59, 206
total field 15, 70, 74, 96, 118, 124, 216
tournament selection 160, 161
transmitted power 243
transmitter 101, 139, 221
transmitting antenna 13, 60, 61, 102, 136, 207, 215, 216, 217, 221, 231
transverse electric 27
transverse magnetic 25, 28, 29, 41, 42, 59, 167, 221
traveling-wave structure 214
trigonometric series 166
truncated Landweber method 84, 125, 127, 129
truncated singular value decomposition 125, 240
trunk 208, 238, 240
tumor 242, 243, 244, 245, 247, 250
tunnel 173, 252
two-dimensional problem 24, 27, 156

ultrawideband technique 101
unbounded medium 10
uniqueness 10, 21, 80, 81, 88

vacuum 8, 48, 96
V-antenna 214
vascularization 242
vector wave equation 10, 68
velocity vector 179
vertical plane 206
Vivaldi antenna 211, 253
void 196, 232, 233, 240, 252
volume equivalence principle 16

waste 253
water 206, 207, 242, 244, 245, 253
waveguide 206, 217, 220, 231, 232, 240
wavelength 13, 73, 95, 119, 167, 221, 223
wavenumber 10, 43
well-posed problem 82

well-posedness 21, 90
wideband system 211
window function 107
wood 208, 238, 240

X-ray 103, 241, 242

Yeh's solution 42

WILEY SERIES IN MICROWAVE AND OPTICAL ENGINEERING

KAI CHANG, Editor
Texas A&M University

FIBER-OPTIC COMMUNICATION SYSTEMS, Third Edition • *Govind P. Agrawal*

ASYMMETRIC PASSIVE COMPONENTS IN MICROWAVE INTEGRATED CIRCUITS • *Hee-Ran Ahn*

COHERENT OPTICAL COMMUNICATIONS SYSTEMS • *Silvello Betti, Giancarlo De Marchis, and Eugenio Iannone*

PHASED ARRAY ANTENNAS: FLOQUET ANALYSIS, SYNTHESIS, BFNs, AND ACTIVE ARRAY SYSTEMS • *Arun K. Bhattacharyya*

HIGH-FREQUENCY ELECTROMAGNETIC TECHNIQUES: RECENT ADVANCES AND APPLICATIONS • *Asoke K. Bhattacharyya*

RADIO PROPAGATION AND ADAPTIVE ANTENNAS FOR WIRELESS COMMUNICATION LINKS: TERRESTRIAL, ATMOSPHERIC, AND IONOSPHERIC • *Nathan Blaunstein and Christos G. Christodoulou*

COMPUTATIONAL METHODS FOR ELECTROMAGNETICS AND MICROWAVES • *Richard C. Booton, Jr.*

ELECTROMAGNETIC SHIELDING • *Salvatore Celozzi, Rodolfo Araneo, and Giampiero Lovat*

MICROWAVE RING CIRCUITS AND ANTENNAS • *Kai Chang*

MICROWAVE SOLID-STATE CIRCUITS AND APPLICATIONS • *Kai Chang*

RF AND MICROWAVE WIRELESS SYSTEMS • *Kai Chang*

RF AND MICROWAVE CIRCUIT AND COMPONENT DESIGN FOR WIRELESS SYSTEMS • *Kai Chang, Inder Bahl, and Vijay Nair*

MICROWAVE RING CIRCUITS AND RELATED STRUCTURES, Second Edition • *Kai Chang and Lung-Hwa Hsieh*

MULTIRESOLUTION TIME DOMAIN SCHEME FOR ELECTROMAGNETIC ENGINEERING • *Yinchao Chen, Qunsheng Cao, and Raj Mittra*

DIODE LASERS AND PHOTONIC INTEGRATED CIRCUITS • *Larry Coldren and Scott Corzine*

EM DETECTION OF CONCEALED TARGETS • *David J. Daniels*

RADIO FREQUENCY CIRCUIT DESIGN • *W. Alan Davis and Krishna Agarwal*

MULTICONDUCTOR TRANSMISSION-LINE STRUCTURES: MODAL ANALYSIS TECHNIQUES • *J. A. Brandão Faria*

PHASED ARRAY-BASED SYSTEMS AND APPLICATIONS • *Nick Fourikis*

FUNDAMENTALS OF MICROWAVE TRANSMISSION LINES • *Jon C. Freeman*

OPTICAL SEMICONDUCTOR DEVICES • *Mitsuo Fukuda*

MICROSTRIP CIRCUITS • *Fred Gardiol*

HIGH-SPEED VLSI INTERCONNECTIONS, Second Edition • *Ashok K. Goel*

FUNDAMENTALS OF WAVELETS: THEORY, ALGORITHMS, AND APPLICATIONS • *Jaideva C. Goswami and Andrew K. Chan*

HIGH-FREQUENCY ANALOG INTEGRATED CIRCUIT DESIGN • *Ravender Goyal (ed.)*

ANALYSIS AND DESIGN OF INTEGRATED CIRCUIT ANTENNA MODULES • K. C. Gupta and Peter S. Hall

PHASED ARRAY ANTENNAS, Second Edition • R. C. Hansen

STRIPLINE CIRCULATORS • Joseph Helszajn

THE STRIPLINE CIRCULATOR: THEORY AND PRACTICE • Joseph Helszajn

LOCALIZED WAVES • Hugo E. Hernández-Figueroa, Michel Zamboni-Rached, and Erasmo Recami (eds.)

MICROSTRIP FILTERS FOR RF/MICROWAVE APPLICATIONS • Jia-Sheng Hong and M. J. Lancaster

MICROWAVE APPROACH TO HIGHLY IRREGULAR FIBER OPTICS • Huang Hung-Chia

NONLINEAR OPTICAL COMMUNICATION NETWORKS • Eugenio Iannone, Francesco Matera, Antonio Mecozzi, and Marina Settembre

FINITE ELEMENT SOFTWARE FOR MICROWAVE ENGINEERING • Tatsuo Itoh, Giuseppe Pelosi, and Peter P. Silvester (eds.)

INFRARED TECHNOLOGY: APPLICATIONS TO ELECTROOPTICS, PHOTONIC DEVICES, AND SENSORS • A. R. Jha

SUPERCONDUCTOR TECHNOLOGY: APPLICATIONS TO MICROWAVE, ELECTRO-OPTICS, ELECTRICAL MACHINES, AND PROPULSION SYSTEMS • A. R. Jha

OPTICAL COMPUTING: AN INTRODUCTION • M. A. Karim and A. S. S. Awwal

INTRODUCTION TO ELECTROMAGNETIC AND MICROWAVE ENGINEERING • Paul R. Karmel, Gabriel D. Colef, and Raymond L. Camisa

MILLIMETER WAVE OPTICAL DIELECTRIC INTEGRATED GUIDES AND CIRCUITS • Shiban K. Koul

ADVANCED INTEGRATED COMMUNICATION MICROSYSTEMS • Joy Laskar, Sudipto Chakraborty, Manos Tentzeris, Franklin Bien, and Anh-Vu Pham

MICROWAVE DEVICES, CIRCUITS AND THEIR INTERACTION • Charles A. Lee and G. Conrad Dalman

ADVANCES IN MICROSTRIP AND PRINTED ANTENNAS • Kai-Fong Lee and Wei Chen (eds.)

SPHEROIDAL WAVE FUNCTIONS IN ELECTROMAGNETIC THEORY • Le-Wei Li, Xiao-Kang Kang, and Mook-Seng Leong

ARITHMETIC AND LOGIC IN COMPUTER SYSTEMS • Mi Lu

OPTICAL FILTER DESIGN AND ANALYSIS: A SIGNAL PROCESSING APPROACH • Christi K. Madsen and Jian H. Zhao

THEORY AND PRACTICE OF INFRARED TECHNOLOGY FOR NONDESTRUCTIVE TESTING • Xavier P. V. Maldague

METAMATERIALS WITH NEGATIVE PARAMETERS: THEORY, DESIGN, AND MICROWAVE APPLICATIONS • Ricardo Marqués, Ferran Martín, and Mario Sorolla

OPTOELECTRONIC PACKAGING • A. R. Mickelson, N. R. Basavanhally, and Y. C. Lee (eds.)

OPTICAL CHARACTER RECOGNITION • Shunji Mori, Hirobumi Nishida, and Hiromitsu Yamada

ANTENNAS FOR RADAR AND COMMUNICATIONS: A POLARIMETRIC APPROACH • Harold Mott

INTEGRATED ACTIVE ANTENNAS AND SPATIAL POWER COMBINING • Julio A. Navarro and Kai Chang

ANALYSIS METHODS FOR RF, MICROWAVE, AND MILLIMETER-WAVE PLANAR TRANSMISSION LINE STRUCTURES • Cam Nguyen

FREQUENCY CONTROL OF SEMICONDUCTOR LASERS • Motoichi Ohtsu (ed.)

WAVELETS IN ELECTROMAGNETICS AND DEVICE MODELING • George W. Pan

OPTICAL SWITCHING • Georgios Papadimitriou, Chrisoula Papazoglou, and Andreas S. Pomportsis

SOLAR CELLS AND THEIR APPLICATIONS • Larry D. Partain (ed.)

MICROWAVE IMAGING • Matteo Pastorino

ANALYSIS OF MULTICONDUCTOR TRANSMISSION LINES • Clayton R. Paul

INTRODUCTION TO ELECTROMAGNETIC COMPATIBILITY, Second Edition • Clayton R. Paul

ADAPTIVE OPTICS FOR VISION SCIENCE: PRINCIPLES, PRACTICES, DESIGN AND APPLICATIONS • Jason Porter, Hope Queener, Julianna Lin, Karen Thorn, and Abdul Awwal (eds.)

ELECTROMAGNETIC OPTIMIZATION BY GENETIC ALGORITHMS • Yahya Rahmat-Samii and Eric Michielssen (eds.)

INTRODUCTION TO HIGH-SPEED ELECTRONICS AND OPTOELECTRONICS • Leonard M. Riaziat

NEW FRONTIERS IN MEDICAL DEVICE TECHNOLOGY • Arye Rosen and Harel Rosen (eds.)

ELECTROMAGNETIC PROPAGATION IN MULTI-MODE RANDOM MEDIA • Harrison E. Rowe

ELECTROMAGNETIC PROPAGATION IN ONE-DIMENSIONAL RANDOM MEDIA • Harrison E. Rowe

HISTORY OF WIRELESS • Tapan K. Sarkar, Robert J. Mailloux, Arthur A. Oliner, Magdalena Salazar-Palma, and Dipak L. Sengupta

PHYSICS OF MULTIANTENNA SYSTEMS AND BROADBAND PROCESSING • Tapan K. Sarkar, Magdalena Salazar-Palma, and Eric L. Mokole

SMART ANTENNAS • Tapan K. Sarkar, Michael C. Wicks, Magdalena Salazar-Palma, and Robert J. Bonneau

NONLINEAR OPTICS • E. G. Sauter

APPLIED ELECTROMAGNETICS AND ELECTROMAGNETIC COMPATIBILITY • Dipak L. Sengupta and Valdis V. Liepa

COPLANAR WAVEGUIDE CIRCUITS, COMPONENTS, AND SYSTEMS • Rainee N. Simons

ELECTROMAGNETIC FIELDS IN UNCONVENTIONAL MATERIALS AND STRUCTURES • Onkar N. Singh and Akhlesh Lakhtakia (eds.)

ANALYSIS AND DESIGN OF AUTONOMOUS MICROWAVE CIRCUITS • Almudena Suárez

ELECTRON BEAMS AND MICROWAVE VACUUM ELECTRONICS • Shulim E. Tsimring

FUNDAMENTALS OF GLOBAL POSITIONING SYSTEM RECEIVERS: A SOFTWARE APPROACH, Second Edition • James Bao-yen Tsui

RF/MICROWAVE INTERACTION WITH BIOLOGICAL TISSUES • André Vander Vorst, Arye Rosen, and Youji Kotsuka

InP-BASED MATERIALS AND DEVICES: PHYSICS AND TECHNOLOGY • Osamu Wada and Hideki Hasegawa (eds.)

COMPACT AND BROADBAND MICROSTRIP ANTENNAS • Kin-Lu Wong

DESIGN OF NONPLANAR MICROSTRIP ANTENNAS AND TRANSMISSION LINES • Kin-Lu Wong

PLANAR ANTENNAS FOR WIRELESS COMMUNICATIONS • Kin-Lu Wong

FREQUENCY SELECTIVE SURFACE AND GRID ARRAY • T. K. Wu (ed.)

ACTIVE AND QUASI-OPTICAL ARRAYS FOR SOLID-STATE POWER COMBINING • Robert A. York and Zoya B. Popović (eds.)

OPTICAL SIGNAL PROCESSING, COMPUTING AND NEURAL NETWORKS • Francis T. S. Yu and Suganda Jutamulia

ELECTROMAGNETIC SIMULATION TECHNIQUES BASED ON THE FDTD METHOD • Wenhua Yu, Xiaoling Yang, Yongjun Liu, and Raj Mittra

SiGe, GaAs, AND InP HETEROJUNCTION BIPOLAR TRANSISTORS • *Jiann Yuan*

PARALLEL SOLUTION OF INTEGRAL EQUATION-BASED EM PROBLEMS • *Yu Zhang and Tapan K. Sarkar*

ELECTRODYNAMICS OF SOLIDS AND MICROWAVE SUPERCONDUCTIVITY • *Shu-Ang Zhou*